Lecture Notes in Physics

T0224107

The Lecture Notes in Physics

The series Lecture Notes in Physics (LNP), founded in 1969, reports new developments in physics research and teaching – quickly and informally, but with a high quality and the explicit aim to summarize and communicate current knowledge in an accessible way. Books published in this series are conceived as bridging material between advanced graduate textbooks and the forefront of research and to serve three purposes:

- to be a compact and modern up-to-date source of reference on a well-defined topic
- to serve as an accessible introduction to the field to postgraduate students and nonspecialist researchers from related areas
- to be a source of advanced teaching material for specialized seminars, courses and schools

Both monographs and multi-author volumes will be considered for publication. Edited volumes should, however, consist of a very limited number of contributions only. Proceedings will not be considered for LNP.

Volumes published in LNP are disseminated both in print and in electronic formats, the electronic archive being available at springerlink.com. The series content is indexed, abstracted and referenced by many abstracting and information services, bibliographic networks, subscription agencies, library networks, and consortia.

Proposals should be sent to a member of the Editorial Board, or directly to the managing editor at Springer:

Christian Caron
Springer Heidelberg
Physics Editorial Department I
Tiergartenstrasse 17
69121 Heidelberg / Germany
christian.caron@springer.com

F. Benatti
M. Fannes
R. Floreanini
D. Petritis (Eds.)

Quantum Information, Computation and Cryptography

An Introductory Survey of Theory, Technology and Experiments

 Springer

Fabio Benatti
Università Trieste
Dipto. Fisica Teorica
Strada Costiera, 11
34014 Trieste
Miramare
Italy
benatti@ts.infn.it

Mark Fannes
Afdeling Theoretische Fysica
Celestijnenlaan 200d
B-3001 Heverlee
Belgium
mark.fannes@fys.kuleuven.be

Roberto Floreanini
Dipartimento di Fisica Teorica
Strada Costiera 11
I-34151 Trieste
Italy
florean@ts.infn.it

Dimitri Petritis
Institut de Recherche Mathématique de Rennes
Université de Rennes 1
Campus de Beaulieu
35042 Rennes Cedex
France
Dimitri.Petritis@univ-rennes1.fr

Benatti F. et al. (Eds.): *Quantum Information, Computation and Cryptography: An Introductory Survey of Theory, Technology and Experiments*, Lect. Notes Phys. 808 (Springer, Berlin Heidelberg 2010), DOI 10.1007/978-3-642-11914-9

Lecture Notes in Physics ISSN 0075-8450   e-ISSN 1616-6361
ISBN 978-3-642-11913-2   e-ISBN 978-3-642-11914-9
DOI 10.1007/978-3-642-11914-9
Springer Heidelberg Dordrecht London New York

Library of Congress Control Number: 2010926043

Cover design: Integra Software Services Pvt. Ltd., Pondicherry

Printed on acid-free paper

Springer is part of Springer Science+Business Media (www.springer.com)

Preface

This book is intended for undergraduate students with a minimal physical and mathematical background; its purpose is to provide them with an introduction to the basic tools of the theory and technology of quantum information theory which comprehends quantum information proper, quantum communication, quantum computation, and quantum cryptography. The structure and contents of the book have been suggested by the following two observations:

- the rich interdisciplinary context that has resulted from 20 years of joint efforts of researchers from as different fields as quantum mechanics, both theoretical and experimental, classical information theory, computer science, and cryptography;
- the lack of a textbook for undergraduate students that need to be guided step by step through their many interconnections; indeed, the many existing books address readers already actively researching in these fields.

Complying with the above two points, the book consists of the integrated contributions of various experts and must not be taken as an overview of the results so far obtained in their fields, rather as a textbook specifically thought for the class of readers specified above. With this in mind, all the authors have concentrated on those aspects they deemed important for an undergraduate student to be instructed about. An effort has thus been made to clarify with theory, examples and practical applications a series of carefully selected and correlated aspects of the theory, rather than trying to comprehensively cover the whole of quantum information theory.

The book benefits from the past and present scientific activity of the authors both as researchers and as teachers and from their personal viewpoints on the subjects treated in the chapters. In particular, some important issues are approached from different and stimulating perspectives that may only improve their comprehension.

Also, an extra flavor comes to the book from the participation of some of the authors in the construction of what was known as "quantum probability," a theory which predated quantum information and has provided the latter with useful technical tools and ideas. Their chapters reflect the authors' long acquaintance with the intriguing relations between quantum mechanics and probability, thus offering an opportunity to fully appreciate the subsequent developments and achievements.

The book consists of ten chapters; the first one by D. Petz, provides an introduction to the mathematical tools of quantum mechanics including the von Neumann

entropy and the quantum relative entropy. The second one by Y. Suhov is an introduction to classical probability and information theory, with particular emphasis
on Shannon's coding theorems. In chapter "Quantum Probability and Quantum
Information Theory" H. Maassen offers an introduction to the physics of quantum
mechanics with particular attention to its most puzzling aspects as entanglement and
the Bell's inequalities. The presentation of the quantum entanglement phenomenon
is further developed by F. Benatti in chapter "Bipartite Quantum Entanglement"
with the use of entropic tools and techniques from quantum open system theory.
While chapter "Quantum Probability and Information Theory" and "Bipartite Quantum Entanglement" mainly treat the quantum mechanics of finite-level systems,
in chapter "Field-Theoretical Methods" by R. Alicki concentrates instead on the
quantum mechanics of systems with infinitely many degrees of freedom. Quantum
statistical mechanics and quantum field theory are indeed becoming more and more
important in the recent developments of quantum information both theoretically and
experimentally.

Quantum information transmission with the quantum versions of Shannon's classical coding theorems is the theme of chapter "Quantum Entropy and Information"
by N. Datta, while chapter "Photonic Realization of Quantum Information Protocols" by M. Genovese and chapter "Physical Realizations of Quantum Information" by F. De Melo and A. Buchleitner deal with the experimental achievements in
photonics, respectively, atomic physics that turned the puzzling aspects of quantum
mechanics into actual physical resources; finally, chapter "Quantum Cryptography"
by D. Bruss and T. Meyer and the last chapter by J. Kempe and T. Widick address
those applications of quantum mechanics to cryptography, respectively, computation
theory that so greatly contributed to the ever growing interest in the theoretical and
technological issues presented in this book.

All chapters contain examples, problems, and exercises whose aim is to make
the students actively interact with the text. The references are provided not with the
purpose of being exhaustive, an almost impossible task that is much better accomplished by the many existing advanced books and reviews. Rather, in view of the
specific class of readers addressed by this book, those references to the literature
have been selected that may help to integrate the authors viewpoints and to suggest
the reader a path toward the latest advances of quantum information theory.

Contents

Hilbert Space Methods for Quantum Mechanics

D. Petz

The mathematical backbone of quantum physics is provided by the theory of linear operators on Hilbert spaces. In fact, quite a few concepts related to linear operators were motivated by quantum theory.

1 Hilbert Spaces

The starting point of the quantum mechanical formalism is the *Hilbert space*. The Hilbert space is a mathematical concept; it is a space in the sense that it is a complex vector space which is endowed with an *inner* or *scalar product* $\langle \cdot, \cdot \rangle$. The linear space $\mathbf{C}^n$ of all n-tuples of complex numbers becomes a Hilbert space with the inner product

$$\langle x, y \rangle = \sum_{i=1}^{n} x_i^* y_i = \left(x_1^*, x_2^*, \ldots, x_n^* \right) \begin{pmatrix} y_1 \\ y_2 \\ \vdots \\ y_n \end{pmatrix}, \tag{1}$$

where z^* denotes the complex conjugate of the complex number $z \in \mathbf{C}$. Another example is the space of square-integrable complex-valued function on the real Euclidean space $\mathbb{R}^n$. If f and g are such functions then

$$\langle f, g \rangle = \int_{\mathbb{R}^n} f^*(x) g(x) \, dx \tag{2}$$

gives the inner product. The latter space is denoted by $L^2(\mathbb{R}^n)$ and it is infinite dimensional contrary to the n-dimensional space $\mathbf{C}^n$. Below we are mostly satisfied with finite-dimensional spaces. The inner product of the vectors $|x\rangle$ and $|y\rangle$ will be

D. Petz (✉)
Alfréd Rényi Institute of Mathematics, Hungarian Academy of Sciences, Budapest, Hungary,
petz@renyi.hu

Petz, D.: *Hilbert Space Methods for Quantum Mechanics*. Lect. Notes Phys. **808**, 1–31 (2010)
DOI 10.1007/978-3-642-11914-9_1   © Springer-Verlag Berlin Heidelberg 2010

often denoted as $\langle x|y\rangle$; this notation, sometimes called bra and ket, is popular in physics. On the other hand, $|x\rangle\langle y|$ is a linear operator which acts on the vector $|z\rangle$ as

$$\big(|x\rangle\langle y|\big)\,|z\rangle = |x\rangle\,\langle y|z\rangle \equiv \langle y|z\rangle\,|x\rangle\,. \tag{3}$$

Therefore,

$$|x\rangle\langle y| = \begin{pmatrix} x_1 \\ x_2 \\ \vdots \\ x_n \end{pmatrix} \big(y_1^*, y_2^*, \ldots, y_n^*\big) \tag{4}$$

is conjugate linear in $|y\rangle$, while $\langle x|y\rangle$ is linear.

1.1 Orthogonal Expansions in a Hilbert Space

Let $\mathcal{H}$ be a complex vector space. A functional $\langle\,\cdot\,,\,\cdot\,\rangle : \mathcal{H}\times\mathcal{H} \to \mathbf{C}$ of two variables is called *inner product* if

(1) $\langle x + y, z\rangle = \langle x, z\rangle + \langle y, z\rangle$ $(x, y, z \in \mathcal{H})$,
(2) $\langle \lambda x, y\rangle = \lambda^*\langle x, y\rangle$, $(\lambda \in \mathbf{C},\ x, y \in \mathcal{H})$,
(3) $\langle x, y\rangle = (\langle y, x\rangle)^*$ $(x, y \in \mathcal{H})$,
(4) $\langle x, x\rangle \geq 0$ for every $x \in \mathcal{H}$ and $\langle x, x\rangle = 0$ only for $x = 0$.

These conditions imply the *Schwartz inequality*

$$\big|\langle x, y\rangle\big|^2 \leq \langle x, x\rangle\,\langle y, y\rangle\,. \tag{5}$$

The inner product determines a *norm*

$$\|x\| := \sqrt{\langle x, x\rangle}, \tag{6}$$

which has the property

$$\|x + y\| \leq \|x\| + \|y\|\,. \tag{7}$$

$\|x\|$ is interpreted as the length of the vector x. A further requirement in the definition of a Hilbert space that every Cauchy sequence must be convergent, that is, the space is *complete*.

Exercise 1 *Show that*

$$\|x - y\|^2 + \|x + y\|^2 = 2\|x\|^2 + 2\|y\|^2 \tag{8}$$

which is called the parallelogram law.

If $\langle x, y \rangle = 0$ for the vectors x and y of a Hilbert space $\mathcal{H}$, then x and y are called *orthogonal*, in notation $x \perp y$. For any subset $H \subset \mathcal{H}$, the orthogonal subset

$$H^\perp := \left\{ x \in \mathcal{H} : x \perp h \quad \forall h \in H \right\} \tag{9}$$

is a closed subspace.

Example 1 Let $L^2[a, b]$ be the set of square-integrable (complex-valued) functions on the interval $[a, b]$. This is a Hilbert space with the inner product

$$\langle f, g \rangle := \int_a^b f^*(x)\, g(x)\, dx \tag{10}$$

and with the norm

$$\|f\| := \sqrt{\int_a^b \|f(x)\|^2\, dx}\,. \tag{11}$$

A family $\{x_i\}$ of vectors is called *orthonormal* if $\langle x_i, x_i \rangle = 1$ and $\langle x_i, x_j \rangle = 0$ if $i \neq j$. A maximal orthonormal system is called *basis*. The cardinality of a basis is called the dimension of the Hilbert space. (The cardinality of any two bases is the same.)

Example 2 The infinite dimensional analogue of $\mathbf{C}^n$ is the space $\ell^2(\mathbb{N})$:

$$\ell^2(\mathbb{N}) := \left\{ x = (x_1, x_2, \dots) : x_n \in \mathbf{C}, \sum_n |x_n|^2 < +\infty \right\}. \tag{12}$$

The inner product is

$$\langle x, x' \rangle := \sum_n x_n^* x_n'\,. \tag{13}$$

The canonical basis in this space is the sequence δ_n ($n = 1, 2, \dots$):

$$\delta_n = (0, 0, \dots, 1, 0, \dots) \qquad \text{(1 is at the nth place).} \tag{14}$$

Theorem 1 *Let $x_1, x_2, \dots$ be a basis in a Hilbert space $\mathcal{H}$. Then for any vector $x \in \mathcal{H}$ the expansion*

$$x = \sum_n \langle x_n, x \rangle x_n \tag{15}$$

holds.

Example 3 In the space $L^2[0, \pi]$ the functions

$$f_n(x) = \sqrt{\frac{2}{\pi}} \sin(n\,x) \qquad (16)$$

form a basis. Any function $g \in L^2[0, \pi]$ has an expansion $g = \sum_n a_n f_n$. The convergence is in the L^2-norm. (It is known from the theory of Fourier series that for a continuous g the expansion is convergent point-wise as well.)

Theorem 2 (Projection theorem) *Let $\mathcal{M}$ be a closed subspace of a Hilbert space $\mathcal{H}$. Any vector $x \in \mathcal{H}$ can be written in a unique way in the form $x = x_0 + y$, where $x_0 \in \mathcal{M}$ and $y \perp \mathcal{M}$.*

The mapping $P : x \mapsto x_0$ defined in the context of the previous theorem is called *orthogonal projection* onto the subspace $\mathcal{M}$. This mapping is linear:

$$P(\lambda x + \mu y) = \lambda P x + \mu P y. \qquad (17)$$

Moreover, $P^2 = P$.

Let $A : \mathcal{H} \to \mathcal{H}$ be a linear mapping and $e_1, e_2, \ldots, e_n$ be a basis in the Hilbert space $\mathcal{H}$. The mapping A is determined by the vectors $A e_k$, $k = 1, 2, \ldots, n$. Furthermore, the vector $A e_k$ is determined by its coordinates:

$$A e_k = c_{k,1} e_1 + c_{k,2} e_2 + \cdots + c_{k,n} e_n. \qquad (18)$$

The numbers $c_{i,j}$ form an $n \times n$ matrix; it is called the *matrix* of the linear transformation A in the basis $e_1, e_2, \ldots, e_n$. When $B : \mathcal{H} \to \mathcal{H}$ is another linear transformation, the matrix of the composition $A \circ B$ is the usual matrix product of the matrix of A and that of B. If a basis is fixed, then it induces a one-to-one correspondence between linear transformations and $n \times n$ matrices.

The *norm* of a linear operator $A : \mathcal{H} \to \mathcal{K}$ is defined as

$$\|A\| := \sup\{\|Ax\| : x \in \mathcal{H}, \|x\| = 1\}. \qquad (19)$$

Exercise 2 *Show that $\|AB\| \leq \|A\| \|B\|$.*

Exercise 3 *Let f be a continuous function on the interval $[a, b]$. Define a linear operator $M_f : L^2[a, b] \to L^2[a, b]$ as*

$$M_f g = f g. \qquad (20)$$

(This is the multiplication by the function f.) Show that

$$\|M_f\| = \sup\{|f(x)| : x \in [a, b]\}. \qquad (21)$$

1.2 The Adjoint of a Linear Operator

Let $\mathcal{H}$ and $\mathcal{K}$ be Hilbert spaces. If $T : \mathcal{H} \to \mathcal{K}$ is a bounded linear operator, then its adjoint $T^\dagger : \mathcal{K} \to \mathcal{H}$ is determined by the formula

$$\langle x, Ty \rangle_\mathcal{K} = \langle T^\dagger x, y \rangle_\mathcal{H} \qquad (x \in \mathcal{K}, y \in \mathcal{H}). \tag{22}$$

$T \in B(\mathcal{H})$ is called self-adjoint if $T^\dagger = T$. T is self-adjoint if and only if $\langle x, Tx \rangle$ is real for every vector $x \in \mathcal{H}$.

Exercise 4 *Show that any orthogonal projection is self-adjoint.*

Example 4 Let $S : \ell^2(\mathbb{N}) \to \ell^2(\mathbb{N})$ be the right shift defined as $S\delta_n = \delta_{n+1}$ in the canonical basis. Then

$$S^\dagger(x_1, x_2, x_3, \ldots) = (x_2, x_3, x_4, \ldots). \tag{23}$$

In another way,

$$S^\dagger \delta_1 = 0, \qquad S^\dagger \delta_{n+1} = \delta_n . \tag{24}$$

$S^\dagger$ is called the left shift.

Theorem 3 *The properties of the adjoint are as follows:*

(1) $(A + B)^\dagger = A^\dagger + B^\dagger$, $(\lambda A)^\dagger = \lambda^* A^\dagger$ $\qquad (\lambda \in \mathbf{C})$,
(2) $(A^\dagger)^\dagger = A$, $(AB)^\dagger = B^\dagger A^\dagger$,
(3) $(A^{-1})^\dagger = (A^\dagger)^{-1}$ *if A is invertible,*
(4) $\|A\| = \|A^\dagger\|$.

Example 5 Let $A : \mathcal{H} \to \mathcal{H}$ be a linear mapping and $e_1, e_2, \ldots, e_n$ be a basis in the Hilbert space $\mathcal{H}$. The i, j element of the matrix of A is $\langle e_i, Ae_j \rangle$. Since

$$\langle e_i, Ae_j \rangle = (\langle e_j, A^\dagger e_i \rangle)^*, \tag{25}$$

this is the complex conjugate of the j, i element of the matrix of $A^\dagger$.

Example 6 For any $A \in B(\mathcal{H})$, the operator $A^\dagger A$ is self-adjoint.

An invertible operator $U \in B(\mathcal{H})$ is called a *unitary* if $U^{-1} = U^\dagger$.

Example 7 For any $A = -A^\dagger \in B(\mathcal{H})$, the operator

$$e^A := \sum_{n=0}^{\infty} \frac{A^n}{n!} \tag{26}$$

is a unitary.

Exercise 5 *Show that the product of any two unitary operators is a unitary.*

1.3 Tensor Product of Hilbert Spaces and Operators

Let $\mathcal{H}$ and $\mathcal{K}$ be Hilbert spaces. Their *algebraic tensor product* consists of the formal finite sums

$$\sum_{i,j} x_i \otimes y_j \qquad (x_i \in \mathcal{H}, y_i \in \mathcal{K}). \tag{27}$$

Computing with these sums, one should use the following rules:

$$(x_1 + x_2) \otimes y = x_1 \otimes y + x_2 \otimes y, \quad (\lambda x) \otimes y = \lambda(x \otimes y),$$
$$x \otimes (y_1 + y_2) = x \otimes y_1 + x \otimes y_2, \quad x \otimes (\lambda y) = \lambda(x \otimes y). \tag{28}$$

The inner product is defined as

$$\left\langle \sum_{i,j} x_i \otimes y_j, \sum_{k,l} z_k \otimes w_l \right\rangle = \sum_{i,j,k,l} \langle x_i, z_k \rangle \langle y_j, w_l \rangle. \tag{29}$$

When $\mathcal{H}$ and $\mathcal{K}$ are finite dimensional spaces, we arrive at the *tensor product* Hilbert space $\mathcal{H} \otimes \mathcal{K}$, otherwise the algebraic tensor product must be completed in order to get a Hilbert space.

Example 8 If $f \in \mathcal{H} := L^2(X, \mu)$ and $g \in \mathcal{K} := L^2(Y, \nu)$, then $f \otimes g$ can be interpreted as a function of two variables: $f(x)g(y)$.

The tensor product of finitely many Hilbert spaces is defined similarly.

If $e_1, e_2, \ldots$ and $f_1, f_2, \ldots$ are bases in $\mathcal{H}$ and $\mathcal{K}$, respectively, then $\{e_i \otimes e_j : i, j\}$ is a basis in the tensor product space. This shows that

$$\dim(\mathcal{H} \otimes \mathcal{K}) = \dim(\mathcal{H}) \times \dim(\mathcal{H}). \tag{30}$$

Example 9 In the Hilbert space $L^2(\mathbb{R}^2)$ we can get a basis if the space is considered as $L^2(\mathbb{R}) \otimes L^2(\mathbb{R})$. In the space $L^2(\mathbb{R})$ the Hermite functions

$$\varphi_n(x) = \exp(-x^2/2) H_n(x) \tag{31}$$

form a good basis, where $H_n(x)$ is the appropriately normalized Hermite polynomial. Therefore, the two variable Hermite functions

$$\varphi_{nm}(x, y) := e^{-(x^2+y^2)/2} H_n(x) H_m(y) \qquad (n, m = 0, 1, \ldots) \tag{32}$$

form a basis in $L^2(\mathbb{R}^2)$.

Example 10 Let $\{e_1, e_2, e_3\}$ be a basis in $\mathcal{H}$ and $\{f_1, f_2\}$ be a basis in $\mathcal{K}$. If $[A_{ij}]$ is the matrix of $A \in B(\mathcal{H}_1)$ and $[B_{kl}]$ is the matrix of $B \in B(\mathcal{H}_2)$, then

$$(A \otimes B)(e_j \otimes f_l) = \sum_{i,k} A_{ij} B_{kl} e_i \otimes f_k . \tag{33}$$

It is useful to order the tensor product bases lexicographically: $e_1 \otimes f_1, e_1 \otimes f_2, e_2 \otimes f_1, e_2 \otimes f_2, e_3 \otimes f_1, e_3 \otimes f_2$. Fixing this ordering, we can write down the matrix of $A \otimes B$ and we have

$$\begin{pmatrix} A_{11}B_{11} & A_{11}B_{12} & A_{12}B_{11} & A_{12}B_{12} & A_{13}B_{11} & A_{13}B_{12} \\ A_{11}B_{21} & A_{11}B_{22} & A_{12}B_{21} & A_{12}B_{22} & A_{13}B_{21} & A_{13}B_{22} \\ A_{21}B_{11} & A_{21}B_{12} & A_{22}B_{11} & A_{22}B_{12} & A_{23}B_{11} & A_{23}B_{12} \\ A_{21}B_{21} & A_{21}B_{22} & A_{22}B_{21} & A_{22}B_{22} & A_{23}B_{21} & A_{23}B_{22} \\ A_{31}B_{11} & A_{31}B_{12} & A_{32}B_{11} & A_{32}B_{12} & A_{33}B_{11} & A_{33}B_{12} \\ A_{31}B_{21} & A_{31}B_{22} & A_{32}B_{21} & A_{32}B_{22} & A_{33}B_{21} & A_{33}B_{22} \end{pmatrix} . \tag{34}$$

Let $\mathcal{H}$ be a Hilbert space. The k-fold tensor product $\mathcal{H} \otimes \cdots \otimes \mathcal{H}$ is called the kth tensor power of $\mathcal{H}$, in notation $\mathcal{H}^{\otimes k}$. When $A \in B(\mathcal{H})$, $A^{(1)} \otimes A^{(2)} \cdots \otimes A^{(k)}$ is a linear transformation on $\mathcal{H}^{\otimes k}$ and it is denoted by $A^{\otimes k}$.

$\mathcal{H}^{\otimes k}$ has two important subspaces, the symmetric and the antisymmetric ones. If $v_1, v_2, \ldots, v_k \in \mathcal{H}$ are vectors, then their *antisymmetric* tensor product is the linear combination

$$v_1 \wedge v_2 \wedge \cdots \wedge v_k := \frac{1}{\sqrt{k!}} \sum_{\pi} (-1)^{\sigma(\pi)} v_{\pi(1)} \otimes v_{\pi(2)} \otimes \cdots \otimes v_{\pi(k)} \tag{35}$$

where the summation is over all permutations π of the set $\{1, 2, \ldots, k\}$ and $\sigma(\pi)$ is the number of inversions in π. The terminology "antisymmetric" comes from the property that an antisymmetric tensor changes its sign if two elements are exchanged. In particular, $v_1 \wedge v_2 \wedge \cdots \wedge v_k = 0$ if $v_i = v_j$ for different i and j.

The computational rules for the antisymmetric tensors are similar to (28):

$$\lambda(v_1 \wedge v_2 \wedge \cdots \wedge v_k) = v_1 \wedge v_2 \wedge \cdots \wedge v_{\ell-1} \wedge (\lambda v_\ell) \wedge v_{\ell+1} \wedge \cdots \wedge v_k \tag{36}$$

for every ℓ and

$$\begin{aligned} & (v_1 \wedge v_2 \wedge \cdots \wedge v_{\ell-1} \wedge v \wedge v_{\ell+1} \wedge \cdots \wedge v_k) \\ & + (v_1 \wedge v_2 \wedge \cdots \wedge v_{\ell-1} \wedge v' \wedge v_{\ell+1} \wedge \cdots \wedge v_k) \\ & = v_1 \wedge v_2 \wedge \cdots \wedge v_{\ell-1} \wedge (v + v') \wedge v_{\ell+1} \wedge \cdots \wedge v_k . \end{aligned} \tag{37}$$

The subspace spanned by the vectors $v_1 \wedge v_2 \wedge \cdots \wedge v_k$ is called the kth antisymmetric tensor power of $\mathcal{H}$, in notation $\wedge^k \mathcal{H}$. So $\wedge^k \mathcal{H} \subset \otimes^k \mathcal{H}$. If $A \in B(\mathcal{H})$, then the transformation $\otimes^k A$ leaves the subspace $\wedge^k \mathcal{H}$ invariant. Its restriction is denoted by $\wedge^k A$ which is equivalently defined as

$$\wedge^k A(v_1 \wedge v_2 \wedge \cdots \wedge v_k) = Av_1 \wedge Av_2 \wedge \cdots \wedge Av_k. \tag{38}$$

If $e_1, e_2, \ldots, e_n$ is a basis in $\mathcal{H}$, then

$$\left\{ e_{i(1)} \wedge e_{i(2)} \wedge \cdots \wedge e_{i(k)} \; : \; 1 \le i(1) < i(2) < \cdots < i(k)) \le n \right\} \tag{39}$$

is a basis in $\wedge^k \mathcal{H}$. It follows that the dimension of $\wedge^k \mathcal{H}$ is

$$\binom{n}{k} \quad \text{if} \quad k \le n , \tag{40}$$

otherwise for $k > n$ the power $\wedge^k \mathcal{H}$ has dimension 0. Consequently, $\wedge^n \mathcal{H}$ has dimension 1 and for any operator $A \in B(\mathcal{H})$, we have

$$\wedge^n A = \lambda \times \mathbb{1} . \tag{41}$$

Exercise 6 *Show that $\lambda = \det A$ in (41). Use this to prove that $\det(AB) = \det A \times \det B$.*
Hint: Show that $\wedge^k(AB) = (\wedge^k A)(\wedge^k B)$.

The *symmetric* tensor product of the vectors $v_1, v_2, \ldots, v_k \in \mathcal{H}$ is

$$v_1 \vee v_2 \vee \cdots \vee v_k := \frac{1}{\sqrt{k!}} \sum_{\pi} v_{\pi(1)} \otimes v_{\pi(2)} \otimes \cdots \otimes v_{\pi(k)} , \tag{42}$$

where the summation is over all permutations π of the set $\{1, 2, \ldots, k\}$ again. The linear span of the symmetric tensors is the symmetric tensor power $\vee^k \mathcal{H}$. It has the basis

$$\left\{ e_{i(1)} \vee e_{i(2)} \vee \cdots \vee e_{i(k)} \; : \; 1 \le i(1) \le i(2) \le \cdots \le i(k) \le n \right\} . \tag{43}$$

Exercise 7 *Give the dimension of $\vee^k \mathcal{H}$ if $\dim(\mathcal{H}) = n$.*

1.4 Positive Operators

The *spectrum*, $\mathrm{sp}(T)$, of an operator $T \in B(\mathcal{H})$ consists of all numbers $\lambda \in \mathbf{C}$ such that the operator $\lambda \mathbb{1} - T$ does not have a bounded inverse. If $\mathcal{H}$ is finite dimensional, then $\lambda \mathbb{1} - T$ does not have a bounded inverse if and only if there is a vector $x \neq 0$ such that $\lambda x - Tx = 0$. In this case x is an eigenvector and λ is the corresponding eigenvalue.

The eigenvalues of a self-adjoint matrix are real and the eigenvectors corresponding to different eigenvalues are orthogonal. Therefore, the matrix (or the corresponding Hilbert space operator) can be written in the form

$$\sum_{i=1}^{k} \lambda_i E_i , \qquad (44)$$

where $\lambda_1, \lambda_2, \ldots, \lambda_k$ are the different eigenvalues and E_i is the orthogonal projection onto the subspace spanned by the eigenvectors corresponding to the eigenvalue λ_i, $1 \le i \le k$.

Exercise 8 *Let $A \in B(\mathcal{H})$ and $B \in B(\mathcal{H})$ be operators on the finite dimensional spaces $\mathcal{H}$ and $\mathcal{K}$. Show that*

$$\det(A \otimes B) = (\det A)^m (\det B)^n, \qquad (45)$$

where $n = \dim \mathcal{H}$, $m = \dim \mathcal{K}$.
Hint: The determinant is the product of the eigenvalues.

Exercise 9 *Show that $\| A \otimes B \| = \| A \| \cdot \| B \|$.*

$T \in B(\mathcal{H})$ is called *positive* if $\langle x, Tx \rangle \ge 0$ for every vector $x \in \mathcal{H}$, in notation $T \ge 0$. A positive operator is self-adjoint.

Exercise 10 *Show that $T^{\dagger}T$ is positive for all $T \in B(\mathcal{H})$. Show that an orthogonal projection is positive.*

Theorem 4 *Let $T \in B(\mathcal{H})$ be a self-adjoint operator and $e_1, e_2, \ldots, e_n$ be a basis in the Hilbert space $\mathcal{H}$. T is positive if and only if for any $1 \le k \le n$ the determinant of the $k \times k$ matrix*

$$(\langle e_i, T e_j \rangle)_{ij=1}^{k} \qquad (46)$$

is positive.

The spectrum, in particular the eigenvalues of a positive operator, lies in $\mathbb{R}^+$. Conversely, if all the eigenvalues are positive for a self-adjoint operator acting on a finite dimensional space, then it is positive. Positive matrices are also called positive semi-definite.

Let $A, B \in B(\mathcal{H})$ be self-adjoint operators. $A \le B$ if $B - A$ is positive. Of positive operators $B(\mathcal{H}) \ni T \ge 0$ one can consider the square root $\sqrt{T}$; the square root of $T^{\dagger}T$, $|T| = \sqrt{T^{\dagger}T}$, is called the modulus of T.

Let $T \in B(\mathcal{H})$ and $e_1, e_2, \ldots, e_n$ be a basis in the Hilbert space $\mathcal{H}$; the sum of the diagonal matrix elements of T is called the trace of T,

$$\operatorname{Tr} T = \sum_{i=1}^{n} \langle e_i | T e_i \rangle . \qquad (47)$$

It is independent of the orthonormal basis which is chosen to compute it and satisfies the cyclicity property

$$\mathrm{Tr}\,(A\,B) = \mathrm{Tr}\,(B\,A) \qquad \forall A, B \in B(\mathcal{H})\,. \tag{48}$$

Example 11 Let $f : \mathbb{R}^+ \to \mathbb{R}$ be a smooth function. f is called *matrix monotone* if

$$0 \leq A \leq B \quad \text{implies that} \quad f(A) \leq f(B)\,. \tag{49}$$

f is matrix monotone if and only for every positive operator A, X and for the real parameter $t \geq 0$,

$$\frac{\partial}{\partial t} \langle x, f(A + tX)x \rangle \geq 0 \tag{50}$$

holds for every vector x which means that

$$\frac{\partial}{\partial t} f(A + tX) \geq 0. \tag{51}$$

We want to show that the square root function is matrix monotone. Let

$$F(t) := \sqrt{A + tX}\,. \tag{52}$$

It is enough to see that the eigenvalues of $F'(t)$ are positive. Differentiating the equality $F(t)F(t) = A + tX$, we get

$$F'(t)F(t) + F(t)F'(t) = X. \tag{53}$$

If $F'(t) = \sum_i \lambda_i E_i$ is the spectral decomposition, then

$$\sum_i \lambda_i (E_i F(t) + F(t)E_i) = X \tag{54}$$

and after multiplication by E_j from the left and from the right, we have for the trace

$$2\lambda_j \mathrm{Tr}\left(E_j\,F(t)\,E_j\right) = \mathrm{Tr}\left(E_j\,X\,E_j\right)\,. \tag{55}$$

Since both traces are positive, λ_j must be positive as well.

Exercise 11 *Show that the square function is not matrix monotone.*
Hint: Choose A to be diagonal and

$$X = \begin{pmatrix} 1 & 1 \\ 1 & 1 \end{pmatrix}\,. \tag{56}$$

Use the argument of the previous example for 2×2 matrices.

Exercise 12 *Show that the function*

$$f(x) = -\frac{1}{x} \tag{57}$$

is matrix monotone.

Exercise 13 *Use the formula*

$$\log x = \int_0^\infty \left(\frac{1}{1+t} - \frac{1}{x+t} \right) dt \tag{58}$$

to show that the function $f(t) = \log t$ is matrix monotone.

Example 12 Let $f : \mathbb{R}^+ \to \mathbb{R}$ be a smooth function. f is called *matrix convex* if

$$0 \le A, B \quad \text{implies that} \quad f(\lambda A + (1-\lambda)B) \le \lambda f(A) + (1-\lambda)f(B) \tag{59}$$

for any number $0 < \lambda < 1$. f is matrix convex if and only for every positive operator A and B

$$f\left(\frac{A+B}{2} \right) \le \frac{f(A) + f(B)}{2} \tag{60}$$

holds.

We want to show that the square root function is matrix concave. This follows by taking the square root of the inequality

$$\frac{A+B}{2} \ge \left(\frac{\sqrt{A} + \sqrt{B}}{2} \right)^2. \tag{61}$$

1.5 The Spectral Theorem

The spectral theorem extends the results about discrete spectra (see Sect. 1.4) to arbitrary self-adjoint operator A. Then the spectrum is not necessarily discrete and the finite sum is replaced by an integral.

Let $\mathcal{X}$ be a complete separable metric space and $\mathcal{H}$ be a Hilbert space. In such a case one considers the smallest σ-algebra (see chapter "Quantum Probability and Quantum Information Theory", Sect. 4) containing all open subsets relative to the given metric: its measurable subsets are called Borel subsets. Assume that for each Borel set $B \subset \mathcal{X}$ a positive operator $E(B) \in B(\mathcal{H})$ is given such that

(1) $0 \le E(B) \le I$, $E(\emptyset) = 0$, $E(\mathbf{C}) = I$,
(2) If (B_i) is a sequence of pairwise disjoint Borel subsets of $\mathcal{X}$ and $B = \cup_{i=1}^\infty B_i$, then

$$E(B)\,|e\rangle = \left(\sum_{i=1}^{\infty} E(B_i)\right)|e\rangle \qquad (62)$$

for every vector $e \in \mathcal{H}$.

In this case E is called a *positive operator-valued measure*, shortly *POVM*. In the most important examples $\mathcal{X}$ is a finite set, the real line $\mathbb{R}$, or the unit circle $\mathbb{T}$.

We want to integrate a function $f : \mathcal{X} \to \mathbf{C}$ with respect to a POVM on $\mathcal{X}$. When $\mathcal{X}$ is a finite set,

$$\int_{\mathcal{X}} f(x)\,\mathrm{d}E(x) = \sum_{x \in \mathcal{X}} f(x)E(\{x\}) \qquad (63)$$

is a finite sum. In the general case, the definition of the integral can be reduced to many integrals with respect to common measures. Given a vector $e \in \mathcal{H}$,

$$\mu_e(B) = \langle e, E(B)e\rangle \qquad (64)$$

gives us a positive measure on the Borel sets of $\mathcal{X}$.

We say that the integral $\int_{\mathcal{X}} f(x)\,\mathrm{d}E(x) = T \in B(\mathcal{H})$, if

$$\langle e, Te\rangle = \int_{\mathcal{X}} f(x)\,\mathrm{d}\mu_e(x) \qquad (65)$$

holds for every $e \in \mathcal{X}$.

A POVM E is called a *projection-valued measure* if $E(B)$ is a projection operator for every Borel set B, that is $E(B) = E(B)^2$.

Exercise 14 *Let E be a projection-valued measure and let B_1, B_2 be disjoint Borel set. Show that if a vector e is in the range of $E(B_1)$, then $E(B_2)e = 0$. (Therefore, $E(B_1)$ and $E(B_2)$ are orthogonal.)*

The next theorem is the *spectral theorem* for a bounded self-adjoint operator.

Theorem 5 *Let $A = A^{\dagger} \in B(\mathcal{H})$. Then there exists a unique projection-valued measure on the real line such that*

$$A = \int \lambda\,\mathrm{d}E(\lambda). \qquad (66)$$

Moreover, if $B \subset \mathbb{R}$ and the spectrum of A is disjoint, then $E(B) = 0$ and

$$f(A) = \int f(\lambda)\,\mathrm{d}E(\lambda) \qquad (67)$$

for every continuous function defined on the spectrum of A.

The projection-valued measure in the theorem is called the *spectral measure* of the operator A. A similar result holds for unbounded self-adjoint operator A but in this case A and $f(A)$ are not defined everywhere. A similar theorem holds for unitary operators, then the spectral measure is on the unit circle.

2 Postulates of Quantum Mechanics

The first postulate of quantum mechanics tells that to each quantum mechanical system a Hilbert space $\mathcal{H}$ is associated. The (pure) physical states of the system correspond to unit vectors of the Hilbert space. This correspondence is not one-to-one. When f_1 and f_2 are unit vectors, the corresponding states are identical if $f_1 = z f_2$ for a complex number z of modulus 1. Such z is often called *phase*.

2.1 n-Level Quantum Systems

A *pure physical state* of a system determines a corresponding state vector up to a phase.

Example 13 The two-dimensional Hilbert space $\mathbf{C}^2$ is used to describe a two-level quantum system called qubit. The canonical basis vectors $(1, 0)$ and $(0, 1)$ are usually denoted by $|\uparrow\rangle$ and $|\downarrow\rangle$, respectively. (An alternative notation is $|1\rangle$ for $(0, 1)$ and $|0\rangle$ for $(1, 0)$.) Since the polarization of a photon is an important example of a qubit, the state $|\uparrow\rangle$ may have the interpretation that the "polarization is vertical") and $|\downarrow\rangle$ means that the "polarization is horizontal."

To specify a state of a qubit we need to give a positive number r and a complex number s such that $r^2 + |s|^2 = 1$. Then the state vector is

$$r |\uparrow\rangle + s |\downarrow\rangle. \tag{68}$$

(Indeed, multiplying a unit vector $s_1 |\uparrow\rangle + s_2 |\downarrow\rangle$, $|s_1|^2 + |s_2|^2 = 1$, by an appropriate phase, we can make the coefficient of $|\uparrow\rangle$ positive and the corresponding state remains the same, in the sense that it gives the same statistical predictions of the old one.)

Splitting s into real and imaginary parts as $s = a + ib$, we have the constraint $r^2 + a^2 + b^2 = 1$ for the parameters $(r, a, b) \in \mathbb{R}^3$.

We shall see that the space of all pure states of a qubit is conveniently visualized as the unit sphere in the three-dimensional Euclidean space; it is called the *Bloch sphere*.

Traditional quantum mechanics distinguishes between pure states and *mixed states*. Mixed states are described by *density matrices*. A density matrix or statistical operator is a positive operator of trace 1 on the Hilbert space. This means that the space has a basis consisting of eigenvectors of the statistical operator and the sum of eigenvalues is 1. (In the finite dimensional case the first condition is automatically fulfilled.) The pure states represented by unit vectors of the Hilbert space are among the density matrices under an appropriate identification. If $x = |x\rangle$ is a unit vector, then $|x\rangle\langle x|$ is a density matrix. Geometrically $|x\rangle\langle x|$ is the orthogonal projection onto the linear subspace generated by x. Note that $|x\rangle\langle x| = |y\rangle\langle y|$ if the vectors x and y differ in a phase.

(A1) The physical states of a quantum mechanical system are described by statistical operators acting on the Hilbert space.

Example 14 A state of the spin (of $1/2$) can be represented by the 2×2 matrix

$$\frac{1}{2}\begin{pmatrix} 1 + x_3 & x_1 - ix_2 \\ x_1 + ix_2 & 1 - x_3 \end{pmatrix}. \tag{69}$$

This is a density matrix if and only if $x_1^2 + x_2^2 + x_3^2 \le 1$.

The second axiom is about observables.

(A2) The observables of a quantum mechanical system are described by self-adjoint operators acting on the Hilbert space.

A *self-adjoint operator* A on a Hilbert space $\mathcal{H}$ is a linear operator $\mathcal{H} \to \mathcal{H}$ which satisfies

$$\langle Ax,\, y\rangle = \langle x,\, Ay\rangle \tag{70}$$

for $x, y \in \mathcal{H}$. Self-adjoint operators on a finite dimensional Hilbert space $\mathbf{C}^n$ are $n \times n$ self-adjoint matrices. A self-adjoint matrix admits a *spectral decomposition* $A = \sum_i \lambda_i E_i$, where λ_i are the different eigenvalues of A and E_i is the orthogonal projection onto the subspace spanned by the eigenvectors corresponding to the eigenvalue λ_i. The multiplicity of λ_i is exactly the rank of E_i.

Example 15 For a quantum spin $1/2$ the matrices

$$\sigma_1 = \begin{pmatrix} 0 & 1 \\ 1 & 0 \end{pmatrix}, \qquad \sigma_2 = \begin{pmatrix} 0 & -i \\ i & 0 \end{pmatrix}, \qquad \sigma_3 = \begin{pmatrix} 1 & 0 \\ 0 & -1 \end{pmatrix} \tag{71}$$

are used to describe the spin of directions x, y, z (with respect to a coordinate system). They are called *Pauli matrices*. Any 2×2 self-adjoint matrix is of the form

$$A_{(x_0, x)} := x_0\sigma_0 + x_1\sigma_1 + x_2\sigma_2 + x_3\sigma_3 \tag{72}$$

if σ_0 stands for the unit matrix I. The density matrix (69) can be written as

$$\tfrac{1}{2}(\sigma_0 + \mathbf{x} \cdot \boldsymbol{\sigma}), \qquad \mathbf{x} = (x_1, x_2, x_3), \qquad \boldsymbol{\sigma} = (\sigma_1, \sigma_2, \sigma_3), \tag{73}$$

where $\|\mathbf{x}\| \leq 1$. Formula (73) makes an affine correspondence between 2×2 density matrices and the unit ball in the Euclidean three-space. The extreme points of the ball correspond to pure state and any mixed state is the convex combination of pure states in infinitely many different ways (see the discussion in chapter "Quantum Probability and Quantum Information Theory", Sect. 3.5). In higher dimension the situation is much more complicated.

Any density matrix can be written in the form

$$\rho = \sum_i \lambda_i |x_i\rangle\langle x_i| \tag{74}$$

by means of unit vectors $|x_i\rangle$ and coefficients $\lambda_i \geq 0$, $\sum_i \lambda_i = 1$. Since ρ is self-adjoint such a decomposition is deduced from the spectral theorem and the vectors $|x_i\rangle$ may be chosen as pairwise orthogonal eigenvectors and λ_i are the corresponding eigenvalues. Under this condition (74) is called *Schmidt decomposition*. It is unique if the spectrum of ρ is non-degenerate, that is, if there is no multiple eigenvalue.

2.2 Measurements

Quantum mechanics is not deterministic. If we prepare two identical systems in the same state, and we measure the same observable on each, then the result of the *measurement* may not be the same. This indeterminism or stochastic feature is fundamental.

(A3) Let $\mathcal{X}$ be a finite set and for $x \in \mathcal{X}$ an operator $V_x \in B(\mathcal{H})$ be given such that

$$\sum_{x \in \mathcal{X}} V_x^\dagger V_x = \mathbb{1} . \tag{75}$$

Such an indexed family of operators is a model of a measurement with values in $\mathcal{X}$. If the measurement is performed in a state ρ, then the outcome $x \in \mathcal{X}$ appears with probability $\mathrm{Tr}\left(V_x \rho V_x^\dagger\right)$ and after the measurement the state of the system is

$$\frac{V_x \rho V_x^\dagger}{\mathrm{Tr}\left(V_x \rho V_x^\dagger\right)} . \tag{76}$$

A particular case is the measurement of an observable described by a self-adjoint operator A with spectral decomposition $\sum_i \lambda_i E_i$. In this case $\mathcal{X} = \{\lambda_i\}$ is the set of eigenvalues and $V_i = E_i$. One computes easily that the expectation of the random outcome is $\mathrm{Tr}(\rho A)$. The functional $A \mapsto \mathrm{Tr}(\rho A)$ is linear and has two important properties:

1. if $A \geq 0$, then $\mathrm{Tr}\,(\rho\, A) \geq 0$ and
2. $\mathrm{Tr}\,(\rho\, \mathbb{1}) = 1$.

These two properties allow to see quantum states in a different way. If $\varphi :$ $B(\mathcal{H}) \to \mathbf{C}$ is a linear functional such that

$$\varphi(A) \geq 0 \quad \text{if} \quad A \geq 0 \quad \text{and} \quad \varphi(\mathbb{1}) = 1, \tag{77}$$

then there exists a density matrix ρ_φ such that

$$\varphi(A) = \mathrm{Tr}\left(\rho_\varphi\, A\right). \tag{78}$$

The functional φ associates expectation values with the observables.

2.3 Composite Systems

According to axiom (A1), a Hilbert space is associated with any quantum mechanical system. Assume that a *composite system* consists of subsystems (1) and (2); they are described by the Hilbert spaces $\mathcal{H}_1$ and $\mathcal{H}_2$. (Each subsystem could be a particle or a spin, for example.) Then we have

(A4) The composite system is described by the tensor product Hilbert space $\mathcal{H}_1 \otimes \mathcal{H}_2$.

When $\{e_j : j \in J\}$ is a basis in $\mathcal{H}_1$ and $\{f_i : i \in I\}$ is a basis in $\mathcal{H}_2$, then $\{e_j \otimes f_j : j \in J, i \in I\}$ is a basis of $\mathcal{H}_1 \otimes \mathcal{H}_2$. Therefore, the dimension of $\mathcal{H}_1 \otimes \mathcal{H}_2$ is $\dim\mathcal{H}_1 \times \dim\mathcal{H}_2$. If $A_i \in B(\mathcal{H}_i)$ $(i = 1, 2)$, then the action of the tensor product operator $A_1 \otimes A_2$ is determined by

$$(A_1 \otimes A_2)(\eta_1 \otimes \eta_2) = A_1\eta_1 \otimes A_2\eta_2 \tag{79}$$

since the vectors $\eta_1 \otimes \eta_2$ span $\mathcal{H}_1 \otimes \mathcal{H}_2$.

When $A = A^\dagger$ is an observable of the first system, its expectation value in the vector state $\psi \in \mathcal{H}_1 \otimes \mathcal{H}_2$ is

$$\langle \psi, (A \otimes \mathbb{1}_2)\psi \rangle, \tag{80}$$

where $\mathbb{1}_2$ is the identity operator on $\mathcal{H}_2$.

Example 16 The Hilbert space of two spins $1/2$ is $\mathbf{C}^2 \otimes \mathbf{C}^2$. In this space, the vectors

$$e_1 := |\uparrow\rangle \otimes |\uparrow\rangle, \quad e_2 := |\uparrow\rangle \otimes |\downarrow\rangle, \quad e_3 := |\downarrow\rangle \otimes |\uparrow\rangle, \quad e_4 := |\downarrow\rangle \otimes |\downarrow\rangle \tag{81}$$

form a basis. The vector state

$$\phi = \frac{|\uparrow\rangle \otimes |\downarrow\rangle - |\downarrow\rangle \otimes |\uparrow\rangle}{\sqrt{2}} \tag{82}$$

has a surprising property. Consider the observable

$$A := \sum_{i=1}^{4} i |e_i\rangle\langle e_i|, \tag{83}$$

which has eigenvalues 1, 2, 3, 4 and the corresponding eigenvectors are just the basis vectors. Measurement of this observable yields the values 1, 2, 3, 4 with probabilities 0, 1/2, 1/2, and 0, respectively. The 0 probability occurs when both spins are up or both are down. Therefore in the vector state ϕ the spins are anti-correlated.

We consider now the composite system $\mathcal{H}_1 \otimes \mathcal{H}_2$ in a state $\phi \in \mathcal{H}_1 \otimes \mathcal{H}_2$. Let $A \in B(\mathcal{H}_1)$ be an observable which is localized at the first subsystem. If we want to consider A as an observable of the total system, we have to define an extension to the space $\mathcal{H}_1 \otimes \mathcal{H}_2$. The tensor product operator $A \otimes I$ will do, I is the identity operator of $\mathcal{H}_2$.

Lemma 1 *Assume that $\mathcal{H}_1$ and $\mathcal{H}_2$ are finite dimensional Hilbert spaces. Let $\{e_j : j \in J\}$ be a basis in $\mathcal{H}_1$ and $\{f_i : i \in I\}$ be a basis in $\mathcal{H}_2$. Assume that*

$$\phi = \sum_{i,j} w_{ij} \, e_j \otimes f_i \tag{84}$$

is the expansion of a unit vector $\phi \in \mathcal{H}_1 \otimes \mathcal{H}_2$. Set W for the matrix which is determined by the entries w_{kl}. Then $W^\dagger W$ is a density matrix and

$$\langle \phi, (A \otimes \mathbb{1})\phi \rangle = \mathrm{Tr}\left(A \, W^\dagger W \right). \tag{85}$$

Proof Let E_{kl} be an operator on $\mathcal{H}_1$ which is determined by the relations $E_{kl}e_j = \delta_{lj}e_k$ $(k, l \in I)$. As a matrix, E_{kl} is called matrix unit, it is a matrix such that (k, l) entry is 1, all others are 0. Then

$$\langle \phi, (E_{kl} \otimes \mathbb{1})\phi \rangle = \left\langle \sum_{i,j} w_{ij} \, e_j \otimes f_i, \, (E_{kl} \otimes \mathbb{1}) \sum_{t,u} w_{tu} \, e_u \otimes f_t \right\rangle$$

$$= \sum_{i,j} \sum_{t,u} w_{ij}^* w_{tu} \langle e_j, E_{kl}e_u \rangle \langle f_i, f_t \rangle$$

$$= \sum_{i,j} \sum_{t,u} w_{ij}^* w_{tu} \, \delta_{lu}\delta_{jk}\delta_{it} = \sum_{i} w_{ik}^* w_{il}. \tag{86}$$

Then we arrived at the (k, l) entry of $W^\dagger W$. Our computation may be summarized as

$$\langle \phi, (E_{kl} \otimes \mathbb{1})\phi \rangle = \mathrm{Tr}\, E_{kl}(W^\dagger W) \qquad (k, l \in I). \tag{87}$$

Since any linear operator $A \in B(\mathcal{H}_1)$ is of the form $A = \sum a_{kl} E_{kl}$ $(a_{kl} \in \mathbf{C})$, taking linear combinations of the previous equations, we have

$$\langle \phi, (A \otimes \mathbb{1}) \phi \rangle = \mathrm{Tr}\,(A\, W^\dagger W). \tag{88}$$

$W^\dagger W$ is positive (see Exercise 10) and

$$\mathrm{Tr}\,(W^\dagger W) = \sum_{i,j} |w_{ij}|^2 = \|\phi\|^2 = 1. \tag{89}$$

Therefore it is a density matrix.   □

This lemma shows a natural way from state vectors to density matrices. Given a density matrix ρ on $\mathcal{H}_1 \otimes \mathcal{H}_2$ there are density matrices $\rho_i \in B(\mathcal{H}_i)$ such that

$$\mathrm{Tr}\,(A \otimes \mathbb{1})\rho = \mathrm{Tr}\,(A\, \rho_1) \qquad (A \in B(\mathcal{H}_1)) \tag{90}$$

and

$$\mathrm{Tr}\,(\mathbb{1} \otimes B)\rho = \mathrm{Tr}\,(B\, \rho_2) \qquad (B \in B(\mathcal{H}_2)). \tag{91}$$

ρ_1 and ρ_2 are called *reduced density matrices*. (They are the quantum analogue of marginal distributions.)

The proof of Lemma 1 contains the reduced density of $|\phi\rangle\langle\phi|$ on the first system; it is $W^\dagger W$. One computes similarly that the reduced density on the second subsystem is $(WW^\dagger)^t$, where X^t denotes the transpose of the matrix X. Since $W^\dagger W$ and $(WW^\dagger)^t$ have the same non-zero eigenvalues, the two subsystems are very strongly connected if the total system is in a pure state.

Let $\mathcal{H}_1$ and $\mathcal{H}_2$ be Hilbert spaces and let $\dim \mathcal{H}_1 = m$ and $\dim \mathcal{H}_2 = n$. It is well known that the matrix of a linear operator on $\mathcal{H}_1 \otimes \mathcal{H}_2$ has a block matrix form

$$U = \left(U_{ij}\right)_{i,j=1}^{m} = \sum_{i,j=1}^{m} E_{ij} \otimes U_{ij}, \tag{92}$$

relative to the lexicographically ordered product basis, where U_{ij} are $n \times n$ matrices. For example,

$$A \otimes \mathbb{1} = \left(X_{ij}\right)_{i,j=1}^{m}, \quad \text{where} \quad X_{ij} = A_{ij} \mathbb{1}_n \tag{93}$$

and

$$\mathbb{1} \otimes B = \left(X_{ij}\right)_{i,j=1}^{m}, \quad \text{where} \quad X_{ij} = \delta_{ij} B. \tag{94}$$

Assume that

$$\rho = \left(\rho_{ij}\right)_{i,j=1}^{m} \tag{95}$$

is a density matrix of the composite system, then

$$\text{Tr}(A \otimes \mathbb{1}\, \rho) = \sum_{i,j} A_{ij}\, \text{Tr}(\mathbb{1}_n \rho_{ij}) = \sum_{i,j} A_{ij}\, \text{Tr}\, \rho_{ij} \tag{96}$$

and this gives that for the first reduced density matrix we have

$$(\rho_1)_{ij} = \text{Tr}\, \rho_{ij} \ . \tag{97}$$

We can compute similarly the second reduced density ρ_2. Since

$$\text{Tr}(\mathbb{1} \otimes B\, \rho) = \sum_i \text{Tr}(B\, \rho_{ii}) \tag{98}$$

we obtain

$$\rho_2 = \sum_{i=1}^{m} \rho_{ii}. \tag{99}$$

The reduced density matrices might be expressed by the *partial traces*. Tr_2 : $B(\mathcal{H}_1) \otimes B(\mathcal{H}_2) \to B(\mathcal{H}_1)$ and $\text{Tr}_1 : B(\mathcal{H}_1) \otimes B(\mathcal{H}_2) \to B(\mathcal{H}_2)$ are defined as

$$\text{Tr}_2(A \otimes B) = A\, (\text{Tr}\, B)\ , \qquad \text{Tr}_1(A \otimes B) = (\text{Tr}\, A)\, B\ . \tag{100}$$

We have

$$\rho_1 = \text{Tr}_2 \rho \qquad \text{and} \qquad \rho_2 = \text{Tr}_1 \rho\ . \tag{101}$$

Axiom (A4) tells about a composite quantum system consisting of two quantum components. In the case of more quantum components, the formalism is similar and more tensor factors appear. It may happen that the quantum system under study has a classical and a quantum component; assume that the first component is classical. Then the description by tensor product Hilbert space is still possible. A basis $(|e_i\rangle)_i$ of $\mathcal{H}_1$ can be fixed and the possible density matrices of the joint system are of the form

$$\sum_i p_i |e_i\rangle\langle e_i| \otimes \rho_i^{(2)}\ , \tag{102}$$

where $(p_i)_i$ is a probability distribution and $\rho_i^{(2)}$ are densities on $\mathcal{H}_2$. Then the reduced state on the first component is the probability density $(p_i)_i$ (which may be regarded as a diagonal density matrix) and $\sum_i p_i \rho_i^{(2)}$ is the second reduced density.

Another postulate of quantum mechanics tells about the *time development* of a closed quantum system. If the system is not subject to any measurement in the time interval $I \subset \mathbb{R}$ and ρ_t denotes the statistical operator at time t, then

(A5) $\rho_t = U(t,s)\rho_s U(t,s)^\dagger$ $(t,s \in I)$,

where the *unitary propagator* $U(t,s)$ is a family of unitary operators such that

(i) $U(t,s)U(s,r) = U(t,r)$,
(ii) $(s,t) \mapsto U(s,t) \in B(\mathcal{H})$ is strongly continuous.

The first-order approximation of the unitary $U(s,t)$ is the *Hamiltonian*:

$$U(t + \Delta t, t) = \mathbb{1} - \frac{\mathrm{i}}{\hbar} H(t)\Delta t, \qquad (103)$$

where $H(t)$ is the Hamiltonian at time t. If the Hamiltonian is time independent, then

$$U(s,t) = \exp\left(-\frac{\mathrm{i}}{\hbar}(s - t)H\right). \qquad (104)$$

In the approach followed here the density matrices are transformed in time; this is the so-called *Schrödinger picture* of quantum mechanics [22]. When discrete time development is considered, a single unitary U gives the transformation of the vector state in the form $\psi \mapsto U\psi$ or in the density matrix formalism $\rho \mapsto U\rho U^\dagger$. When the unitary time development is viewed as a quantum algorithm in connection with quantum computation, the term *gate* is used instead of unitary. So the gates constituting an algorithm are simply unitary operators.

2.4 State Transformations

Assume that $\mathcal{H}$ is the Hilbert space of our quantum system which initially has a statistical operator ρ (acting on $\mathcal{H}$). When the quantum systems are not closed, they are coupled to another system, called *environment*. The environment has a Hilbert space $\mathcal{H}_e$ and statistical operator ρ_e. Before interaction the total system has density $\rho_e \otimes \rho$. The dynamical change caused by the interaction is implemented by a unitary and $U(\rho_e \otimes \rho)U^\dagger$ is the new statistical operator and the reduced density $\tilde{\rho}$ is the new statistical operator of the quantum system we are interested in. The affine change $\rho \mapsto \tilde{\rho}$ is typical for quantum mechanics and called *state transformation*. In this way the map $\rho \mapsto \tilde{\rho}$ is defined on density matrices but it can be extended by linearity to all matrices. In this way we obtain a trace-preserving and positivity-preserving linear transformation.

The above-defined state transformation can be described in several other forms and reference to the environment could be omitted completely. Assume that ρ is an $n \times n$ matrix and ρ_e is of the form $(z_k z_l^*)_{kl}$ where $(z_1, z_2, \ldots, z_m)$ is a unit vector in the m-dimensional space $\mathcal{H}_e$. (ρ_e is pure state.) All operators acting on $\mathcal{H}_e \otimes \mathcal{H}$ are written in a block matrix form; they are $m \times m$ matrices with $n \times n$ matrix entries. In particular, $U = (U_{ij})_{i,j=1}^{m}$ and $U_{ij} \in M_n$. If U is a unitary, then $U^\dagger U$ is the identity and this implies that

$$\sum_i U_{ik}^\dagger U_{il} = \delta_{kl} \mathbb{1}_n. \tag{105}$$

Formula (99) for the reduced density matrix gives

$$\tilde{\rho} = \sum_i (U(\rho_e \otimes \rho)U^\dagger)_{ii} = \sum_{i,k,l} U_{ik}(\rho_e \otimes \rho)_{kl}(U^\dagger)_{li}$$

$$= \sum_{i,k,l} U_{ik}(z_k z_l^* \rho)(U_{il})^\dagger = \sum_i \left(\sum_k z_k U_{ik}\right) \rho \left(\sum_l z_l U_{il}\right)^\dagger$$

$$= \sum_i A_i \rho A_i^\dagger, \tag{106}$$

where the operators $A_i := \sum_k z_k U_{ik}$ satisfy

$$\sum_p A_p^\dagger A_p = \mathbb{1} \tag{107}$$

due to (105) and $\sum_k |z_k|^2 = 1$.

Theorem 6 *Any state transformation $\rho \mapsto \mathcal{E}(\rho)$ can be written in the form*

$$\mathcal{E}(\rho) = \sum_p A_p \rho A_p^\dagger, \tag{108}$$

where the operator coefficients satisfy (107). Conversely, all linear mappings of this form are state transformations.

Proof The first part of the theorem was obtained above. To prove the converse part, we need to solve the equations

$$A_i := \sum_k z_k U_{ik} \qquad (i = 1, 2, \ldots, m). \tag{109}$$

Choose simply $z_1 = 1$ and $z_2 = z_3 = \cdots = z_m = 0$ and the equations reduce to $U_{p1} = A_p$. This means that the first column is given from the block matrix U and we need to determine the other columns in such a way that U should be a unitary. Thanks to condition (107) this is possible. Condition (107) tells us that the first column of our block matrix determines an isometry which extends to a unitary. $\square$

The coefficients A_p in the *operator sum representation* are called the *operation elements* of the state transformation. The terms quantum (state) operation and channeling transformation are also often used instead of state transformation. The state transformations form a convex subset of the set of all positive trace-preserving linear transformations. (It is not known what the extremal points of this set are.)

$\mathcal{E}$ is called *completely positive* if $\mathcal{E} \otimes \mathrm{id}_n$ is positivity preserving for the identical mapping $\mathrm{id}_n : M_n(\mathbf{C}) \to M_n(\mathbf{C})$ on any matrix algebra.

Theorem 7 *Let $\mathcal{E} : M_n(\mathbf{C}) \to M_k(\mathbf{C})$ be a linear mapping. Then $\mathcal{E}$ is completely positive if and only if it admits a representation*

$$\mathcal{E}(A) = \sum_u V_u \, A \, V_u^\dagger \tag{110}$$

by means of some linear operators $V_u : \mathbf{C}^n \to \mathbf{C}^k$.

This result was first proven by Kraus. It follows that stochastic mappings are completely positive and the operator sum representation is also called *Kraus representation*. Note that this representation is not unique. Let $\mathcal{E} : M_n(\mathbf{C}) \to M_k(\mathbf{C})$ be a linear mapping. $\mathcal{E}$ is determined by the block matrix $(X_{ij})_{1 \le i,j \le k}$, where

$$X_{ij} = \mathcal{E}(E_{ij}). \tag{111}$$

(Here E_{ij} denote the matrix units.) This is the *block matrix representation* of $\mathcal{E}$.

Theorem 8 *Let $\mathcal{E} : M_n(\mathbf{C}) \to M_k(\mathbf{C})$ be a linear mapping. Then $\mathcal{E}$ is completely positive if and only if the representing block matrix $(X_{ij})_{1 \le i,j \le k} \in M_k(\mathbf{C}) \otimes M_n(\mathbf{C})$ is positive.*

Example 17 Consider the transpose mapping $A \mapsto A^t$ on 2×2 matrices:

$$\begin{pmatrix} x & y \\ z & w \end{pmatrix} \mapsto \begin{pmatrix} x & z \\ y & w \end{pmatrix}. \tag{112}$$

The representing block matrix is

$$X = \begin{pmatrix} 1 & 0 & 0 & 0 \\ 0 & 0 & 1 & 0 \\ 0 & 1 & 0 & 0 \\ 0 & 0 & 0 & 1 \end{pmatrix}. \tag{113}$$

This is not positive, so the transpose mapping is not completely positive.

Example 18 Consider a positive trace-preserving transformation of the form $\mathcal{E} : M_n(\mathbf{C}) \to M_m(\mathbf{C})$ such that its range consists of commuting operators. We show that $\mathcal{E}$ is automatically a state transformation.

Since a commutative subalgebra of $M_m(\mathbf{C})$ is the linear span of some pairwise orthogonal projections P_k, one can see that $\mathcal{E}$ has the form

$$\mathcal{E}(A) = \sum_k P_k \operatorname{Tr}(F_k A) \,, \tag{114}$$

where F_k is a positive operator in $M_n(\mathbf{C})$; it induces the coefficient of P_k as a linear functional on $M_n(\mathbf{C})$.

We want to show the positivity of the representing block matrix:

$$\sum_{ij} E_{ij} \otimes \left(\sum_k P_k \operatorname{Tr}(F_k E_{ij}) \right) = \sum_k \left(\sum_{ij} E_{ij} \otimes P_k \right) \circ \left(\sum_{ij} E_{ij} \operatorname{Tr}(F_k E_{ij}) \otimes \mathbb{1} \right), \tag{115}$$

where ∘ denotes the Hadamard (or entry-wise product) of $nm \times nm$ matrices. Recall that according to Schur's theorem the *Hadamard product* of positive matrices is positive. The first factor is

$$[P_k, P_k, \dots, P_k]^\dagger [P_k, P_k, \dots, P_k] \tag{116}$$

and the second factor is $F_k \otimes I$; both are positive. Consider the particular case of (114) where each P_k is of rank 1 and $\sum_{k=1}^r F_k = \mathbb{1}$. Such a family of F_k's describe a measurement which associates the r-tuple $(\operatorname{Tr} \rho F_1, \operatorname{Tr} \rho F_2, \dots, \operatorname{Tr} \rho F_r)$ with the density matrix ρ. Therefore a measurement can be formulated as a state transformation with diagonal outputs.

The Kraus representation and the block matrix representation are convenient ways to describe a state transformation in any finite dimension. In the 2×2 case we have the possibility to expand the mappings in the basis $\sigma_0, \sigma_1, \sigma_2, \sigma_3$.

Any trace-preserving mapping $\mathcal{E} : M_2(\mathbf{C}) \to M_2(\mathbf{C})$ has a matrix

$$T = \begin{pmatrix} 1 & 0 \\ t & T_3 \end{pmatrix} \tag{117}$$

with respect to this basis, where $T_3 \in M_3$ and

$$\mathcal{E}(w_0 \sigma_0 + \mathbf{w} \cdot \boldsymbol{\sigma}) = w_0 \sigma_0 + (t + T_3 \mathbf{w}) \cdot \boldsymbol{\sigma} \,. \tag{118}$$

The following examples of state transformations are given in terms of the T-representation:

Example 19 (Pauli channels) $t = 0$ and $T_3 = \operatorname{Diag}(\alpha, \beta, \gamma)$. Density matrices are sent to density matrices if and only if

$$-1 \le \alpha, \beta, \gamma \le 1 \tag{119}$$

for the real parameters α, β, γ.

It is not difficult to compute the representing block matrix; we have

$$X = \begin{pmatrix} \dfrac{1+\gamma}{2} & 0 & 0 & \dfrac{\alpha+\beta}{2} \\ 0 & \dfrac{1-\gamma}{2} & \dfrac{\alpha-\beta}{2} & 0 \\ 0 & \dfrac{\alpha-\beta}{2} & \dfrac{1-\gamma}{2} & 0 \\ \dfrac{\alpha+\beta}{2} & 0 & 0 & \dfrac{1+\gamma}{2} \end{pmatrix}. \tag{120}$$

X is unitarily equivalent to the matrix

$$\begin{pmatrix} \dfrac{1+\gamma}{2} & \dfrac{\alpha+\beta}{2} & 0 & 0 \\ \dfrac{\alpha+\beta}{2} & \dfrac{1+\gamma}{2} & 0 & 0 \\ 0 & 0 & \dfrac{1-\gamma}{2} & \dfrac{\alpha-\beta}{2} \\ 0 & 0 & \dfrac{\alpha-\beta}{2} & \dfrac{1-\gamma}{2} \end{pmatrix}. \tag{121}$$

This matrix is obviously positive if and only if

$$|1 \pm \gamma| \geq |\alpha \pm \beta|. \tag{122}$$

For $\alpha = \beta = \gamma = p$ the positivity condition holds when $\dfrac{1}{3} \leq p \leq 1$.

Exercise 15 *Consider the linear transformation* $\mathcal{E} : M_2(\mathbf{C}) \to M_2(\mathbf{C})$ *defined as*

$$\mathcal{E}(w_0\sigma_0 + \mathbf{w} \cdot \sigma) = w_0\sigma_0 + \left(\frac{w_1}{\sqrt{3}}, 0, \frac{1}{3} + \frac{w_3}{3} \right) \cdot (\sigma_1, \sigma_2, \sigma_3). \tag{123}$$

Show that this is a state transformation.
Hint: Compute the representing block matrix.

Example 20 The *depolarizing channel*

$$\mathcal{E}_{p,n} : M_n \to M_n, \qquad \mathcal{E}_{p,n}(A) = pA + (1-p)\frac{\mathbb{1}}{n} \operatorname{Tr}(A) \tag{124}$$

is trivially completely positive for $0 \leq p \leq 1$. The representing block matrix is

$$X = p \sum_{ij} E_{ij} \otimes E_{ij} + \frac{1-p}{n} \mathbb{1} \otimes \mathbb{1}. \tag{125}$$

The matrix

$$\frac{1}{n} \sum_{ij} E_{ij} \otimes E_{ij} \tag{126}$$

is a self-adjoint idempotent (that is, a projection), so its spectrum is $\{0, 1\}$. Consequently, the eigenvalues of X are

$$pn + \frac{1-p}{n}, \frac{1-p}{n}. \tag{127}$$

They are positive when $-1/(n^2 - 1) \leq p \leq 1$.

3 Some Applications

In the traditional approach to quantum mechanics, a physical system is described in a Hilbert space: observables correspond to self-adjoint operators and statistical operators are associated with the states. von Neumann associated an entropy quantity with a statistical operator in 1927 [15] and the discussion was extended in his book [16].

3.1 von Neumann Entropy

von Neumann's argument was a "gedanken experiment" on the ground of phenomenological thermodynamics which is not repeated here, only the conclusion is [16, 20]. Assume that the density ρ is the mixture of orthogonal densities ρ_1 and ρ_2, $\rho = p\rho_1 + (1 - p)\rho_2$. Then

$$pS(\rho_1) + (1 - p)S(\rho_2) = S(\rho) + \kappa p \log p + \kappa (1 - p) \log(1 - p), \tag{128}$$

where S is a certain thermodynamical entropy quantity, relative to the fixed temperature and molecule density and κ a suitable proportionality constant, the Boltzman constant of thermodynamics. (Remember that the orthogonality of states has a particular meaning in quantum mechanics.) From the two-component mixture, we can easily move to an arbitrary density matrix $\rho = \sum_i \lambda_i |\varphi_i\rangle\langle\varphi_i|$ and we have

$$S(\rho) = \sum_i \lambda_i S(|\varphi_i\rangle\langle\varphi_i|) - \kappa \sum_i \lambda_i \log \lambda_i. \tag{129}$$

This formula reduces the determination of the (thermodynamical) entropy of a mixed state to that of pure states. The so-called *Schatten decomposition* $\sum_i \lambda_i |\varphi_i\rangle\langle\varphi_i|$ of a statistical operator is not unique although $\langle\varphi_i, \varphi_j\rangle = 0$ is assumed for $i \neq j$. When λ_i is an eigenvalue with multiplicity, the corresponding eigenvectors can be chosen in many ways. If we expect the entropy $S(\rho)$ to be independent of the Schatten decomposition, then we are led to the conclusion that $S(|\varphi\rangle\langle\varphi|)$ must be independent of the state vector $|\varphi\rangle$. This argument assumes that there are no super-selection sectors, that is, any vector of the Hilbert space can be a state vector. (von Neumann's argument was somewhat different; see the original paper [15, 20].) If the entropy of pure states is defined to be 0 as a kind of normalization, then we have the *von Neumann entropy* formula

$$S(\rho) = -\kappa \sum_i \lambda_i \log \lambda_i = \kappa \operatorname{Tr} \eta(\rho) \tag{130}$$

if λ_i are the eigenvalues of ρ and $\eta(t) = -t \log t$. For the sake of simplicity the multiplicative constant κ will mostly be omitted.

It is worthwhile to note that if $S(\rho)$ is interpreted as the uncertainty carried by the statistical operator ρ, then (128) seems to be natural,

$$S(p\rho_1 + (1-p)\rho_2) = pS(\rho_1) + (1-p)S(\rho_2) + H(p, 1-p) \tag{131}$$

holds for an orthogonal mixture and Shannon's classical information measure is involved. The *mixing property* (131) essentially determines the von Neumann entropy and tells us that the relation of orthogonal quantum states is classical. A detailed axiomatic characterization of the von Neumann entropy is Theorem 2.1 in [18].

Theorem 9 *Let ρ_1 and ρ_2 be density matrices and $0 < p < 1$. The following inequalities hold:*

$$p\,S(\rho_1) + (1-p)S(\rho_2) \leq S(p\rho_1 + (1-p)\rho_2),$$
$$S(p\rho_1 + (1-p)\rho_2) \leq p\,S(\rho_1) + (1-p)S(\rho_2) + H(p, 1-p). \tag{132}$$

Proof The first inequality is an immediate consequence of the concavity of the function $\eta(t) = -t \log t$. In order to obtain the second inequality we benefit from the formula

$$\operatorname{Tr} A\big(\log(A+B) - \log A\big)$$
$$= \int_0^\infty \operatorname{Tr} A(A+t)^{-1} B(A+B+t)^{-1} \, dt \geq 0 \qquad (A, B \geq 0) \tag{133}$$

and infer

$$\operatorname{Tr}\left(p\,\rho_1 \log\left(p\,\rho_1 + (1-p)\,\rho_2\right)\right) \geq \operatorname{Tr}\left(p\,\rho_1 \log p\,\rho_1\right) \tag{134}$$

and

$$\mathrm{Tr}\left((1-p)\rho_2 \log\left(p\,\rho_1+(1-p)\,\rho_2\right)\right) \geq \mathrm{Tr}\left((1-p)\,\rho_2 \log(1-p)\,\rho_2\right). \quad (135)$$

Adding the latter two inequalities we obtain the second inequality of the theorem. □

The von Neumann entropy is the trace of a continuous function of the density matrix, hence it is an obviously continuous functional on the states. From a result due to *Fannes* [7, 18], if an estimate

$$|S(\rho) - S(\sigma)| \leq F(\|\rho - \sigma\|_1) \quad (136)$$

holds classically, that is for measures or for commuting ρ and σ, then the same estimate holds also in the quantum case. This observation yields

Theorem 10 *Let ρ_1 and ρ_2 be densities on a d-dimensional Hilbert space. If $\varepsilon :=$ $\|\rho_1 - \rho_2\|_1/2$, then the inequality*

$$|S(\rho_1) - S(\rho_2)| \leq \varepsilon \log(d-1) + \eta(\varepsilon) + \eta(1-\varepsilon) \quad (137)$$

holds, where $\|X\|_1 := \mathrm{Tr}\,(X^\dagger X)^{1/2}$ is the so-called trace norm.

The proof is found in [21]. Note that on an infinite dimensional Hilbert space the von Neumann entropy is not continuous, unless restricted to a set $\{\rho : S(\rho) \leq c\}$.

Most properties of the von Neumann entropy will be deduced from the behavior of the relative entropy; see [18].

3.2 Relative Entropy

Assume that ρ_1 and ρ_2 are density matrices on a Hilbert space $\mathcal{H}$, then their *relative entropy* is

$$S(\rho_1\|\rho_2) = \begin{cases} \mathrm{Tr}\,\rho_1(\log\rho_1 - \log\rho_2) & \text{if supp}\,\rho_1 \leq \text{supp}\,\rho_2, \\ +\infty & \text{otherwise,} \end{cases} \quad (138)$$

where supp X denotes the support projection of a linear operator X, namely, the projection onto the largest subspace over which $X \neq 0$. Note that it is not a symmetric function of the two arguments.

The relative entropy expresses statistical distinguishability and therefore it decreases under any state transformation $\mathcal{E}$:

$$S(\rho_1\|\rho_2) \geq S(\mathcal{E}(\rho_1)\|\mathcal{E}(\rho_2)). \quad (139)$$

This is Uhlmann's monotonicity theorem [13, 17, 24]. The proof will be sketched.

We need that for a matrix convex function f, the matrix form of the Jensen inequality

$$f(CAC^\dagger + DBD^\dagger) \le Cf(A)C^\dagger + Df(B)D^\dagger \tag{140}$$

holds when $CC^\dagger + DD^\dagger = \mathbb{1}$. In particular, if $f(0) = 0$, then

$$f(CAC^\dagger) \le Cf(A)C^\dagger, \tag{141}$$

see [8]. $f(t) = -\log t$ is matrix monotone decreasing and matrix convex.

The other formula we need is the expression of the relative entropy in terms of the relative modular operator (denoted by Δ). Let $\Delta = LR$, where

$$La = \rho_2 a \quad \text{and} \quad Ra = a\rho_1^{-1} \quad (a \in B(\mathcal{H})). \tag{142}$$

Since $\log \Delta = \log L + \log R$, we have a very useful expression for the relative entropy [10]:

$$\mathrm{Tr}\left(\rho_1\big(\log \rho_1 - \log \rho_2\big)\right) = \langle \rho_1^{1/2}, (-\log \Delta)\rho_1^{1/2}\rangle, \tag{143}$$

where $\langle X, Y\rangle = \mathrm{Tr}(X^\dagger Y)$ denotes the so-called *Hilbert-Schmidt scalar product* of two matrices. The operator Δ_0 is defined similarly from $\mathcal{E}(\rho_1)$ and $\mathcal{E}(\rho_2)$. The relative entropy $S(\mathcal{E}(\rho_1)\|\mathcal{E}(\rho_2))$ is expressed by Δ_0.

The proof of the Uhlmann monotonicity theorem can go as follows:

1. The adjoint map $\mathcal{E}^\dagger : B(\mathcal{K}) \to B(\mathcal{H})$ of $\mathcal{E}$ defined by the Hilbert–Schmidt scalar product is unit preserving and the Schwartz inequality holds:

$$|\langle X, Y\rangle|^2 \le \mathrm{Tr}\left(X^\dagger X\right)\mathrm{Tr}\left(Y^\dagger Y\right). \tag{144}$$

2. The operator

$$V : x\sqrt{\mathcal{E}(\rho_1)} \mapsto \mathcal{E}^\dagger(x)\sqrt{\rho_1} \tag{145}$$

is a contraction with respect to the *Hilbert–Schmidt norm* $\|X\|_{\mathrm{HS}}^2 = \mathrm{Tr}(X^\dagger X)$.

3. $V^\dagger \Delta V \le \Delta_0$. Therefore, $f(\Delta_0) \le f(V^\dagger \Delta V)$ (decreasing), $f(V^\dagger \Delta V) \le V^\dagger f(\Delta)V$ (concavity). Therefore,

$$f(\Delta_0) \le V^\dagger f(\Delta)V. \tag{146}$$

4. Apply $\sqrt{\mathcal{E}(\rho_1)}$ to the vector $\sqrt{\rho_1}$:

$$\left\langle \sqrt{\mathcal{E}(\rho_1)}, f(\Delta_0)\sqrt{\mathcal{E}(\rho_1)}\right\rangle \le \left\langle \sqrt{\rho_1}, f(\Delta)\sqrt{\rho_1}\right\rangle. \tag{147}$$

Theorem 11 *Let ρ_{123} be a density matrix for a system with three components: $B(\mathcal{H}_1 \otimes \mathcal{H}_2 \otimes \mathcal{H}_3)$. Then the strong subadditivity (SSA) inequality*

$$S(\rho_{123}) + S(\rho_2) \leq S(\rho_{12}) + S(\rho_{23}) \tag{148}$$

holds, where ρ_2, ρ_{12}, and ρ_{23} are the reduced density matrices to the subsystems indicated by the subscripts.

Proof An equivalent form of the strong subadditivity can be expressed by relative entropies:

$$S\left(\rho_{12}\|d_1^{-1}\mathbb{1}_1 \otimes \rho_2\right) \leq S\left(\rho_{123}\|d_1^{-1}\mathbb{1}_1 \otimes \rho_{23}\right). \tag{149}$$

The reduced densities of ρ_{123} and $d_1^{-1}\mathbb{1}_1 \otimes \rho_{23}$ are ρ_{12} and $d_1^{-1}\mathbb{1}_1 \otimes \rho_2$. Hence the monotonicity theorem for the relative entropy gives the inequality. $\qquad\square$

3.3 Fidelity

How close are two quantum states? There are many possible answers to this question. Restricting ourselves to pure states, we have to consider two unit vectors $|\varphi\rangle$ and $|\psi\rangle$. Quantum mechanics has used the concept of transition probability $|\langle \varphi \mid \varphi \rangle|^2$ for a long time. This quantity is phase invariant; it lies between 0 and 1. It equals to 1 if and only if the two states coincide, that is, $|\varphi\rangle$ equals to $|\psi\rangle$ up to a phase.

We call *fidelity* the square root of the transition probability, namely

$$F(|\varphi\rangle, |\psi\rangle) := |\langle \varphi \mid \psi \rangle|. \tag{150}$$

Shannon used a nonnegative distortion measure, and we may regard

$$1 - F(|\varphi\rangle, |\psi\rangle) \tag{151}$$

as a distortion function on quantum states. Under a quantum operation pure states could be transformed into mixed states, hence we need extension of the fidelity

$$F\left(|\varphi\rangle\langle\varphi|, \rho\right) = \sqrt{\langle \varphi \mid \rho \mid \varphi \rangle} \tag{152}$$

or in full generality

$$F(\rho_1, \rho_2) = \mathrm{Tr}\left(\sqrt{\rho_1^{1/2}\rho_2\,\rho_1^{1/2}}\right) \tag{153}$$

for positive matrices ρ_1 and ρ_2. This quantity was studied by Uhlmann in a different context [23] and he proved a variational formula:

Theorem 12

$$F(\rho_1, \rho_2) = \inf \left\{ \sqrt{\operatorname{Tr}(\rho_1 G) \operatorname{Tr}(\rho_2 G^{-1})} : 0 \leq G \text{ is invertible} \right\}. \tag{154}$$

From Theorem 12 the symmetry of $F(\rho_1, \rho_2)$ is obvious and we can easily deduce the *monotonicity of the fidelity* under state transformation:

$$F\left(\mathcal{E}(\rho_1), \mathcal{E}(\rho_2)\right)^2 \geq \operatorname{Tr}\left(\mathcal{E}(\rho_1)G\right) \operatorname{Tr}\left(\mathcal{E}(\rho_2)G^{-1}\right) - \varepsilon$$

$$\geq \operatorname{Tr}\left(\rho_1 \mathcal{E}^\dagger(G)\right) \operatorname{Tr}\left(\rho_2 \mathcal{E}^\dagger(G^{-1})\right) - \varepsilon, \tag{155}$$

where $\mathcal{E}^\dagger$ is the adjoint of $\mathcal{E}$ with respect to the Hilbert–Schmidt inner product, $\varepsilon > 0$ is arbitrary, and G is chosen to be appropriate. It is well known that $\mathcal{E}^\dagger$ is unital and positive; hence $\mathcal{E}^\dagger(G)^{-1} \geq \mathcal{E}^\dagger(G^{-1})$:

$$\operatorname{Tr}\left(\rho_1 \mathcal{E}^\dagger(G)\right) \operatorname{Tr}\left(\rho_2 \mathcal{E}^\dagger(G^{-1})\right) \geq \operatorname{Tr}\left(\rho_1 \mathcal{E}^\dagger(G)\right) \operatorname{Tr}\left(\rho_2 \mathcal{E}^\dagger(G)^{-1}\right)$$

$$\geq F(\rho_1, \rho_2)^2. \tag{156}$$

In this way the monotonicity is concluded:

Theorem 13 *For a state transformation $\mathcal{E}$ the inequality*

$$F\left(\mathcal{E}(\rho_1), \mathcal{E}(\rho_2)\right) \geq F(\rho_1, \rho_2) \tag{157}$$

holds.

Another remarkable operational formula is [6]

$$F(\rho_1, \rho_2) = \max \left\{ |\langle \psi_1 | \psi_2 \rangle| : \mathcal{E}(|\psi_1\rangle\langle\psi_1|) = \rho_1, \right. \tag{158}$$

$$\left. \mathcal{E}(|\psi_2\rangle\langle\psi_2|) = \rho_2 \text{ for some state transformation } \mathcal{E} \right\}.$$

This variational expression reduces the understanding of the fidelity of arbitrary states to the case of pure states. The monotonicity property is implied by this formula easily.

Convergence in fidelity is equivalent to convergence in trace norm $F(\rho_n, \rho_n') \to 1$ if and only if $\operatorname{Tr}|\rho_n - \rho_n'| \to 0$. This property of the fidelity is a consequence of the inequalities

$$1 - F(\rho_1, \rho_2) \leq \frac{1}{2}\operatorname{Tr}|\rho_1 - \rho_2| \leq \sqrt{1 - F(\rho_1, \rho_2)}. \tag{159}$$

References

A number of general references have been included in the bibliography: [1–5, 9, 11, 12, 14, 16, 18, 19, 21]. They can be consulted for more detailed information on the topics treated in this lecture.

1. Alberti, P.M., Uhlmann, A.: Stochasticity and Partial Order. Doubly Stochastic Maps and Unitary Mixing. VEB Deutscher Verlag Wiss., Berlin (1981)
2. Alicki, R., Fannes, M.: J. Phys. A **34**, 155 (2004)
3. Bhatia, R.: Matrix Analysis. Springer, New York (1996)
4. Blank, J., Exner, P., Havlíček, M.: Hilbert Space Operators in Quantum Physics. American Institute of Physics, New York (1994)
5. Bratteli, O., Robinson, D.W.: Operator Algebras and Quantum Statistical Mechanics II. Springer, New York (1981)
6. Dodd, J.L., Nielsen, M.A.: Phys. Rev. A **66**, 044301 (2002)
7. Fannes, M.: Commun. Math. Phys. **31**, 291 (1973)
8. Hansen, F., Pedersen, G.K.: Math. Anal. **258**, 229 (1982)
9. Helstrom, C.W.: Quantum Detection and Estimation Theory. Academic Press, New York (1976)
10. Hiai, F., Petz, D.: Commun. Math. Phys. **143**, 99 (1991)
11. Holevo, A.S.: Probabilistic and Statistical Aspects of Quantum Theory. North-Holland, Amsterdam (1982)
12. Holevo, A.S.: Statistical Structure of Quantum Theory. Springer, Berlin (2001)
13. Lindblad, G.: Commun. Math. Phys. **40**, 147 (1975)
14. Marshall, A.W., Olkin, I.: Inequalities: Theory of Majorization and Its Applications. Academic, New York (1979)
15. von Neumann, J.: Gött. Nach. **1**, 273 (1927)
16. von Neumann, J.: Mathematische Grundlagen der Quantenmechanik. Springer, Berlin (1932)
17. Nielsen, M.A., Petz, D.: Quantum Inf. Comput. **5**, 507 (2005)
18. Ohya, M., Petz, D.: Quantum Entropy and Its Use. Springer, Berlin (1993)
19. Petz, D.: Algebra of the Canonical Commutation Relation. Leuven University Press, Leuven (1990)
20. Petz, D.: Entropy, von Neumann and the von Neumann entropy. In: Rédei, M., Stöltzner, M. (eds.) John von Neumann and the Foundations of Quantum Physics. Kluwer, Dordrecht (2001)
21. Petz, D.: Lectures on Quantum Information Theory and Quantum Statistics. Springer, Berlin (2008)
22. Schrödinger, E.: Proc. Camb. Philol. Soc. **31**, 446 (1936)
23. Uhlmann, A.: Rep. Math. Phys. **9**, 273 (1976)
24. Uhlmann, A.: Commun. Math. Phys. **54**, 21 (1977)

Classical Information Theory

Y. Suhov

1 Entropy

There is no rule in the world
but through Chance and Chaos,
and entropies are messengers of Chance
and measures of Chaos.
All variables are random,
but some are more random than others.
From the series 'Thus spoke Supervisor'.

1.1 Definitions and Examples

We begin with the definition of information gained by knowing that an event A has occurred:

$$\iota(A) = -\log_2 \mathbf{P}(A). \tag{1}$$

(A dual point of view is also useful (although more evasive), where $\iota(A)$ is the amount of information needed to specify event A.) Here and below $\mathbf{P}$ stands for the underlying probability distribution. So the rarer an event A, the more information we gain if we know it has occurred. (More broadly, the rarer an event A, the more impact it will have. For example, the unlikely event that occurred in 1938 when fishermen caught a coelacanth – a prehistoric fish believed to be extinct – required a significant change to beliefs about evolution and biology. On the other hand, the likely event of catching a herring or a tuna would hardly imply any change in theories.)

It is obvious from the above equation that for independent events, A_1 and A_2, the information relative to the joint events A_1 and A_2 ($A_1 \cap A_2$) satisfies

$$\iota(A_1 \cap A_2) = \iota(A_1) + \iota(A_2), \tag{2}$$

Y. Suhov (✉)
Statistical Laboratory, DPMMS, University of Cambridge, Cambridge, UK
yms@statslab.cam.ac.uk

Suhov, Y.: *Classical Information Theory.* Lect. Notes Phys. **808**, 33–64 (2010)
DOI 10.1007/978-3-642-11914-9_2

and for event A with $p(A) = 1/2$,

$$\iota(A) = 1. \tag{3}$$

A justification of this definition of information comes from the fact that any function $\iota^*(A)$, which (i) depends on probability $\mathbf{P}(A)$ (i.e., obeys $\iota^*(A) = \iota^*(A')$ if $\mathbf{P}(A) = \mathbf{P}(A')$), (ii) is continuous in $\mathbf{P}(A)$, and (iii) satisfies the two aforementioned properties, coincides with $\iota(A)$.

Definition 1 Let X be a discrete random variable (RV) with finitely many distinct values $x_1, x_2, \ldots, x_m$, taken with probabilities $\mathbf{p} = (p_1, p_2, \ldots, p_m)$. Then the *entropy* of X is the expected amount of information gained on learning the value of X:

$$h(X) = -\sum_{i=1}^{m} p_i \log_2 p_i. \tag{4}$$

It is clear that the entropy $h(X)$ depends on the probability distribution $\mathbf{p}$, but not on the values of X, so we will write interchangeably $h(X)$, $h(p_1, \ldots, p_m)$, or even just $h(\mathbf{p})$. In addition, we will often use notation $p(x_i)$ instead of p_i, writing

$$h(X) = -\sum_{x_i} p(x_i) \log_2 p(x_i) = \mathbb{E} - \log_2 p(X). \tag{5}$$

Here and below, symbol $\mathbb{E}$ stands for the expectation, relative to an underlying probability distribution; in some cases we will need to specify which particular probability distribution we bear in mind.

Thus, if we consider a random variable $\phi(X)$ where g is a one-to-one function then $h(\phi(X)) = h(X)$.

Note an unusual form of a random variable under expectation in (5); it is $\log_2 p(X)$. It takes values $-\log_2 p_1, \ldots, -\log_2 p_m$, with probabilities $p_1, \ldots, p_m$, respectively. It may seem tautological, but such a tautology greatly simplifies further analysis.

As some of the probabilities may be equal to 0, we will adopt the convention that $0 \log 0 = 0$ (by continuity $\lim_{x \to 0+} -x \log_2 x = 0$).

Example 1 If X is a Bernoulli(p) random variable (taking the value 0 with probability $1 - p$ and 1 with probability p), then

$$h(X) = -p \log_2 p - (1 - p) \log_2(1 - p). \tag{6}$$

Many authors write $h(p)$ for this function (further abusing the notation). We too will use this notation on occasions.

For $p = 0$ or 1, X is deterministic and $h(X) = 0$. On the other hand, for $p = 1/2$, X is "as random as it can be" and $h(X) = 1$, the maximum value for random variables taking two values, see Fig. 1. This fits in with our intuition as to how a sensible measure of uncertainty should behave.

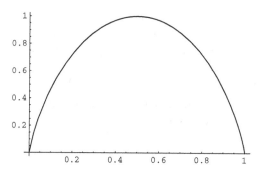

Fig. 1 The entropy of a Bernoulli(p) variable for different p

Further, $h''(p) = -\log_2 e/(p(1-p)) \leq 0$, that is, the function $h(p)$ is concave.

Example 2 If X is a uniform(n) random variable, with $\mathbf{P}(X = x_j) = \dfrac{1}{n}, j = 1, \ldots, n$, then

$$h(X) = \log_2 n. \tag{7}$$

Definition 1 can be extended to the case of random variables taking countably many values $x_1, x_2, \ldots$ by replacing the sum from 1 to n with a sum from 1 to ∞ (here we adopt the convention that if the sum $-\sum_{i=1}^{+\infty} p_i \log_2 p_i$ diverges, then the entropy is infinite).

Example 3 Let X be a geometric(p) random variable, with $\mathbf{P}(X = r) = (1-p)p^r$ for $r = 0, 1, \ldots$. Then

$$
\begin{aligned}
h(X) &= \sum_r \mathbf{P}(X = r)\left(-\log_2(1-p) - r\log_2 p\right)\\
&= -\log_2(1-p) - \frac{p}{1-p}\log_2 p,
\end{aligned}
\tag{8}
$$

since $\mathbb{E}X = p/(1-p)$.

For a Poisson(λ) random variable X, with $\mathbf{P}(X = r) = (\lambda^r/r!)e^{-\lambda r}$ for $r = 0, 1, \ldots$, the entropy $h(X)$ looks rather ugly:

$$
\begin{aligned}
h(X) &= \sum \mathbf{P}(X = r)\left(\frac{\lambda}{\ln 2} - r\log_2 \lambda + \log_2(r!)\right)\\
&= \frac{\lambda}{\ln 2} - \lambda\log_2 \lambda + \sum \mathbf{P}(X = r)\sum_{l=0}^{r}\log_2 l\\
&= \frac{\lambda}{\ln 2} - \lambda\log_2 \lambda + \sum_{r=0}^{\infty}\mathbf{P}(X \geq r)\log_2 r .
\end{aligned}
\tag{9}
$$

An easy observation is that in the discrete case, entropy $h(X)$ is always non-negative: $h(X) \geq 0$. This is because the sum in the right-hand side of (4) and (5) contains only non-negative terms, i.e., $-\log_2 p_i \geq 0$ as $0 \leq p_i \leq 1$. Moreover, $h(X) > 0$ unless X is a deterministic random variable, that is, X takes some value with probability 1 (and hence all other values with probability 0).

Definition 2 Let X be a random variable with a probability density function $p(x)$. Then the differential entropy of X is defined in analogy with the discrete case:

$$h(X) = -\int p(x) \log_2 p(x) dx. \tag{10}$$

Example 4

1. If X is uniform(α, β), and $p(x) = \dfrac{1}{\beta - \alpha} \mathbf{1}_{(\alpha, \beta)}(x)$, with $\mathbf{1}_\Delta$ the characteristic function of the interval Δ and $-\infty < \alpha < \beta < +\infty$, then

$$h(X) = \log_2(\beta - \alpha). \tag{11}$$

2. Let X be a normal(μ, σ) random variable, with $-\infty < \mu < +\infty$, $\sigma > 0$, and

$$p(x) = \frac{1}{\sqrt{2\pi}\sigma} \exp\left[-\frac{(x - \mu)^2}{2\sigma^2}\right] \text{ for } -\infty < x < +\infty.$$

Then

$$h(X) = \frac{1}{2}\left[1 + \log_2(\pi \, e \, \sigma^2)\right]. \tag{12}$$

3. If X is an exponential(λ) random variable, with $\lambda > 0$ and $p(x) = \lambda e^{-\lambda x} \mathbf{1}(x > 0)$, then

$$h(x) = \frac{1}{\ln 2}\left(1 - \log_2 \lambda\right). \tag{13}$$

Unlike the entropy of a discrete random variable, the differential entropy may be negative: you can see it in all examples above.

The entropy and the differential entropy $h(X)$ are used to "measure" an amount of randomness (or uncertainty) in a given random variable X. Despite its rather straightforward appearance (and a good deal of research), $h(X)$ still is a concept full of mystery, and quantum information theory only made it more enigmatic (see chapters "Hilbert Space Methods for Quantum Mechanics," "Bipartite Quantum Entanglement," and "Quantum Entropy and Information").

The entropy and the differential entropy are often called Shannon entropies, in honor of C. Shannon. In fact, Shannon was not the first scientist to use this quantity;

it was used long before him in thermodynamic and statistical physics, by authors like Carnot, Clausius, Boltzmann, Gibbs, and Helmholtz. However, he was the pioneer in using the concept of entropy in the context of information theory (in fact, the whole context of information theory as Shannon saw it was based on the concept of entropy, although, admittedly, the importance of entropy in this area was stressed some 20 years earlier by Hartley). There is an unconfirmed legend that in the beginning of information theory, Shannon went to Princeton to see J. von Neumann who was doing ground-breaking research in (classical) computer science at that time. Shannon presented his preliminary findings to von Neumann and said that he needed a kind of measure of randomness. Von Neumann replied that in his view it should be entropy and wrote down its definition. (Von Neumann himself used the quantum analog of entropy, again 20 or so years earlier, for mixed states in Hilbert space (although to a rather limited extent), hence the term von Neumann entropy in the context of quantum information theory.) Then von Neumann said to Shannon "Try it. Nobody understands what it is in reality, so you may well get away with it."

A profound analogy with statistical physics is that, as maximum entropy states play a significant role within statistical physics, maximum entropy probability distributions naturally arise in probability and statistics.

1.2 Joint Entropy, Conditional Entropy, Relative Entropy, Mutual Entropy

In this section we consider discrete random variables only. The use of entropy in information theory is almost always related to joint random variables. In this regard, the concepts of joint, conditional, relative, and mutual entropies are of particular importance.

From now on log stands for $\log_2$.

Definition 3 Given a pair of random variables, X, Y, with values x_i and y_j, the *joint entropy* $h(X, Y)$ is defined by

$$h(X, Y) = - \sum_{x_i, y_j} p_{X,Y}(x_i, y_j) \log p_{X,Y}(x_i, y_j) = \mathbb{E} - \log p_{X,Y}(X, Y). \quad (14)$$

Here and below, $p_{X,Y}(x_i, y_j) = \mathbf{P}(X = x_i, Y = y_j)$ denotes the joint probability. The expectation in (14) is taken relative to $p_{X,Y}$. Thus, $h(X, Y)$ is the entropy of the random variable (or the random vector) (X, Y) with values (x_i, y_j).

The *conditional entropy*, $h(X|Y)$, of a random variable X, given random variable Y, is defined as the expected amount of information gained from observing X given that a value of Y is known:

$$h(X|Y) = - \sum_{x_i, y_j} p_{X,Y}(x_i, y_j) \log p_{X|Y}(x_i|y_j) = \mathbb{E} - \log p_{X|Y}(X|Y). \quad (15)$$

Here and below, $p_{X|Y}(x_i|y_j)$ denotes the conditional probability $\mathbf{P}(X = x_i|Y = y_j)$. The expectation in (14) is again taken relative to $p_{X,Y}$. As follows from equations (14) and (15),

$$h(X|Y) = h(X, Y) - h(Y). \tag{16}$$

Note that in general $h(X|Y) \neq h(Y|X)$. Also, $h(X|Y)$ is a mixture of entropies $h(\mathbf{p}_{X|Y}(\cdot|y_j))$ of conditional distributions $\mathbf{p}_{X|Y}(\cdot|y_j)$, with coefficients $p_Y(y_j)$. Formally

$$h(X|Y) = \sum_{y_j} p_Y(y_j)h(\mathbf{p}_{X|Y}(\cdot|y_j)), \tag{17}$$

where

$$h(\mathbf{p}_{X|Y}(\cdot|y_j)) = -\sum_{x_i} p_{X|Y}(x_i|y_j) \log p_{X|Y}(x_i|y_j). \tag{18}$$

Next, if random variables X and Y take values from the same (finite or countable) set (say I), and the marginal probability $p_Y(x) = \mathbf{P}(Y = x) > 0 \,\forall x \in I$, then the *relative entropy* $h(X||Y)$ (the entropy of X relative to Y) is defined by

$$h(X||Y) = \sum_{x} p_X(x) \log \frac{p_X(x)}{p_Y(x)} = \mathbb{E} - \log \frac{p_Y(X)}{p_X(X)}. \tag{19}$$

Here and below, p_X and p_Y stand for the probability distributions of X and Y, respectively ($p_X(x) = \mathbf{P}(X = x)$ and $p_Y(x) = \mathbf{P}(Y = x)$, $x \in I$). The expectation in (19) is taken relative to p_X.

Finally, the *mutual information* between X and Y is defined as

$$\begin{aligned}
\iota(X : Y) &= \mathbb{E} \log \frac{p_{X,Y}(X, Y)}{p_X(X)p_Y(Y)} \\
&= \sum_{x,y} p_{X,Y}(x, y) \log \frac{p_{X,Y}(x, y)}{p_X(x)p_Y(y)} \\
&= h(X) + h(Y) - h(X, Y).
\end{aligned} \tag{20}$$

The expectation in (20) is taken relative to $p_{X,Y}$. In terms of the relative entropy, $\iota(X : Y) = h_{X \otimes Y}(X, Y)$. Here $X \otimes Y$ denotes the pair of independent random variables, one of which is distributed as X and the other as Y. Physically, $\iota(X : Y)$ measures the amount of information about X conveyed by Y (and vice versa). Note that the mutual information is symmetric:

$$\iota(X : Y) = \iota(Y : X), \text{ i.e., } h(X) - h(X|Y) = h(Y) - h(Y|X). \tag{21}$$

Of course, if X and Y are independent then $\iota(X : Y) = 0$.

All entropies introduced in Definitions 1 and 3 are related to each other. One entropy is helpful in proving properties of another and vice versa. We begin our analysis with

Theorem 1 (The Gibbs inequality) *For any random variables X and Y, with the same set of values I,*

$$h(X\|Y) \geq 0 . \tag{22}$$

Moreover, equality in (22) is attained if and only if $p_X(i) = p_Y(i)$, $i \in I$, that is, X and Y are identically distributed (have the same distribution).

Proof The bound

$$\log x \leq \frac{x-1}{\ln 2} \tag{23}$$

holds for each $x > 0$, with equality if and only if $x = 1$. Denoting $I' = \{i : p_X(i) > 0\}$, we have

$$\sum_i p_X(i) \log \frac{p_Y(i)}{p_X(i)} = \sum_{i \in I'} p_X(i) \log \frac{p_Y(i)}{p_X(i)} \leq \frac{1}{\ln 2} \sum_{i \in I'} p_X(i) \left(\frac{p_Y(i)}{p_X(i)} - 1 \right)$$

$$= \frac{1}{\ln 2} \left(\sum_{i \in I'} p_Y(i) - \sum_{i \in I'} p_X(i) \right) = \frac{1}{\ln 2} \left(\sum_{i \in I'} p_Y(i) - 1 \right) \leq 0. \tag{24}$$

For equality we need (a) $\sum_{i \in I'} p_Y(i) = 1$, i.e., $p_Y(i) = 0$ when $p_X(i) = 0$, and (b) $p_Y(i)/p_X(i) = 1$ for $i \in I'$.

Theorem 1 will be helpful in the proof of Theorem 2.

Theorem 2

1. *If random variable X takes at most m distinct values, then*

$$0 \leq h(X) \leq \log m . \tag{25}$$

   The left-hand equality occurs if and only if X takes a single value and the right-hand equality if and only if X takes m values with equal probabilities: $\mathbf{P}(X = x_i) = \frac{1}{m}$, $i = 1, \ldots, m$.

2. *For every pair of random variables X, Y,*

$$h(X, Y) \leq h(X) + h(Y) , \tag{26}$$

   with equality if and only if X and Y are independent.

Proof Use the Gibbs inequality (Theorem 1)

1. with $\{p(i)\}$ being the distribution of X and $p'(i) = 1/m$, $1 \le i \le m$ and
2. with i being a pair (i_1, i_2) of values of X and Y, $p(i) = p_{X,Y}(i_1, i_2)$ being the joint distribution of X and Y and $p'(i) = p_X(i_1)p_Y(i_2)$, the product of their marginal distributions.

Then

1. $h(X) = - \sum_i p(i) \log p(i) \le \sum_i p(i) \log m = \log m,$
2. $\forall X, Y,$

$$
\begin{aligned}
h(X, Y) &= - \sum_{(i_1, i_2)} p_{X,Y}(i_1, i_2) \log p_{X,Y}(i_1, i_2) \\
&\le - \sum_{(i_1, i_2)} p_{X,Y}(i_1, i_2) \times \log \left(p_X(i_1) p_Y(i_2) \right) \\
&= - \sum_{i_1} p_X(i_1) \log p_X(i_1) - \sum_{i_2} p_Y(i_2) \log p_Y(i_2) \\
&= h(X) + h(Y).
\end{aligned}
\tag{27}
$$

We used here the fact that $\sum_{i_2} p_{X,Y}(i_1, i_2) = p_X(i_1)$, $\sum_{i_1} p_{X,Y}(i_1, i_2) = p_Y(i_2)$.

Lemma 1 (The pooling inequality) *For any $q_1, q_2 \ge 0$, with $q_1 + q_2 > 0$,*

$$
\begin{aligned}
-(q_1 + q_2) \log (q_1 + q_2) &\le -q_1 \log q_1 - q_2 \log q_2 \\
&\le -(q_1 + q_2) \log \frac{q_1 + q_2}{2};
\end{aligned}
\tag{28}
$$

the left-hand equality occurs if and only if $q_1 q_2 = 0$ (i.e., one of the values q_1, q_2 vanishes) and the right-hand equality if and only if $q_1 = q_2$.

Proof Bound (28) is equivalent to

$$
0 \le h \left(\frac{q_1}{q_1 + q_2}, \frac{q_2}{q_1 + q_2} \right) \le \log 2 \, (= 1) .
\tag{29}
$$

Lemma 1 means that when you "glue" together values of a random variable you diminish the corresponding contribution to the entropy. On the other hand, if you "re-distribute" the probabilities making them equal you increase the contribution.

An immediate corollary of Lemma 1 is the following

Theorem 3 *Suppose that a random variable X is a function of random variable Y, i.e., $X = \phi(Y)$. Then*

$$
h(X) \le h(Y),
\tag{30}
$$

with equality if and only if ϕ is invertible.

Proof Indeed, if ϕ is invertible then the probability distributions of X and Y differ only in the order of probabilities, which does not change the entropy. If ϕ "glues" some values y_j then you can repeatedly use the left-hand side of the pooling inequality.

Theorem 4 (The Fano inequality) *Suppose a random variable X takes $m > 1$ values, and one of them has probability $(1 - \varepsilon)$. Then*

$$h(X) \leq G(\varepsilon) + \varepsilon \log (m - 1), \tag{31}$$

where

$$G(\varepsilon) = -\varepsilon \log \varepsilon - (1 - \varepsilon) \log (1 - \varepsilon). \tag{32}$$

Proof Suppose that $p_1 = p(x_1) = 1 - \varepsilon$. Then

$$h(X) = h(p_1, \ldots, p_m) = -\sum_{i=1}^{m} p_i \log p_i = -p_1 \log p_1 - (1 - p_1) \log (1 - p_1)$$

$$+ (1 - p_1) \log (1 - p_1) - \sum_{i=2}^{m} p_i \log p_i$$

$$= h(p_1, 1 - p_1) + (1 - p_1)h \left(\frac{p_2}{1 - p_1}, \ldots, \frac{p_m}{1 - p_1} \right) ; \tag{33}$$

in the right-hand side the first term is $G(\varepsilon)$ and the second does not exceed $\varepsilon \log (m - 1)$.

The Fano inequality shows how the entropy $h(X)$ grows when X is "near" a constant random variable.

Definition 4 Given a triple of random variables, X, Y, and Z, we say that X and Y are conditionally independent, given Z, if

$$p(X = x, Y = y | Z = z) = p(X = x | Z = z)p(Y = y | Z = z) \tag{34}$$

for any z with $p(Z = z) > 0$ and all x and y.

For the conditional entropy you immediately obtain

Theorem 5

1. For all random variables X and Y,

$$0 \leq h(X|Y) \leq h(X); \tag{35}$$

the left-hand equality occurs if and only if X is a function of Y and the right-hand equality if and only if X and Y are independent.

2. For all random variables X, Y, and Z,

$$h(X|Y, Z) \le h(X|Y) \le h(X|\phi(Y)); \tag{36}$$

the left-hand equality occurs if and only if X and Z are conditionally independent given Y and the right-hand equality if and only if X and Y are conditionally independent given $\phi(Y)$.

Proof

1. The left-hand bound in (31) follows from the definition (see (14)). The right-hand bound follows from (35) and (25). The left-hand equality in (35) is equivalent to $h(X, Y) = h(Y)$; the last equality occurs if and only if, with probability 1, the map $(X, Y) \mapsto Y$ is invertible which means that X is a function of Y. The right-hand equality in (35) occurs if and only if $h(X, Y) = h(X) + h(Y)$, i.e., X and Y are independent.

2. For the left-hand bound, use a formula analogous to (16)

$$h(X|Y, Z) = h(X, Z|Y) - h(Z|Y) \tag{37}$$

and an inequality analogous to (16):

$$h(X, Z|Y) \le h(X|Y) + h(Z|Y), \tag{38}$$

with equality if and only if X and Z are conditionally independent given Y. For the right-hand bound, use (i) a formula that is a particular case of (37)

$$h(X|Y, \phi(Y)) = h(X, Y|\phi(Y)) - h(Y|\phi(Y)), \tag{39}$$

together with the remark that

$$h(X|Y, \phi(Y)) = h(X|Y) \tag{40}$$

and (ii) an inequality which is a particular case of (38):

$$h(X, Y|\phi(Y)) \le h(X|\phi(Y)) + h(Y|\phi(Y)), \tag{41}$$

with equality if and only if X and Y are conditionally independent given $\phi(Y)$.

Theorem 6 (The generalized Fano inequality) *For a pair of random variables, X and Y, with values* $x_1, \ldots, x_m$ *and* $y_1, \ldots, y_m$ *if*

$$\sum_{j=1}^{m} p(X = x_j, Y = y_j) = 1 - \varepsilon, \tag{42}$$

then

$$h(X|Y) \leq G(\varepsilon) + \varepsilon \log(m-1), \tag{43}$$

where $G(\varepsilon)$ is defined in (32).

Proof Denoting $\varepsilon_j = p(X \neq x_j | Y = y_j)$, you can write

$$\sum_j p_Y(y_j)\varepsilon_j = \sum_j p(X \neq x_j, Y = y_j) = \varepsilon. \tag{44}$$

By definition of the conditional entropy, the Fano inequality, and concavity of function $G(\varepsilon)$,

$$\begin{aligned} h(X|Y) &\leq \sum_j p_Y(y_j)\Big(G(\varepsilon_j) + \varepsilon_j \log(m-1)\Big) \\ &\leq \sum_j p_Y(y_j)G(\varepsilon_j) + \varepsilon \log(m-1) \leq G(\varepsilon) + \varepsilon \log(m-1). \end{aligned} \tag{45}$$

Most of the properties listed are extended to the case of random vectors.

Theorem 7

1. *For every pair of random vectors, $\mathbf{X}^{(n)} = (X_1, \ldots, X_n)$ and $\mathbf{Y}^{(n)} = (Y_1, \ldots, Y_n)$,*

$$h(\mathbf{X}^{(n)}) = \sum_{i=1}^n h(X_i|\mathbf{X}^{(i-1)}) \leq \sum_{i=1}^n h(X_i), \tag{46}$$

with equality if and only if $X_1, \ldots, X_n$ are independent.
2. *For every pair of random vectors, $\mathbf{X}^{(n)} = (X_1, \ldots, X_n)$ and $\mathbf{Y}^{(n)} = (Y_1, \ldots, Y_n)$,*

$$h(\mathbf{X}^{(n)}|\mathbf{Y}^{(n)}) \leq \sum_{i=1}^n h(X_i|\mathbf{Y}^{(n)}) \leq \sum_{i=1}^n h(X_i|Y_i), \tag{47}$$

with the left-hand equality if and only if $X_1, \ldots, X_n$ are conditionally independent, given $\mathbf{Y}^{(n)}$, and the right-hand equality if and only if, for each $i = 1, \ldots, n$, X_i and $\{Y_r : 1 \leq r \leq n, r \neq i\}$ are conditionally independent, given Y_i.

The proof repeats the previously used arguments.

An immediate corollary of Theorems 2 and 4 is the following:

Theorem 8 *For all random variables X and Y and any function ϕ,*

$$0 \leq \iota(X : \phi(Y)) \leq \iota(X : Y). \tag{48}$$

The left-hand equality occurs if and only if X and $\phi(Y)$ are independent and the right-hand equality if and only if X and Y are conditionally independent, given $\phi(Y)$.

Note that X and Y in Definition 4 and Theorem 9 may be random vectors. In addition, for a pair of random vectors, $\mathbf{X}^{(n)}$ and $\mathbf{Y}^{(n)}$, Theorem 8 yields the following theorem:

Theorem 9

1. *For all random vectors* $\mathbf{X}^{(n)} = (X_1, \ldots, X_n)$ *and* $\mathbf{Y}^{(n)} = (Y_1, \ldots, Y_n)$,

$$\iota(\mathbf{X}^{(n)} : \mathbf{Y}^{(n)}) \geq h(\mathbf{X}^{(n)}) - \sum_{i=1}^{n} h(X_i|\mathbf{Y}^{(n)}) \geq h(\mathbf{X}^{(n)}) - \sum_{i=1}^{n} h(X_i|Y_i). \quad (49)$$

2. *If $X_1, \ldots, X_n$ are independent, then*

$$\iota(\mathbf{X}^{(n)} : \mathbf{Y}^{(n)}) \leq \sum_{i=1}^{n} \iota(X_i : \mathbf{Y}^{(n)}). \quad (50)$$

Observe that the right-hand side of (50) is always

$$\geq \sum_{i=1}^{n} \iota(X_i : Y_i). \quad (51)$$

An important special case is where the connection between entry X_i and vector $\mathbf{Y}^{(n)}$ is through Y_i, the corresponding entry Y_i, that is, $\forall$ sample values x of X and $y^{(n)} = (y_1, \ldots, y_n)$ of $\mathbf{Y}^{(n)}$:

$$\mathbf{P}(X_i = x|\mathbf{Y}^{(n)} = y^{(n)}) = \mathbf{P}(X_i = x|Y_i = y_i). \quad (52)$$

In this case, $\iota(X_i : \mathbf{Y}^{(n)}) = \iota(X_i : Y_i)$, and bound (50) becomes

$$\iota(\mathbf{X}^{(n)} : \mathbf{Y}^{(n)}) \leq \sum_{i=1}^{n} \iota(X_i : Y_i). \quad (53)$$

2 Source Coding

Encode in the 15th century, decode in the 21st.
(The Da Vinci First Coding Principle)
From the series 'Theorems never learned'

2.1 Introduction to Coding

A typical scheme used in information transmission is as follows:

| A message source | $\to$ | An encoder | $\to$ | A channel | $\to$ | A decoder | $\to$ | A destination |

Consider the following example:

- A message source: A Cambridge college choir.
- An encoder: A BBC recording unit. It translates the sound to a binary array and writes it to a CD or DVD track. A disk is then produced and put on the market.
- A channel: A customer in Australia buys a CD or DVD on Amazon.com; the disk is mailed to him/her and received by the customer who occasionally plays it. The channel is subject to "noise": possible damage (mechanical, electrical, chemical, etc.) incurred during transportation and use.
- A decoder: A CD or DVD player in Australia.
- A destination: An audience in Australia.
- The goal: To ensure high-quality sound despite possible damage.

In fact, a CD (and to a lesser degree a DVD) can sustain damage done by a needle while making a neat hole in it or by a tiny drop of acid (you are not encouraged to make such an experiment!)

In technical terms, typical goals of information transmission are

- fast encoding of information,
- reliable transmission of encoded messages,
- effective use of the channel available (i.e., maximum transfer of information per unit time or space),
- fast decoding, and
- correcting errors (as many as possible) introduced by noise in the channel.

As usual, these goals contradict each other, and one has to find an optimal solution. Shannon's coding theorems provide an effective quantitative assessment of what can and what cannot be achieved.

The first operation in the above scheme is *encoding*. The aims of encoding are

- compressing data to reduce redundant information contained in a message,
- protecting data from unauthorized users, and
- enabling errors to be corrected.

So, we start by studying *sources* and *encoders*. A standard model is where a source emits a sequence of letters,

$$u_1 u_2 \ldots u_n \ldots , \tag{54}$$

where $u_j \in I$, $I(= I_m)$ is an m-element set $\{1, \ldots, m\}$ (a source alphabet). For instance, it may be a literary English text: here $m = 26 + 7$, 26 letters plus 7 punctuation symbols: ., :;–(). (Sometimes one adds ? ! ' ' and ".) A telegraph English corresponds to $m = 27$. In the above example of DVD recording it is an analog signal quantified according to a grid of values of relevant parameters (the spectrum of sound and light harmonics and their amplitudes); the value of m reaches several thousands. The coding process in this example is going in parallel with the process of message emission.

A common approach is to consider (54) as a *sample* from a random source, i.e., a sequence of random variables

$$U_1, U_2, \ldots, U_n, \ldots \qquad (55)$$

and try to develop a theory for a reasonable class of such sequences.

Example 5

1. The simplest example of a random source is a sequence of independent, identically distributed (IID) random variables:

$$\mathbf{P}(U_1 = u_1, U_2 = u_2, \ldots, U_k = u_k) = \prod_{j=1}^{k} p(u_j), \qquad (56)$$

   where $p(u) = \mathbf{P}(U_j = u), u \in I$, is the marginal distribution of a single variable. A random source with IID symbols is often called a Bernoulli source.
   A particular case where $p(u)$ does not depend on $u \in U$ (and hence equals $1/m$) corresponds to the equiprobable Bernoulli source.
   A good intuitive model of a Bernoulli source (which Shannon used a lot) is a sequence of outcomes of a coin toss or the die cast; the coin and the die can be biased or unbiased.

2. The next example is a Markov source where the symbols form a Markov chain (M.c.):

$$\mathbf{P}(U_1 = u_1, U_2 = u_2, \ldots, U_k = u_k) = \lambda(u_1) \prod_{j=1}^{k-1} P(u_j, u_{j+1}), \qquad (57)$$

   where, following a Cambridge tradition, $\lambda(u)$ are initial probabilities, whereas $P(u, u') = P(u'|u)$ are probabilities of transition:

$$\lambda(u) = \mathbf{P}(U_1 = u), \quad P(u, u') = \mathbf{P}(U_{i+1} = u'|U_j = u), \quad u, u' \in I. \qquad (58)$$

   A Markov source is called stationary if

$$\mathbf{P}(U_j = u) = p(u), \quad j \geq 1, \qquad (59)$$

   i.e., $p = \{p(u)\}$ is an invariant vector for matrix $P = \{P(u, u')\}$: $pP = p$.

3. A "degenerate" example of a Markov source is where a source emits repeated symbols. Here,

$$P(U_1 = U_2 = \cdots = U_k = u) = q(u), \ u \in I,$$
$$P(U_k \neq U_{k'}) = 0, \ 1 \leq k < k', \tag{60}$$

where $0 \leq q(u) \leq 1$ and $\sum_{u \in I} q(u) = 1$. $q(u)$ is the probability of string $uu \ldots u \ldots$. This example is in a sense opposite to Bernoulli.

An (initial) piece of sequence (54)

$$\mathbf{u}^{(n)} = u_1 u_2 \ldots u_n \tag{61}$$

is called a (source) sample n-string, or n-word (with digits from I), and is treated as a "message." Correspondingly, one considers a random n-string

$$\mathbf{U}^{(n)} = (U_1, U_2, \ldots, U_n) . \tag{62}$$

An encoder (or coder) uses an alphabet $J(= J_a) = \{0, 1, \ldots, a - 1\}$; typically $a < m$ (or even $a \ll m$); in many cases $a = 2$ (a binary coder).

A *code* (also *coding*, or *encoding*) is a map, f, that takes a string $\mathbf{u}^{(n)}$ into a string $\mathbf{x}^{(N)} = f\left(\mathbf{u}^{(n)}\right)$ where $\mathbf{x}^{(N)} = (x_1, \ldots, x_N)$ and digits $x_j \in J$:

$$f : \mathbf{u}^{(n)} \in I^n \mapsto \mathbf{x}^{(N)} = f\left(\mathbf{u}^{(n)}\right) \in J^N. \tag{63}$$

String $f\left(\mathbf{u}^{(n)}\right) \in J^N$ is called a *codeword* for message $\mathbf{u}^{(n)}$ under code f.

In this definition, a lot of things have been left unspecified, viz.

1. Is the length N of $\mathbf{x}^{(N)}$ fixed for a given n or does it vary with $\mathbf{u}^{(n)}$ (fixed length versus variable length encoding)? In the latter case, (63) should be modified to

$$f : \mathbf{u}^{(n)} \in I^n \mapsto \mathbf{x}^{(N)} = f\left(\mathbf{u}^{(n)}\right) \in J^* = \sqcup_{N \geq 1} J^N , \tag{64}$$

where $\sqcup_{N \geq 1} J^N$ is a disjoint union of Cartesian products J^N.
2. How does f act? Digit-wise (where $f\left(\mathbf{u}^{(n)}\right)$ is a concatenation (denoted by the symbol $\vee$) $\left(f^{(1)}(u_1) \vee \cdots \vee f^{(1)}(u_n)\right)$ of shorter components $f^{(1)}(u_j)$ produced from a single-letter encoding map $f^{(1)} : u \in I \mapsto f^{(1)}(u) \in J^*$) or block-wise, where the same idea is applied to "blocks" (parts of $\mathbf{u}^{(n)}$) rather than to single letters?
3. Do we need f to be one-to-one? Or can we use a "random" map where push-forward images can have several (and possibly many) pull-back inverse images?

All these questions are important but should not be over-stated: It was Shannon's genial insight into both the theory and the practice of the problem that enabled him to carry through. The way out is to ask what do we want? We may go back to the above answers (1)–(3) and then think of examples.

For instance, consider 100 outputs of a very biased coin with probability 0.01 of providing heads (which was not a big deal even in Shannon's time). Here, instead of listing the whole string "$TTTTTTTT...THT...$," we could summarize the output by saying "096881," to represent the fact that the heads occurred in three places 9, 68, and 81. (The initial position is encoded as 00 and the final position as 99.) Or even more boldly, we may refuse to list the positions of all heads and give only two first positions where H occurred. What do we gain? The original task was of being able to list all possible 2^{100} outcomes (which is impossible even for a modern computer; cf. the Avogadro number 6.022×10^{23}). The "reduced" task is to list $\left(100 + 100^2\right)/2 = 5,050$ outcomes: again not a big deal. What do we lose? In the case where there are three or more heads generated, we will be wrong about the whole sequence. What is the probability of this unpleasant event? By what is called the Chebyshev inequality, it does not exceed $1/200$. (In fact, it is considerably smaller.) Are we prepared to take such a risk? This is the essence of *data compression* and suggests an obvious question: given a series of n independent outputs from a random variable, by what factor do we expect to be able to compress the output?

Now it remains to put this into mathematical language, which is done in the next section.

According to an opinion that gains popularity, Shannon gave a brilliant example where knowing too much of math may be not productive. He created his theory at a time where another giant, N. Wiener, actively propagandized his own views on how the academic, technological, and industrial development of present and future should be managed, through what he called cybernetics. An important part of cybernetics is information exchange. Wiener was instinctively a continuous-world thinker, whereas Shannon was distinctly a discrete-world man, and that determined the difference between their approaches. Wiener advocated mainly an analog information processing and control (although at an early stage he wrote a memorandum explaining why digital approach is preferable), whereas Shannon on the whole was firmly on the digital side. Analog principles and devices were a big splash throughout the 1940s to early 1970s but were clearly losing ground to their digital counterparts at the later period (which continues now). As F. Dyson puts it: "Electronic engineers learned information theory, the gospel according to Shannon, as part of their basic training, and cybernetics was forgotten."

2.2 Shannon's First Coding Theorem

To be able to compress data efficiently, we would like to find (relatively) small sets that occur with high probability.

Definition 5 For a given $R > 0$, a source generating a random string $\mathbf{U}^{(n)} = (U_1, \ldots, U_n)$ with digits from an alphabet I is said to be *reliably encodable at rate R* if, for any n you can find a set $A_n \in I^n$ such that

$$\#(A_n) \leq 2^{nR} \quad \text{and} \quad \lim_{n \to \infty} \mathbf{P}(\mathbf{U}^{(n)} \in A_n) = 1, \tag{65}$$

where $\#(X)$ denotes the cardinality of the set X.

The idea here is that we can label the members of the set A_n with a label of length nR, and so since we are "nearly always" in A_n, the average length of a compressed string will be close to nR.

Definition 6 The *information rate H* of a given source is the smallest reliable encoding rate

$$H = \inf[R : R \text{ is reliable}] . \tag{66}$$

Theorem 10 *For a source with alphabet of size m,*

$$0 \leq H \leq \log m , \tag{67}$$

both bounds being attainable.

Proof The left-hand inequality in (67) trivially follows from Definition 6. On the other hand, for any A_n, the size $\#(A_n) \leq \#(I^n) = m^n = 2^{n \log m}$, hence the right-hand side of the inequality. The left-hand equality is attained for a source where always $x_1 = x_2 = x_3 \cdots$. Then A_n contains at most m strings, which is eventually less than 2^{nR} for any $R > 0$.

The right-hand equality is attained for a source with uniformly, independent and identically distributed x_i: in this case for any set of strings $\mathbf{P}(A_n) = (1/m^n)\#(A_n)$, which goes to zero when $\#(A_n) \leq 2^{nR}$ and $R < \log m$.

Although the quantity H seems useful, it also seems hard to write down. We will prove results about it by first introducing a quantity D_n which tells us about H, and then a quantity ξ_n which tells us about D_n. Then, ξ_n is easier to find, and from it we deduce the value of H.

Definition 7 Define the subset maximum by

$$D_n(R) := \max_{A:\#(A)\leq 2^{nR}} \mathbf{P}(\mathbf{U}^{(n)} \in A) . \tag{68}$$

Lemma 2 *For any $\varepsilon > 0$, the information rate H satisfies*

$$\lim_{n\to\infty} D_n(H + \varepsilon) = 1, \quad and, \text{ if } H > 0, \ D_n(H - \varepsilon) \nrightarrow 1 . \tag{69}$$

Proof By definition, $H + \varepsilon$ is a reliable encoding rate. There must therefore exist a sequence of sets $A_n \subset I^n$, with $\#(A_n) \leq 2^{nR}$ and $\mathbf{P}(\mathbf{U}^{(n)} \in A_n) \to 1$, as $n \to \infty$. Since $D_n(R) \geq \mathbf{P}(\mathbf{U}^{(n)} \in A_n)$, $D_n(R) \to 1$.

Alternatively, for any $H > 0$, we can pick ε such that $H - \varepsilon > 0$, but not reliable. Then choosing C_n to be the sets for which the maximum in (68) is attained, the $\#(C_n) \leq 2^{nR}$, but $\mathbf{P}(C_n) \nrightarrow 1$.

Given a string $\mathbf{u}^{(n)} = u_1 \dots u_n$, consider its log-likelihood per source letter:

$$\xi_n(\mathbf{u}^{(n)}) = -\frac{1}{n} \log_+ p_n(\mathbf{u}^{(n)}) , \tag{70}$$

where $p_n(\mathbf{u}^{(n)})$ is the probability assigned to string $\mathbf{u}^{(n)}$:

$$p_n(\mathbf{u}^{(n)}) = \mathbf{P}\left(\mathbf{U}^{(n)} = \mathbf{u}^{(n)}\right). \tag{71}$$

Here, and below, $\log_+ x = \log x$, if $x > 0$, and $= 0$, if $x = 0$. For a random string $\mathbf{u}^{(n)}$,

$$\xi_n(\mathbf{u}^{(n)}) = -\frac{1}{n}\log_+ p_n(\mathbf{u}^{(n)}) \tag{72}$$

is a random variable.

Lemma 3 *For any R and any $\varepsilon > 0$,*

$$\mathbf{P}(\xi_n \leq R) \leq D_n(R) \leq \mathbf{P}(\xi_n \leq R + \varepsilon) + 2^{-n\varepsilon}. \tag{73}$$

Proof For simplicity, we leave out the superscript (n). Define

$$\begin{aligned}B_n &= \{u \in I^n : p_n(u) \geq 2^{-nR}\} = \{u \in I^n : -\log p_n(u) \leq nR\}\\ &= \{u \in I^n : \xi_n(u) \leq R\}\,. \end{aligned} \tag{74}$$

Then

$$1 \geq \mathbf{P}(U \in B_n) = \sum_{u \in B_n} p_n(u) \geq 2^{-nR}(\#(B_n)), \tag{75}$$

and so $\#(B_n) \leq 2^{nR}$. Thus,

$$D_n(R) = \max_{\#(A) \leq 2^{nR}} \mathbf{P}(U \in A) \geq \mathbf{P}(U \in B_n) = \mathbf{P}(\xi_n \leq R), \tag{76}$$

which proves the left-hand side in (73).

On the other hand, there exists $C_n \subseteq I^n$ where the maximum in (68) is attained. For such a C_n,

$$\begin{aligned}D_n(R) = \mathbf{P}(U \in C_n) &= \mathbf{P}(U \in C_n, \xi_n \leq R + \varepsilon) + \mathbf{P}(U \in C_n, \xi_n > R + \varepsilon)\\ &\leq \mathbf{P}(\xi_n \leq R + \varepsilon) + \sum_{u \in C_n} \mathbf{1}\left(p_n(u) < 2^{-n(R+\varepsilon)}\right)p_n(u)\\ &\leq \mathbf{P}(\xi_n \leq R + \varepsilon) + 2^{-n(R+\varepsilon)}\#(C_n)\\ &\leq \mathbf{P}(\xi_n \leq R + \varepsilon) + 2^{-n(R+\varepsilon)}2^{nR} = \mathbf{P}(\xi_n \leq R + \varepsilon) + 2^{-n\varepsilon}, \end{aligned} \tag{77}$$

as required.

Definition 8 We say that a sequence of random variables $\{\eta_n\}$ converge in probability to a constant r if for any $\varepsilon > 0$,

$$\lim_{n \to \infty} \mathbf{P}\left(|\eta_n - r| \geq \varepsilon\right) = 0. \tag{78}$$

Convergence in probability is denoted by $\eta_n \overset{\mathbf{P}}{\longrightarrow} r$.

Convergence in probability is established in the law of large numbers (see Theorem 13).

Theorem 11 (Shannon's first coding theorem) *If ξ_n converges in probability to a constant γ then $\gamma = H$, the information rate of a source.*

Proof Let $\xi_n \overset{\mathbf{P}}{\longrightarrow} \gamma$. Since $\xi_n \geq 0$, its limit $\gamma \geq 0$. By Lemma 3, $\forall \varepsilon > 0$,

$$
\begin{aligned}
D_n(\gamma + \varepsilon) &\geq \mathbf{P}(\xi_n \leq \gamma + \varepsilon) \geq \mathbf{P}(\gamma - \varepsilon \leq \xi_n \leq \gamma + \varepsilon) \\
&= \mathbf{P}(|\xi_n - \gamma| \leq \varepsilon) = 1 - \mathbf{P}(|\xi_n - \gamma| > \varepsilon),
\end{aligned}
\tag{79}
$$

which converges to 1 as $n \to \infty$. We deduce that $H \leq \gamma$.

In particular, if $\gamma = 0$ then $H = 0$. If $\gamma > 0$, by the opposite bound in Lemma 3,

$$
\begin{aligned}
D_n(\gamma - \varepsilon) &\leq \mathbf{P}(\xi_n \leq \gamma - \varepsilon/2) + 2^{-n\varepsilon/2} \\
&\leq \mathbf{P}(|\xi_n - \gamma| \geq \varepsilon/2) + 2^{-n\varepsilon/2},
\end{aligned}
\tag{80}
$$

which tends to $2^{-n\varepsilon/2}$. By Lemma 2, $H \geq \gamma$, and the result follows.

Remark 1 Convergence $\xi_n \overset{\mathbf{P}}{\longrightarrow} \gamma = H$ is equivalent to the following asymptotic equipartition property (AEP): $\forall \varepsilon > 0$,

$$
\lim_{n \to \infty} \mathbf{P}\left(2^{-n(H+\varepsilon)} \leq p_n(\mathbf{u}^{(n)}) \leq 2^{-n(H-\varepsilon)}\right) = 1.
\tag{81}
$$

In fact,

$$
\begin{aligned}
\mathbf{P}\left(2^{-n(H+\varepsilon)} \leq p_n(\mathbf{u}^{(n)})\right. &\left.\leq 2^{-n(H-\varepsilon)}\right) \\
&= \mathbf{P}\left(H - \varepsilon \leq -\frac{1}{n}\log p_n(\mathbf{u}^{(n)}) \leq H + \varepsilon\right) \\
&= \mathbf{P}(|\xi_n - H| \leq \varepsilon) \\
&= 1 - \mathbf{P}(|\xi_n - H| > \varepsilon).
\end{aligned}
\tag{82}
$$

In other words, for any $\varepsilon > 0$ there exists $n_0(\varepsilon)$ such that, for any $n > n_0(\varepsilon)$, we can identify a typical set T_n with the properties that

1. $\mathbf{P}\left(\mathbf{u}^{(n)} \notin T_n\right) < \varepsilon$
2. for any $\mathbf{u}^{(n)} \in T_n$, the $2^{-n(H+\varepsilon)} \leq \mathbf{P}\left(\mathbf{U}^{(n)} = \mathbf{u}^{(n)}\right) \leq 2^{-n(H-\varepsilon)}$

For a source with the AEP, encode the typical strings with codewords of length $n(H + \varepsilon)$, and the rest however you like. You will then have an effective encoding rate $H + o(1)$ bits/source letter.

Theorem 12 *For a Bernoulli source, the information rate equals the entropy of a single letter, that is*

$$H = h(U_j) = - \sum_{u \in I} p(u) \log p(u). \tag{83}$$

Proof For an IID sequence $U_1, U_2, \ldots,$

$$p_n(\mathbf{u}^{(n)}) = \prod_{i=1}^{n} p(u_i), \text{ hence, } -\log p_n(u) = \sum_i -\log p(u_i). \tag{84}$$

For a random string $\mathbf{u}^{(n)} = (U_1, \ldots, U_n),$

$$\xi_n = -\log p_n(\mathbf{u}^{(n)}) = - \sum_{i=1}^{n} \log p(U_i) = \sum_{i=1}^{n} \sigma_i, \tag{85}$$

where random variables $\sigma_i = -\log p(U_i)$ are IID (we refer to this as IID random variables (RVs)). Observe that $\mathbb{E}\sigma_i = -\sum_j p(j) \log p(j) = h$ and so $\mathbb{E}\xi_n = \frac{1}{n} \sum_{i=1}^{n} \mathbb{E}\sigma_i = h$. The following theorem yields convergence in probability $\xi_n \xrightarrow{\mathbf{P}} h$ as an immediate corollary.

Theorem 13 (Weak law of large numbers for IID RVs) *Consider any sequence of IID RVs $X_1, X_2, \ldots$ with finite mean $\mathbb{E}X_i = \mu$ and finite variance $\mathbb{E}(X_i - \mu)^2 = \sigma^2$. Then for any $\varepsilon > 0$,*

$$\lim_{n \to \infty} \mathbf{P} \left(\left| \frac{1}{n} \sum_{i=1}^{n} X_i - \mu \right| \geq \varepsilon \right) = 0. \tag{86}$$

Proof By Chebyshev, for any random variable η and any $\varepsilon > 0$,

$$\mathbf{P}(\eta \geq \varepsilon) \leq \frac{1}{\varepsilon^2} \mathbb{E}\eta^2. \tag{87}$$

Applying this to the left-hand side of (86) gives

$$\mathbf{P} \left(\left| \frac{1}{n} \sum_{i=1}^{n} X_i - \mu \right| \geq \varepsilon \right) \leq \frac{1}{\varepsilon^2} \mathbb{E} \left(\frac{1}{n} \sum_{i=1}^{n} X_i - \mu \right)^2 = \frac{1}{\varepsilon^2 n^2} \mathbb{E} \left(\sum_{i=1}^{n} (X_i - \mu) \right)^2. \tag{88}$$

Since the sum $\sum_{i=1}^{n} (X_i - \mu)$ has mean zero, the right-hand side of (88) is nothing but the variance:

$$\frac{1}{\varepsilon^2 n^2} \text{Var} \left(\sum_{i=1}^{n} (X_i - \mu) \right) = \frac{1}{\varepsilon^2 n^2} \sum_{i=1}^{n} \text{Var}(X_i - \mu) = \frac{1}{n\varepsilon^2} \text{Var } X_1, \tag{89}$$

which goes to 0 when $n \to \infty$. Theorem 13 is proved, and so is Theorem 12.

Note that we refer to this as the weak law of large numbers (WLLNs), because of the (relatively) weak sense in which convergence is proved. Stronger laws are possible: for example, if $\mathbb{E}|X_i| < \infty$, then for IID RVs $X_1, X_2, \ldots$

$$\frac{\sum_{i=1}^{n} X_i}{n} \to (\mathbb{E}X_i) \text{ almost surely.} \tag{90}$$

("Almost surely" means "with probability 1"). However, for our purposes, the weak law will be enough.

We finish this section with the following theorem which is given without a proof:

Theorem 14 *For a Markov source which is an irreducible aperiodic MC, the information rate equals the conditional entropy in equilibrium, that is*

$$
\begin{aligned}
H = h^{\text{eq}}(U_{j+1}|U_j) &= - \sum_{u,v \in I} \pi^{\text{eq}}(u) P(u, v) \log P(u, v) \\
&= - \sum_{u,v \in I} \pi^{\text{eq}}(u) P(v|u) \log P(v|u).
\end{aligned}
\tag{91}
$$

Here $\pi^{\text{eq}}(x)$ are the (uniquely determined) invariant (stationary) probabilities forming an invariant vector $\pi = (\pi(u), u \in I)$: $\pi P = \pi$.

3 Channel Coding

3.1 Introduction to Channels

In this section, we discuss problems of information transmission. That is, we consider whether we can reliably send information through a noisy communication channel. For example, consider two characters, Alice and Bob, A talking to B on their mobile phones.

1. A source (A's voice) generates a signal converted by the microphone into a series of 0s and 1s. This is an initial stage of encoding (or compressing) information, which is highly accurate. We consider this binary sequence as a message $\mathbf{u}^{(n)}$. The source alphabet $I = \{0, 1\}$ and n is about 10,000–30,000.
2. We (a transmitter) use an encoder (a function f, random or deterministic, introduced in formulae (63) and (64)) to convert these 0s and 1s into another series of 0s and 1s. The key will be to make a clever choice of function, to somehow "pad out" the information in the signal and protect it from errors.
3. Next, the output of the function is sent to B's mobile phone, and errors are introduced in transmission. For instance, the mobile phone signal will be sent via masts and base stations to another mobile phone. Errors can be introduced by a variety of processes, but from our point of view, that will not matter – we will

not differentiate between errors introduced by cosmic rays, software errors at the base station, pigeons flying into the mast, and so on). This is what we refer to as the communication channel.

4. B's phone now attempts to decode the signal. The decoder will, roughly speaking, attempt to apply the inverse of the encoding function f, to recover something close to the original message. However, we will need a decoding function that is robust to the introduction of these errors.

Many other processes can be regarded in the same way, and the model of a channel is in fact very abstract.

When we use a channel we hope to

- correct as many errors as possible (deal with as noisy a channel as possible),
- make effective use of the channel available (maximum transfer of information per unit time), and
- quickly encode and decode information (a mobile phone conversation must take place in real time).

As was agreed, the source emits a random text $U_1, U_2, \ldots$, where $U_i \in I$, and we encode a message $u^{(n)}$ by a codeword $x^{(N)}$, by using a code $f(= f_n) : I^n \to J^N$ (often $J = \{0, 1\}$). We stress that the code is known to both the transmitter and the receiver. The key is to understand the relationship between n and N.

A channel is understood as any physical process subject to "noise" which distorts the messages transmitted: a message at the output differs in general from the message at the input. Formally, a channel is characterized by a conditional distribution

$$\mathbf{P}_{\text{channel}} \left(\text{receive word } \mathbf{y}^{(N)} \middle| \text{codeword } \mathbf{x}^{(N)} \text{ sent} \right) , \tag{92}$$

which we again suppose is known to both sender and receiver. These conditional probabilities define a channel. We will often deal with the simplest case.

Definition 9 We say that a channel is memoryless if

$$\mathbf{P}_{\text{channel}} \left(\mathbf{y}^{(N)} \middle| \mathbf{x}^{(N)} \right) = \prod_{i=1}^{N} P_{\text{channel}}(y_i | x_i), \tag{93}$$

where $\mathbf{y}^{(N)} = y_1 \ldots y_N$, $\mathbf{x}^{(N)} = x_1 \ldots x_N$ (the brackets and the commas in the notation $(x_1, \ldots, x_N)$ and $(y_1, \ldots, y_N)$ will henceforth be omitted. Here, $P_{\text{channel}}(y|x)$ are symbol-to-symbol channel probabilities ($P_{\text{channel}}(y|x)$ is the conditional probability to have symbol y at the output of the channel given that symbol x has been sent). We think of P_{channel} as a matrix, often called the channel matrix.

If the rows of the channel matrix P are permutations of each other, we say that the channel is symmetric.

In particular, we often consider the memoryless binary symmetric channel, where $x, y \in \{0, 1\}$, so that $\{P_{\text{channel}}(y|x)\}$ is a 2×2 transition probability matrix of the form

$$P_{\text{channel}} = \begin{pmatrix} 1-p & p \\ p & 1-p \end{pmatrix} \tag{94}$$

and p is called the row error probability (or the symbol error probability).

3.2 Transmission Rates and Capacity

We want to introduce a decoding rule $\widehat{f}_N : J^N \to I^n$ so that the overall probability of error

$$
\begin{aligned}
\varepsilon &= \sum_{\mathbf{u}^{(n)}} \mathbf{P}\left(\widehat{f}(y^{(N)}) \neq \mathbf{u}^{(n)}, \mathbf{u}^{(n)} \text{ emitted by the source}\right) \\
&= \sum_{\mathbf{u}^{(n)}} \mathbf{P}_{\text{source}}\left(\mathbf{U}^{(n)} = \mathbf{u}^{(n)}\right) \mathbf{P}_{\text{channel}}\left(\widehat{f}_N(\mathbf{y}^{(N)}) \neq \mathbf{u}^{(n)} \middle| f_n(\mathbf{u}^{(n)}) \text{ sent}\right)
\end{aligned}
\tag{95}
$$

is small. In fact, we will try to have quantity (95) tending to zero as $n \to \infty$. The construction is based on two facts:

1. For a source with the AEP the number of distinct n-strings emitted is $2^{n(H+o(1))}$ where $H \leq \log m$ is the information rate of the source. That is, we do not need to encode $m^n = 2^{n \log m}$ messages, but rather only $2^{n(H+o(1))}$ which may be considerably less. That implies a codeword length of $\lceil nH \rceil$.

2. We increase the length of the codewords used from $\lceil nH \rceil$ to $\lceil \overline{R}^{-1} nH \rceil$, for some constant $\overline{R}$. That is, we introduce redundancy in code f_n, hoping to reduce the overall error probability (95). We want to have the factor of increase $\overline{R}^{-1}$ as small as possible, that is, $\overline{R}$ as large as possible.

As the codeword length is a crucial parameter, write N instead of $\overline{R}^{-1}Hn$ and $\overline{R}N$ instead of Hn. It is convenient to consider a "typical" set $\mathcal{U}_N$ of distinct strings emitted by the source, with $\#\left(\mathcal{U}_N\right) = 2^{N(\overline{R}+o(1))}$.

Definition 10 Value $\overline{R} \in (0, 1)$ is a reliable transmission rate (for a given channel) if, given that the source strings take equiprobable values from a set $\mathcal{U}_N$ of size $2^{N(\overline{R}+o(1))}$, there exists an encoding rule $f_N : \mathcal{U}_N \to \mathcal{X}_N \subseteq \{0, 1\}^N$ and a decoding rule $\widehat{f}_N : \{0, 1\}^N \to \mathcal{U}_N$ with error probability

$$
\sum_{\mathbf{u}^{(n)} \in \mathcal{U}_N} \frac{1}{\#\left(\mathcal{U}_N\right)} \mathbf{P}_{\text{channel}}\left(\widehat{f}_N(\mathbf{y}^{(N)}) \neq \mathbf{u}^{(n)} \middle| f_N(\mathbf{u}^{(n)}) \text{ sent}\right) \tag{96}
$$

tending to zero as $N \to \infty$.

Equivalently, that is, we look for

$$\overline{R} \sim \frac{\log \#\left(\mathcal{U}_N\right)}{N} \, . \tag{97}$$

Definition 11 The channel capacity is defined as the supremum

$$C = \sup [\, \overline{R} : \text{ is a reliable transmission rate }] \, . \tag{98}$$

The reason for considering the equiprobable distribution on $\mathcal{U}_N$ in Definition 10 is that it is the worst case. The argument uses the idea of randomizing, which proves to be very powerful in channel calculations.

In what follows, whenever appropriate, we will omit the subscript N and the superscript (N).

Theorem 15 *Fix a channel, by choosing $\mathbf{P}_{\text{channel}}$ in (92). Fix a set $\mathcal{U}$ of source strings and denote by $\varepsilon(\mathbf{P})$ the overall error probability (96) for the source with probability distribution $\mathbf{P}$ on $\mathcal{U}$, minimized over all encoding and decoding rules. Then*

$$\varepsilon(\mathbf{P}) \le \varepsilon(\mathbf{P}^{\text{eq}}) \, , \tag{99}$$

where $\mathbf{P}^{\text{eq}}$ is equidistribution over $\mathcal{U}$.

Proof Consider particular encoding and decoding rules, f and $\widehat{f}$, and let a string $\mathbf{u} \in \mathcal{U}$ have probability $\mathbf{P}(\mathbf{u})$. Define the error probability when $\mathbf{u}$ is emitted to be

$$\beta(\mathbf{u}) = \sum_{\mathbf{y}: \widehat{f}(\mathbf{y}) \ne \mathbf{u}} \mathbf{P}_{\text{channel}}(\mathbf{y} | f(\mathbf{u})) \, . \tag{100}$$

The overall error probability therefore equals $\varepsilon(\mathbf{P}, f, \widehat{f}) = \sum_{\mathbf{u} \in \mathcal{U}} \mathbf{P}(\mathbf{u})\beta(\mathbf{u})$.

Now, consider a permutation π of the codewords (that is, instead encode $\mathbf{u}$ by $f(\mathbf{v})$ where $\mathbf{v} = \pi(\mathbf{u})$). The new overall error probability $\varepsilon(\pi) = \sum_{\mathbf{u} \in \mathcal{U}} \mathbf{P}(\mathbf{u})\beta(\pi(\mathbf{u}))$. In the case of $\mathbf{P} = \mathbf{P}^{\text{eq}}$ (equidistribution) $\varepsilon(\pi)$ does not depend on π and equals $\overline{\varepsilon} = \dfrac{1}{\#\left(\mathcal{U}\right)} \sum_{\mathbf{u} \in \mathcal{U}} \beta(\mathbf{u}) = \varepsilon(\mathbf{P}^{\text{eq}}, f, \widehat{f})$.

For each probability distribution $\mathbf{P}$ there exists π such that $\varepsilon(\pi) \le \overline{\varepsilon}$. In fact, take a random permutation, Π, equidistributed among all permutations of codewords. Then

$$\min_{\pi} \varepsilon(\pi) \le \mathbb{E}\varepsilon(\Pi) = \mathbb{E}\sum_{\mathbf{u} \in \mathcal{U}} \mathbf{P}(\mathbf{u})\beta(\Pi\mathbf{u})$$

$$= \sum_{\mathbf{u} \in \mathcal{U}} \mathbf{P}(\mathbf{u})\mathbb{E}\beta(\Pi\mathbf{u}) = \sum_{\mathbf{u} \in \mathcal{U}} \mathbf{P}(\mathbf{u})\frac{1}{\#\left(\mathcal{U}\right)}\sum_{\mathbf{v} \in \mathcal{U}} \beta(\mathbf{v}) = \overline{\varepsilon} \, . \tag{101}$$

Hence, given any f and $\widehat{f}$, you can find new encoding and decoding rules with overall error probability $\le \varepsilon(\mathbf{P}^{\text{eq}}, f, \widehat{f})$. Minimizing over f and $\widehat{f}$ leads to (99).

3.3 Simple Decoding Rules

It is now time to discuss possible decoding rules. As was noted before, a decoding rule (or a decoder) is a map $\widehat{f}_N : \{0, 1\}^N \to \mathcal{U}$.

Physically speaking, a decoding rule is given by fixing, for each codeword $\mathbf{x}_i \in \{0, 1\}^N$, a set $A(\mathbf{x}_i) \subset \{0, 1\}^N$, so that $A(\mathbf{x}_i)$ and $A(\mathbf{x}_j)$ are disjoint for distinct codewords $\mathbf{x}_i$ and $\mathbf{x}_j$, and the union $\bigcup_i A(\mathbf{x}_i)$ gives the whole $\{0, 1\}^N$. Given that $\mathbf{y} \in A(\mathbf{x})$, we decode by $\widehat{f}_N(\mathbf{y}) = \mathbf{x}$. See Fig. 2 for a schematic illustration of this.

Although in the definition of the channel capacity we assume that the source messages are equidistributed (which gives the worst case) of course the source does not always follow this assumption. We may need to distinguish between two situations:

1. The receiver knows the probability distribution $\mathbf{P}_{source}$ and hence the probabilities $p_N(\mathbf{x})$, $\mathbf{x} \in \{0, 1\}^N$, of the codewords.
2. The receiver does not know $p_N(\mathbf{x})$.

Definition 12 Two natural decoding rules are as follows

1. The ideal observer rule: Decode a received word $\mathbf{y}^{(N)}$ by a codeword $\mathbf{x}^\star$ that maximizes the posterior probability

$$\mathbf{P}\left(\mathbf{x} \text{ sent} \,\middle|\, \mathbf{y} \text{ received}\right) = \frac{p_N(\mathbf{x})\mathbf{P}_{channel}(\mathbf{y}|\mathbf{x})}{p_\mathbf{y}(\mathbf{y})}, \tag{102}$$

where

$$p_\mathbf{y}(\mathbf{y}) = \sum_{\mathbf{u} \in \mathcal{U}_N} \mathbf{P}_{source}(\mathbf{u})\mathbf{P}_{channel}(\mathbf{y}|f_N(\mathbf{u})). \tag{103}$$

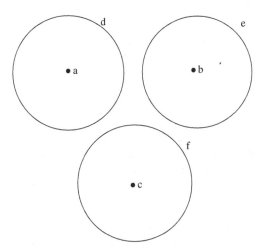

Fig. 2 Decoding rule described in terms of sets around each codeword

2. The maximum likelihood rule: Decode a received word $\mathbf{y}$ by codeword $\mathbf{x}^*$ that maximizes the prior probability

$$\mathbf{P}_{channel}(\mathbf{y}|\mathbf{x}). \qquad (104)$$

Theorem 16 *Suppose that an encoding rule f is defined for all messages that occur with positive probability and is one-to-one:*

1. *For any encoding rule, the ideal observer decoder minimizes the overall error probability among all decoders.*
2. *If the source message $\mathbf{U}$ is equiprobable on a set $\mathcal{U}$ and encoding rule f is $\mathcal{U} \to \mathcal{X}$ then codeword $\mathbf{X} = f(\mathbf{U})$ is equiprobable on $\mathcal{X}$, and the ideal observer and maximum likelihood decoders coincide.*

Proof

1. Note that, given a received word $\mathbf{y}$, the denominator in (100) is fixed, so the ideal observer rule maximizes the joint probability $p(\mathbf{x})\mathbf{P}_{channel}(\mathbf{y}|\mathbf{x})$. Suppose you use an encoding rule f and decoding rule $\widehat{f}$. Then the overall error probability (see (95)) is

$$\sum_{\mathbf{u}} \mathbf{P}_{source}(\mathbf{U} = \mathbf{u})\mathbf{P}_{channel}\left(\widehat{f}(\mathbf{y}) \neq \mathbf{u} \,\middle|\, f(\mathbf{u}) \text{ sent}\right)$$

$$= \sum_{\mathbf{x}} p(\mathbf{x}) \sum_{\mathbf{y}:\widehat{f}(\mathbf{y}) \neq \mathbf{x}} \mathbf{P}_{channel}(\mathbf{y}|\mathbf{x})$$

$$= \sum_{\mathbf{y}} \sum_{\mathbf{x}:\mathbf{x} \neq \widehat{f}(\mathbf{y})} p(\mathbf{x})\mathbf{P}_{channel}(\mathbf{y}|\mathbf{x}) \qquad (105)$$

$$= \sum_{\mathbf{y}} \sum_{\mathbf{x}} p(\mathbf{x})\mathbf{P}_{channel}(\mathbf{y}|\mathbf{x}) - \sum_{\mathbf{y}} p\left(\widehat{f}(\mathbf{y})\right) \mathbf{P}_{channel}\left(\mathbf{y}\,\middle|\,\widehat{f}(\mathbf{y})\right)$$

$$= 1 - \sum_{\mathbf{y}} p\left(\widehat{f}(\mathbf{y})\right) \mathbf{P}_{channel}\left(\mathbf{y}\,\middle|\,\widehat{f}(\mathbf{y})\right).$$

   Now each term in the sum $\sum_{\mathbf{y}} p\left(\widehat{f}(\mathbf{y})\right) \mathbf{P}_{channel}\left(\mathbf{y}\,\middle|\,\widehat{f}(\mathbf{y})\right)$ is maximized when $\widehat{f}$ coincides with the ideal observer rule. Hence, the whole sum is maximized, and the overall error probability minimized.
2. The probabilities being constant means that they can be canceled, and the result is clear.

As we assume, in the definition of the channel capacity, that the source messages are equidistributed, it is natural to use the maximum likelihood decoder. We are going to do so throughout the rest of this chapter.

3.4 Shannon's Second Coding Theorem

In this section, we discuss Shannon's second coding theorem (in the case of a memoryless channel). It turns out that a key quantity in the theorem is the mutual entropy $\iota(X : Y)$ defined in (20). The result is as follows:

Theorem 17 (Shannon's second coding theorem) *For a memoryless channel, the capacity equals the maximum of the mutual entropy between a single input and a single output symbol. That is*

$$C = \sup_{p_X} \iota(X : Y)$$

$$= \sup_{p_X} \sum_{x,y} p_X(x) P_{\text{channel}}(y|x) \log \frac{p_X(x) P_{\text{channel}}(y|x)}{p_X(x) p_Y(y)}. \tag{106}$$

Here p_X stands for the distribution of input symbol X and p_Y for the distribution of input symbol Y:

$$p_Y(y) = \sum_{\widetilde{x}} p_X(\widetilde{x}) P_{\text{channel}}(y|\widetilde{x}). \tag{107}$$

Idea of the proof The proof will be divided into two parts, presented separately as Theorems 18 and 19 that establish, respectively, an upper and a lower bound for C. Traditionally, Theorem 18 is called the inverse part and Theorem 19 the direct part of Shannon's second coding theorem.

Before we immerse in the formal proof, consider the following example of a "noisy typewriter."

Example 6 Pick integers k and l. Consider a channel which has the input and output symbols $\{0, 1, 2, \ldots, kl - 1\}$. The channel takes symbol i to $i, i + 1, \ldots, i + (k - 1)$ with equal probability. Then, for every input distribution p_X

$$\iota(X : Y) = h(Y) - h(Y|X)$$

$$= h(Y) - \sum_{x} p_X(x) h(\mathbf{p}_Y(\cdot|X = x)) = h(Y) - \log k. \tag{108}$$

The largest that $h(Y)$ could be is $\log(kl)$ – one way that we can achieve this is to take X equidistributed on all the possible inputs. Then Y is equidistributed on all the possible outputs. This means that $\sup_{p_X} \iota(X : Y) = \log l$. Now, we can achieve error probability of zero, with this amount of redundancy introduced. That is, if we only use input symbols $\{0, k, 2k, \ldots, (l - 1)k\}$, then the outputs of the channel will be distinct for distinct symbols. That is, we can send (and guarantee to decode) l messages with each symbol. Hence, in (97), this corresponds to $\#\left(\mathcal{U}_1\right) = l$ (and indeed $\#\left(\mathcal{U}_N\right) = l^N$), so that

$$\overline{R} = \frac{\log \#\left(\mathcal{U}_N\right)}{N} = \log l \tag{109}$$

is a reliable transmission rate.

This argument does not really prove anything yet, but we can motivate our proof by saying that the property of joint typicality introduced in Definition 14 equates to saying that the source and channel are roughly like the noisy typewriter.

Theorem 18 (The inverse part of Shannon's second coding theorem) *For a memoryless channel, the channel capacity C obeys*

$$C \leq \sup_{p_X} \iota(X:Y). \tag{110}$$

Proof Suppose we have a source which produces equidistributed symbols on a set $\mathcal{U}_N$, of size 2^{NR_N}. We must show that for a sequence of codes with error probability ε_N converging to zero, then

$$R = \lim_{N \to \infty} R_N \leq \sup_{p_X} \iota(X:Y). \tag{111}$$

Now, suppose that the encoding rule f is 1–1 (if not, the bounds will be even tighter, since the error probability will be higher). Then

$$\begin{aligned}
NR_N = \log(\#\left(\mathcal{U}_N\right)) &= h\left(\mathbf{X}^{(N)}\right) \\
&= h\left(\mathbf{X}^{(N)}|\mathbf{Y}^{(N)}\right) + i\left(\mathbf{X}^{(N)}:\mathbf{Y}^{(N)}\right) \\
&\leq (1 + (NR_N)\varepsilon_N) + N\iota(X:Y),
\end{aligned} \tag{112}$$

where the inequality follows by (43) and (53).

Hence, dividing through by N, since $R_N \to R$ and $\varepsilon_N \to 0$, we can deduce the result.

Prior to proceeding further with formal proofs, consider how Theorem 17 is used to calculate the capacity of a symmetric memoryless channel. A useful formula is $\iota(X:Y) = h(Y) - h(Y|X)$. Then, in the case of a symmetric channel, the entropy $h\left(\mathbf{p}_{Y|X}(\cdot|x)\right)$ of the conditional distribution $\mathbf{p}_{Y|X}(\cdot|x)$ does not depend on x, and so $h(Y|X)$ just equals the entropy of one row. Hence, we only need to maximize $h(Y)$ over input distributions X – sometimes we will have a free choice of achievable Y, sometimes it will be constrained, and we have to use other tricks.

Example 7 For a memoryless binary symmetric channel, for each x,

$$\begin{aligned}
h\left(\mathbf{p}_{Y|X}(\cdot|x)\right) &= -\sum_{y=0,1} P_{\text{channel}}(y|x) \log P_{\text{channel}}(y|x) \\
&= -p \log p - (1-p) \log(1-p) = h(p, 1-p).
\end{aligned} \tag{113}$$

Hence $h(Y|X) = h(p, 1 - p)$ does not depend on input distribution p_X. Thus, in this case

$$C = \sup_{p_X} h(Y) - h(p, 1 - p) . \qquad (114)$$

But $\sup_{p_X} h(Y)$ is equal to $\log 2 = 1$: it is attained at $p_X(0) = p_X(1) = 1/2$, because then

$$p_Y(0) = p_Y(1) = 1/2(p + (1 - p)) = 1/2. \qquad (115)$$

Therefore, in this example,

$$C = 1 - h(p, 1 - p) . \qquad (116)$$

Next, consider the idea of jointly typical sequences, an idea which extends the definition of typical sequences and AEP of Remark 1.

Definition 13 Given a probability distribution $p(x, y)$, we say that the sequences $(\mathbf{x}^{(N)}, \mathbf{y}^{(N)})$ are jointly typical if the following three properties hold:

$$\left| -\frac{1}{N} \log p\left(\mathbf{x}^{(N)}\right) - h(X) \right| < \varepsilon,$$

$$\left| -\frac{1}{N} \log p\left(\mathbf{y}^{(N)}\right) - h(Y) \right| < \varepsilon, \qquad (117)$$

$$\left| -\frac{1}{N} \log p\left(\mathbf{x}^{(N)}, \mathbf{y}^{(N)}\right) - h(X, Y) \right| < \varepsilon,$$

where $p\left(\mathbf{x}^{(N)}\right) = \prod_{i=1}^{N} p(x_i)$ and so on.

We can establish a joint version of the AEP:

Theorem 19 Writing $\mathbf{A}_{\varepsilon}^{(N)}$ for the set of jointly typical sequences

1. The probability of the set $\mathbf{P}\left(\mathbf{A}_{\varepsilon}^{(N)}\right)$ converges to 1 as $N \to \infty$.
2. The size $\#\left(\mathbf{A}_{\varepsilon}^{(N)}\right) \leq 2^{N(h(X,Y)+\varepsilon)}$.
3. Given random variables $\tilde{\mathbf{X}}^{(N)}, \tilde{\mathbf{Y}}^{(N)}$ independent and with the same marginals as $\mathbf{x}^{(N)}$ and $\mathbf{y}^{(N)}$ then

$$\mathbf{P}\left((\tilde{\mathbf{X}}^{(N)}, \tilde{\mathbf{Y}}^{(N)}) \in \mathbf{A}_{\varepsilon}^{(N)}\right) \leq 2^{-N(\iota(X:Y)-3\varepsilon)}.$$

Proof As before, the law of large numbers will give that there exists a number N' such that for $N \geq N'$,

$$\mathbf{P}\left(\left|-\frac{1}{N}\log p\left(\mathbf{x}^{(N)}\right) - h(X)\right| > \varepsilon\right) < \varepsilon/3,$$

$$\mathbf{P}\left(\left|-\frac{1}{N}\log p\left(\mathbf{y}^{(N)}\right) - h(Y)\right| > \varepsilon\right) < \varepsilon/3, \qquad (118)$$

$$\mathbf{P}\left(\left|-\frac{1}{N}\log p\left(\mathbf{x}^{(N)}, \mathbf{y}^{(N)}\right) - h(X, Y)\right| > \varepsilon\right) < \varepsilon/3,$$

so that the probability of the union of these three sets is less than ε. Thus, the first part of Theorem 19 holds.

To show the second part of Theorem 19, note that

$$1 \geq \sum_{\mathbf{A}_\varepsilon^{(N)}} p\left(\mathbf{x}^{(N)}, \mathbf{y}^{(N)}\right) \geq \left(\#\left(\mathbf{A}_\varepsilon^{(N)}\right)\right) 2^{-N(h(X,Y)+\varepsilon)}.$$

The third part follows by considering

$$\begin{aligned}
\mathbf{P}&\left(\left(\widetilde{\mathbf{X}}^{(N)}, \widetilde{\mathbf{Y}}^{(N)}\right) \in \mathbf{A}_\varepsilon^{(N)}\right) \\
&= \sum_{\mathbf{x}^{(N)}, \mathbf{y}^{(N)} \in \mathbf{A}_\varepsilon^{(N)}} \mathbf{P}\left(\left(\widetilde{\mathbf{X}}^{(N)}, \widetilde{\mathbf{Y}}^{(N)}\right) = \left(\mathbf{x}^{(N)}, \mathbf{y}^{(N)}\right)\right) \\
&= \sum_{\mathbf{x}^{(N)}, \mathbf{y}^{(N)} \in \mathbf{A}_\varepsilon^{(N)}} \mathbf{P}\left(\left(\widetilde{\mathbf{X}}^{(N)} = \mathbf{x}^{(N)}\right)\mathbf{P}(\widetilde{\mathbf{Y}}^{(N)}) = \mathbf{y}^{(N)}\right) \qquad (119) \\
&\leq \sum_{\mathbf{x}^{(N)}, \mathbf{y}^{(N)} \in \mathbf{A}_\varepsilon^{(N)}} 2^{-N(h(X) - \varepsilon)} 2^{-N(h(Y) - \varepsilon)} \\
&\leq 2^{N(h(X, Y) + \varepsilon)} 2^{-N(h(X) - \varepsilon)} 2^{-N(h(Y) - \varepsilon)} \\
&= 2^{-N(\iota(X : Y) - 3\varepsilon)}.
\end{aligned}$$

Now, notice that there are about $2^{Nh(X)}$ typical X sequences, about $2^{Nh(Y)}$ typical Y sequences, but only around $2^{Nh(X,Y)}$ jointly typical sequences. This means that if we pick a random typical X sequence, and a random typical Y sequence, the probability that sequence (X, Y) is jointly typical is about $2^{-N\iota(X:Y)}$.

Theorem 20 (The direct part of Shannon's second coding theorem) *For a memoryless channel, the capacity*

$$C \geq \sup_{px} \iota(X : Y). \qquad (120)$$

Proof To prove Theorem 20, we take any probability distribution $p(x)$ of an input symbol and generate a code *randomly*, according to it. That is, we make a list of 2^{NR} codewords, where each symbol of each codeword is chosen independently, according to the same distribution $p(x)$.

Now, we can send 2^{NR} messages, so choose uniformly from the set of 2^{NR} possible codewords one codeword $\mathbf{x}^{(N)} = (x_1, \ldots, x_N)$. The receiver will receive a sequence $\mathbf{y}^{(N)}$, distributed according to the channel probability matrix. Now, we introduce yet another decoding rule, typical set decoding.

Definition 14 On receiving word $\mathbf{y}^{(N)}$, look for a word $\mathbf{x}^{(N)}$ such that $(\mathbf{x}^{(N)}, \mathbf{y}^{(N)})$ forms a jointly typical sequence. If there is a unique such word, return it as the best guess of the transmitted codeword. If there is more than one, or none at all, return an error.

We will show that typical set decoding is asymptotically optimal. We do this by taking a probability distribution p and considering the average performance of the random codes generated as above, averaged across all possible codewords. This significantly simplifies the analysis. That is, we calculate the average across all random codes $\mathcal{C}$: the overall error probability equals

$$\sum_{\mathcal{C}} \mathbf{P}(\mathcal{C})\, \mathbf{P}(\text{error} \mid \mathcal{C})$$

$$= \sum_{\mathcal{C}} \mathbf{P}(\mathcal{C} \text{ used}) \frac{1}{2^{NR}} \sum_{w} \mathbf{P}(\text{error} \mid \mathcal{C} \text{ used, word } w \text{ sent}) \qquad (121)$$

$$= \frac{1}{2^{NR}} \sum_{w} \sum_{\mathcal{C}} \mathbf{P}(\mathcal{C} \text{ used})\, \mathbf{P}(\text{error} \mid \mathcal{C} \text{ used, word } w \text{ sent}) \,.$$

Now notice that since we average across all random codes, it does not matter which codeword was actually sent, so without loss of generality, we can assume $w = 1$.

Note that two types of errors are possible: first if $\mathbf{x}^{(N)}$ and $\mathbf{y}^{(N)}$ are not jointly typical and second if there exists some other codeword which is jointly typical with $\mathbf{y}^{(N)}$. Now, the trick will be to use the joint AEP, which shows that the first type of error has probability tending to zero and that the probability that a given codeword and received word are jointly typical is about 2^{-Ni}, so we can afford to use about 2^{Ni} codewords. That is, if we write

$$E_i = \{\text{codeword } i \text{ is jointly typical with received } \mathbf{y}^{(N)}\} \,, \qquad (122)$$

then the probability of an error is

$$\mathbf{P}(E_1^c \cup E_2 \cup E_3 \cup \cdots \cup E_{2^{NR}}) \leq \mathbf{P}(E_1^c) + \sum_{i=2}^{2^{NR}} \mathbf{P}(E_i) \,. \qquad (123)$$

Now, by the first part of Theorem 19, for N sufficiently large, the first event has probability less than ε. Similarly, the third part of Theorem 19 shows that for $i \neq 1$, $\mathbf{P}(E_i) \leq 2^{-N(\iota(X\,:\,Y) - 3\varepsilon)}$, so that the error probability is less than

$$\varepsilon + \exp_2(N(R - \iota(X\,:\,Y) + 3\varepsilon)) \,, \qquad (124)$$

so that if $R < \iota(X : Y) - 3\varepsilon$, this quantity goes to zero.

Now, notice that we have a free choice of probability distributions p, so we can choose the one which makes $\iota(X : Y)$ as large as possible. Further, since the average code has error probability tending to zero, we can guarantee that we will be able to choose some code which has its error probability tending to zero.

4 Bibliographical Notes

For the bulk material of these lecture notes refer to [1–3] and for basics of probability [4–6].

References

1. Cover, T.M., Thomas, Y.A.: Elements of Information Theory. Wiley, New York (1991)
2. Csiszár, I., Körner, J.: Information Theory: Coding Theorems for Discrete Memoryless Channels. Academic Press, New York (1981)
3. Goldie, C., Pinch, R.G.E.: Communication Theory. Cambridge University Press, Cambridge (1991)
4. Grimmett, G., Stirzaker, D.: Probability and Random Processes. Oxford University Press, New York (2001)
5. Stirzaker, D.: Elementary Probability. Cambridge University Press, Cambridge (2003)
6. Suhov, Y., Kelbert, M.: Probability and Statistics by Example. Cambridge University Press, Cambridge (2005)

Quantum Probability and Quantum Information Theory

H. Maassen

1 Introduction

From its very birth in the 1920s, quantum theory has been characterized by a certain strangeness: It seems to run counter to the intuitions that we humans have about the world we live in.

According to these "realistic" intuitions all things have their definite place and sharply determined qualities, such as speed, color, and weight. Quantum theory, however, refuses to precisely pinpoint them. With respect to this apparent shortcoming of the theory different points of view can be taken. It could be suspected that quantum theory is incomplete, in that it gives a coarse description of a reality that is actually more refined. This is the viewpoint once taken by Einstein, and it still has adherents today. It calls for a search for finer mathematical models of physical reality, based on classical probability, often referred to as "hidden variable models" (see chapter "Photonic Realization of Quantum Information Protocols"). One such attempt is Bohm's theory of non-relativistic quantum mechanics.

However, the work of John Bell in the 1960s and of Alain Aspect in the 1970s and 1980s strongly favors the opposite point of view: Their work has made clear that such models with a classical probabilistic structure are necessarily afflicted with a certain weakness; they must at least allow *action at a distance*. This we regard as a bad property for a theory which aims to describe a physical world where no signals have been observed to travel faster than light. Apart from that, the hidden variable theories which have been found so far are highly artificial and cannot be tested against quantum mechanics since they do not predict any new phenomena.

It is for these reasons that we decide to accept quantum theory with its inherent strangeness and are prepared to modify probability theory accordingly.

H. Maassen (✉)
Radboud University, Nijmegen, The Netherlands, `maassen@math.ru.nl`

Maassen, H.: *Quantum Probability and Quantum Information Theory*. Lect. Notes Phys. **808**, 65–108 (2010)
DOI 10.1007/978-3-642-11914-9_3

1.1 Quantum Probability

So quantum mechanics does not predict the results of physical experiments with certainty, but calculates probabilities for their possible outcomes.

Now, the classical mathematical theory of probability obtained a unified formulation in the 1930s, when Kolmogorov introduced his axioms, defining the universal structure $(\Omega, \Sigma, \mathbf{P})$ of a probability space. For a long time this theory of probability (dealing with probability distributions, stochastic processes, Markov chains, martingales, etc.) remained completely separate from the mathematical development of quantum mechanics (involving vectors in a Hilbert space, Hermitian operators, unitary transformations, and such like).

In the 1970s and 1980s people around Accardi, Lewis, Davies, Kümmerer, building on ideas of von Neumann's and Segal's, developed a unified framework, a generalized, "non-commutative," probability theory, in which classical probability theory and quantum mechanics can be discussed in unison. It consists of ordinary Hilbert space quantum theory, with the emphasis moved toward operators on Hilbert space, and the algebras which they generate. The main objective of this chapter is to sketch the outlines of this framework and show its usefulness for information theory.

1.2 Quantum Information

In Shannon's (classical) information theory (see chapter "Classical Information Theory"), a single unit, the *bit*, serves to quantify all forms of information, be it in print, computer memory, CD-ROM, or strings of DNA. Such a single unit suffices, because different forms of information can be converted into each other by copying, according to fixed "exchange rates." The physical states of quantum systems, however, cannot be copied into such "classical" information, but *can* be converted into each other. This leads to a new unit of information: the *qubit*.

Quantum information theory studies the handling of this new form of information by information-carrying channels. We shall treat the basic properties of these channels and some impossibilities as well as new possibilities connected with quantum information. The impossibility of copying makes quantum information an ideal means to establish secrecy (see chapter "Quantum Cryptography").

1.3 Quantum Computing

It was Richard Feynman who first thought of *employing* the strangeness of quantum mechanics to do things that would be impossible in a classical world.

The idea was developed in the 1980s and 1990s by David Deutsch, Peter Shor, and many others into a flourishing branch of science called "quantum computing": How to make quantum mechanical systems perform calculations more efficiently than ordinary computers can do. This research is still in a predominantly theoretical stage: The quantum computers actually built are as yet extremely primitive and can by no means compete with even the simplest pocket calculator, but expectations are high (see chapter "Quantum Algorithms").

1.4 This Chapter

We start with an introduction to quantum probability. In Sect. 2 we demonstrate the "strangeness" of quantum phenomena by very simple polarization experiments, culminating in Bell's famous inequality, tested in Aspect's experiment. Bell's inequality is a statement in classical probability that is violated in quantum probability and in reality.

Taking polarizers as our starting point, in Sects. 3 and 4 we build up the new probability theory in terms of algebras of operators on a Hilbert space. In Sect. 5 operations on these algebras will be characterized, and some aspects will be discussed in which they differ from classical physical operations. They are subject to certain strange limitations: The impossibility of copying, of coding information into bits, of jointly measuring incompatible observables, of observation without perturbing the object (cf. Sect. 6). But they also open up surprising possibilities: entangling remote systems, teleportation of this entanglement, sending two bits in a single qubit (cf. Sect. 7). Further luring perspectives as highly efficient algorithms for sorting, Fourier transformation, and factoring very large numbers will be treated in chapter "Quantum Algorithms".

2 Why Classical Probability Does Not Suffice

(This section is based on [8].)

2.1 An Experiment with Polarizers

To start with, we consider a simple experiment. In a beam of light of a fixed color we put a pair of polarizing filters, each of which can be rotated around the axis formed by the beam. As is well known, the light falling through both filters changes in intensity when the filters are rotated relative to each other. Starting from the orientation where the resulting intensity is maximal and rotating one of the filters through an angle α, the light intensity decreases with α, vanishing for $\alpha = \frac{1}{2}\pi$. If we call the intensity of the beam before the filters I_0, after the first I_1, and after the second I_2, then $I_1 = \frac{1}{2}I_0$ (we assume the original beam to be unpolarized), and (Fig. 1)

$$I_2 = I_1 \cos^2 \alpha \, . \tag{1}$$

Fig. 1 Two polarizers in conjunction

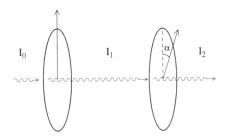

So far the phenomenon is described well by classical physics. During the last century, however, it has been observed that for very low intensities (monochromatic) light comes in small packages, which were called *photons*, whose energy depends on the color, but not on the total intensity. So the intensity must be proportional to the *number* of these photons, and formula (1) must be given a statistical meaning: A photon passing through the first filter has a probability $\cos^2 \alpha$ to pass through the second. Formula (1) then holds only on the average, for large numbers of photons.

Thinking along the lines of classical probability, we may associate with a polarization filter in the direction α a random variable P_α, taking the value $P_\alpha(\omega) = 0$ if the photon ω is absorbed by the filter and $P_\alpha(\omega) = 1$ if it passes through. For two filters in the directions α and β these random variables then should be correlated as follows:

$$\mathbb{E}(P_\alpha P_\beta) \;=\; \mathbf{P}[P_\alpha = 1, P_\beta = 1] \;=\; \tfrac{1}{2}\cos^2(\alpha - \beta). \qquad (2)$$

Here we hit on a difficulty: The function on the right-hand side is not a possible correlation function! This can be seen as follows. Take three polarizing filters, having polarization directions α_1, α_2, and α_3, respectively. Put them on the optical bench in pairs. They should give rise to random variables P_1, P_2, and P_3 satisfying

$$\mathbb{E}(P_i P_j) = \tfrac{1}{2}\cos^2(\alpha_i - \alpha_j) . \qquad (3)$$

Proposition 1 (Bell's three-variable inequality) *[2] For any three 0–1-valued random variables P_1, P_2, and P_3 on a probability space $(\Omega, \mathbf{P})$ the following inequality holds:*

$$\mathbf{P}[P_1 = 1, P_3 = 0] \;\le\; \mathbf{P}[P_1 = 1, P_2 = 0] + \mathbf{P}[P_2 = 1, P_3 = 0]. \qquad (4)$$

Proof

$$\mathbf{P}[P_1 = 1, P_3 = 0] = \mathbf{P}[P_1 = 1, P_2 = 0, P_3 = 0] + \mathbf{P}[P_1 = 1, P_2 = 1, P_3 = 0]$$
$$\le \mathbf{P}[P_1 = 1, P_2 = 0] + \mathbf{P}[P_2 = 1, P_3 = 0]. \qquad \square$$
$$(5)$$

In our example, we have

$$\mathbf{P}[P_i = 1, P_j = 0] = \mathbf{P}[P_i = 1] - \mathbf{P}[P_i = 1, P_j = 1]$$
$$= \tfrac{1}{2} - \tfrac{1}{2}\cos^2(\alpha_i - \alpha_j) = \tfrac{1}{2}\sin^2(\alpha_i - \alpha_j). \tag{6}$$

Bell's inequality thus reads

$$\tfrac{1}{2}\sin^2(\alpha_1 - \alpha_3) \leq \tfrac{1}{2}\sin^2(\alpha_1 - \alpha_2) + \tfrac{1}{2}\sin^2(\alpha_2 - \alpha_3), \tag{7}$$

which is clearly violated for the choices $\alpha_1 = 0$, $\alpha_2 = \tfrac{1}{6}\pi$, and $\alpha_3 = \tfrac{1}{3}\pi$, where it says that

$$\frac{3}{8} \leq \frac{1}{8} + \frac{1}{8}. \tag{8}$$

This example suggests that classical probability cannot even describe this simple experiment!

Remark 1 The above calculation could be summarized as follows: we are in fact looking for a family of 0–1-valued random variables $(P_\alpha)_{0 \leq \alpha < \pi}$ with $\mathbf{P}[P_\alpha = 1] = \tfrac{1}{2}$, satisfying the requirement that

$$\mathbf{P}[P_\alpha \neq P_\beta] = \sin^2(\alpha - \beta). \tag{9}$$

Now, on the space of 0–1-valued random variables on a probability space the function $(X, Y) \mapsto \mathbf{P}[X \neq Y]$ equals the L^1-distance of X and Y:

$$\mathbf{P}[X \neq Y] = \int_\Omega |X(\omega) - Y(\omega)| \, \mathbf{P}(d\omega) = \| X - Y \|_1. \tag{10}$$

On the other hand, the function $(\alpha, \beta) \mapsto \sin^2(\alpha - \beta)$ does not satisfy the triangle inequality for a distance function on the interval $[0, \pi)$. Therefore no family $(P_\alpha)_{0 \leq \alpha < \pi}$ exists which meets requirement (9).

2.2 An Improved Experiment

On closer inspection the above example is not very convincing. Indeed, when two polarizers are arranged on the optical bench, why should not the random variable for the second polarizer depend on the angle of the first? The correlation in (2) would then read

$$\mathbb{E}(P_\alpha P_{\alpha,\beta}) = \mathbf{P}[P_\alpha = 1, P_{\alpha,\beta} = 1] = \tfrac{1}{2}\cos^2(\alpha - \beta), \tag{11}$$

which can easily be satisfied, and the whole refutation collapses.

Fig. 2 Photon pair
production

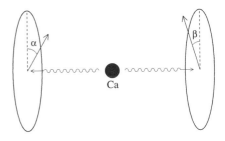

So we should do a better experiment. We must let the filters act on the photons without influence on each other. Maybe we can separate them spatially?

Here a clever technique from quantum optics comes to our aid. It is possible to build a device that produces *pairs* of photons, such that the members of each pair move in opposite directions and show opposite behavior toward parallel polarization filters: If one passes the filter, then the other is surely absorbed. The device contains calcium atoms, which are excited by a laser to a state they can only leave under emission of such a pair (Fig. 2).

With these photon pairs, the very same experiment can be performed, but this time the polarizers are far apart, each one acting on its own photon. The same correlations are measured, say first between P_{α_1} on the left and P_{α_2} on the right, then between P_{α_1} on the left and P_{α_3} on the right, and finally between P_{α_2} on the left and P_{α_3} on the right. The same outcomes are found, violating Bell's three-variable inequality, thus strengthening the case against classical probability.

2.3 The Decisive Experiment

Advocates of classical probability could still find serious fault with the argument given so far. Indeed, do we really *have to* assume that we are measuring the same random variable P_{α_2} on the right as later on the left? Is it really true that the polarizations in these pairs are exactly opposite? There could exist a probabilistic explanation of the phenomena without this assumption.

So the argument has to be tightened still further. This brings us to the experiment which was actually performed by A. Aspect in Orsay (near Paris) in 1982 [1]. In this experiment a random choice out of two different polarization measurements was performed on each side of the pair-producing device, say in the direction α_1 or α_2 on the left and in the direction β_1 or β_2 on the right, giving rise to *four* random variables $P_1 := P(\alpha_1)$, $P_2 := P(\alpha_2)$ and $Q_1 := Q(\beta_1)$, $Q_2 := Q(\beta_2)$, two of which are measured and compared at each trial (see chapter "Photonic Realization of Quantum Information Protocols" for more details).

Proposition 2 (Bell's four-variable inequality) *For any quadruple P_1, P_2, Q_1, and Q_2 of 0–1-valued random variables on $(\Omega, \mathbf{P})$ the following inequality holds:*

$$\mathbf{P}[P_1 = Q_1] \leq \mathbf{P}[P_1 = Q_2] + \mathbf{P}[Q_2 = P_2] + \mathbf{P}[P_2 = Q_1]. \qquad (12)$$

(In fact, by symmetry, neither of these four probabilities is larger than the sum of the other three.)

Proof It is easy to see that for all ω

$$P_1(\omega) = Q_1(\omega) \Rightarrow P_1(\omega) = Q_2(\omega) \quad \text{or} \quad Q_2(\omega) = P_2(\omega) \quad \text{or} \quad P_2(\omega) = Q_1(\omega).$$
(13)

□

Bell's four-variable inequality can be viewed as a "quadrangle inequality" with respect to the metric $(X, Y) \mapsto \| X - Y \|_1$ on random variables X, Y.

On the other hand, quantum mechanics predicts (cf. Sect. 3.6), and the experiment of Aspect showed, that one has,

$$\mathbf{P}[P(\alpha) = Q(\beta) = 1] = \tfrac{1}{2} \sin^2(\alpha - \beta).$$
(14)

Similarly, $\mathbf{P}[P(\alpha) = Q(\beta) = 0] = \tfrac{1}{2} \sin^2(\alpha - \beta)$. Hence

$$\mathbf{P}[P(\alpha) = Q(\beta)] = \sin^2(\alpha - \beta).$$
(15)

So Bell's four-variable inequality reads in this example:

$$\sin^2(\alpha_1 - \beta_1) \le \sin^2(\alpha_1 - \beta_2) + \sin^2(\alpha_2 - \beta_1) + \sin^2(\alpha_2 - \beta_2),$$
(16)

which is clearly violated for the choices $\alpha_1 = 0$, $\alpha_2 = \tfrac{\pi}{3}$, $\beta_1 = \tfrac{\pi}{2}$, and $\beta_2 = \tfrac{\pi}{6}$, see Fig. 3, in which case it reads

$$1 \le \frac{1}{4} + \frac{1}{4} + \frac{1}{4}.$$
(17)

Now we are finished: There does not exist, on any classical probability space, a quadruple P_1, P_2, Q_1, and Q_2 of random variables with the correlations measured in this experiment.

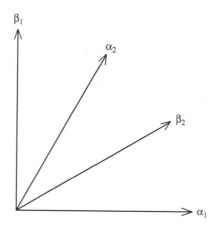

Fig. 3 Directions violating Bell's inequality

Discussion

1. A crucial assumption that goes into Bell's inequality is that it makes sense to compare (cf. (5)) the (possibly random) reactions which a given photon *would* show to different filters, including those it does not actually meet. This assumption is called *realism*; it is made in all classical probabilistic physical theories, but is abandoned in quantum mechanics.

2. A second important assumption, necessary for the validity of Bell's inequality, was mentioned before: The *outcome* on the right (described by $Q(\beta)$ for some β) should not depend on the *angle* α of the polarizer on the left. This assumption is called "locality." In order to justify this assumption, Aspect has made considerable efforts. In his (third) experiment [1], the choice of what to measure on the left (α_1 or α_2) and on the right (β_1 or β_2) was made *during the flight of the photons*, so that any influence which each of these choices might have on *the outcome* on the opposite end would have to travel faster than light. By the causality principle of relativity theory such influences are excluded.

3. The Orsay experiment refutes all imaginable physical theories which are both *local* and *realistic* (cf. 1 and 2 above). Quantum mechanics is local, but not realistic. Its great successes lead us to believe that realism fails for the description of nature. Some prefer to adhere to realism, and so they must give up locality, and hence Einstein causality [4, 9].

4. In our opinion, the phrase "quantum non-locality," which is often heard in the context of Bell's inequalities, signals a misconception. It suggests giving up *both* realism *and* locality. This is too much of a defeat and unnecessary. Quantum mechanics is local. But it describes phenomena which *in a classical theory* could only be explained using some action at a distance.

2.4 The Orsay Experiment as a Card Game

To illustrate the above refutation of local realism more vividly, we shall present the experiment in the form of a card game. Nature can win this game. Can you?

Two players, P and Q, are sitting at a table. They are cooperating to achieve a single goal. There is an arbiter present to deal cards and to count points. On the table there is a board consisting of four squares as drawn in Fig. 4. There are dice and an ordinary deck of playing cards. The deck of cards is shuffled well. (In fact we shall assume that the deck of cards is an infinite sequence of independent cards, chosen fully at random.) First the players are given some time to make agreements on the strategy they are going to follow. Then the game starts, and from this moment on they are no longer allowed to communicate. The following sequence of actions is then repeated many times:

1. The dealer hands a card to P and one to Q. Both look at their own card, but not at the other one's. (The only feature of the card that matters is its color: red or black.)

Fig. 4 Board for the Bell game

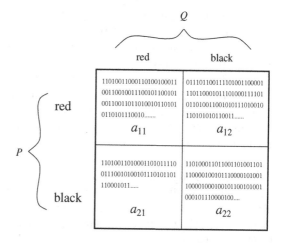

2. The dice are thrown.
3. P and Q simultaneously say "yes" or "no," according to their own choice. They are free to make their answer depending on any information they possess, such as the color of their own card, the agreements made in advance, the numbers shown by the dice, the weather, the time.
4. The cards are laid out on the table. The pair of colors of the cards determines one of the four squares on the board: These are labeled (red, red), (red, black), (black, red), and (black, black).
5. In the square so determined a 0 or a 1 is written: a 0 when the answers of P and Q have been different, a 1 if they have been the same.

In the course of time, the squares on the board get filled with 0s and 1s. The arbiter keeps track of the percentage of 1s in proportion to the total number of bits in each square; we shall call the time limits of these percentages as the game proceeds: a_{11}, a_{12}, a_{21}, and a_{22}. The aim of the game, for both P and Q, is to get a_{11} larger than the sum of the other three limiting percentages. So P and Q must try to give identical answers as often as they can when both their cards are red, but different answers otherwise.

Proposition 3 (Bell's inequality for the game) *P and Q cannot win the game by classical means, namely*

$$a_{11} \leq a_{12} + a_{21} + a_{22} . \tag{18}$$

Proof The best P and Q can do, in order to win the game, is to agree upon some (possibly random) strategy for each turn. For instance, they may agree that P will always say "yes" (i.e., $P_{\text{red}} = P_{\text{black}} =$"yes") and that Q will answer the question "Is my card red?" (i.e., $Q_{\text{red}} =$ "yes" and $Q_{\text{black}} =$"no"). This will lead to a 1 in the (red, red) square or the (black, red) square or to a 0 in one of the other two.

So if the players repeat this strategy indefinitely, on the long run they would get $a_{11} = a_{12} = 1$ and $a_{21} = a_{22} = 0$, disappointingly satisfying Bell's inequality.

The above example is an extremal strategy. There are many (in fact, 16) strategies like this. By the point-wise version (13) of Bell's four-variable inequality, none of these 16 extremal strategies wins the game. Inclusion of the randomness coming from the dice yields a full polytope of random strategies, having the above 16 as its extremal points. But since the inequalities are linear, this averaging procedure does not help. This "proves" our "proposition." Disbelievers are challenged to find a winning strategy.   □

Strangely enough, however, nature does provide us with a strategy to win the game, still essentially based on the $\cos^2$ law 2 for photon absorption! Instead of the dice, put a calcium atom on the table. When the cards have been dealt, P and Q put their polarizers in the direction indicated by their cards. If P has a red card, then he chooses the direction $\alpha_1 = 0$ (cf. Fig. 3). If his card is black, then he chooses $\alpha_2 = \frac{\pi}{3}$. If Q has a red card, then he chooses $\beta_1 = \frac{\pi}{2}$. If his card is black, then he chooses $\beta_2 = \frac{\pi}{6}$. No information on the colors of the cards needs to be exchanged. When the calcium atom has produced its photon pair, each player looks whether his own photon passes his own polarizer, and then says "yes" if it does, "no" if it does not. On the long run they will get $a_{11} = 1$, $a_{12} = a_{21} = a_{22} = \frac{1}{4}$, and thus they win the game.

So the calcium atom, the quantum mechanical die, makes possible what could not be done with the classical die.

3 Toward a Mathematical Model

Coerced by the foregoing considerations, we give up trying to make a classical probabilistic model in order to explain polarization experiments. Instead, we take these experiments as a paradigm for an alternative type of "quantum" probability, to be developed now.

3.1 A Mathematical Description of Polarization

We have discussed (linear) polarization of a light beam. This is completely characterized by a direction in the plane perpendicular to the light beam. So we simply describe states of polarization by different directions in a two-dimensional real plane $\mathbb{R}^2$, or equivalently by unit vectors $\psi \in \mathbb{R}^2$, $\| \psi \| = 1$, pointing in this direction. Actually, since we cannot distinguish between two states which differ by a rotation of π, we shall describe states of polarization by one-dimensional subspaces of $\mathbb{R}^2$. Given two directions of polarization with an angle α between them, spanned by two unit vectors $\psi, \theta \in \mathbb{R}^2$, the probability to find polarization ϑ when a photon is in the state ψ can be expressed as

$$\cos^2 \alpha = \langle \psi, \theta \rangle^2, \tag{19}$$

where $\langle \psi, \theta \rangle$ denotes the scalar product between ψ and θ.

In the mathematical model we should distinguish between the physical state of polarization of a photon on the one hand and the filter on the other hand, i.e., the 0–1-valued random variable which asks whether a photon is polarized in a certain direction. This can be done by identifying the random variable with the orthogonal projection P onto the one-dimensional subspace. We can then write

$$\cos^2 \alpha = \langle \psi, \theta \rangle^2 = \langle \psi, P\psi \rangle . \tag{20}$$

Since P is 0–1-valued (a photon passes or is absorbed), this probability is equal to the expectation of this random variable:

$$\langle \psi, P\psi \rangle = \mathbb{E}(P) . \tag{21}$$

3.2 The Full Truth About Polarization: The Qubit

In the foregoing description of polarization things were presented somewhat simpler than they are: We considered only linear polarization, thus disregarding circular polarization. The full description of polarization leads to the quantum mechanics of a two-level system or *qubit*:

State of polarization of a photon $\triangleq$ one-dimensional subspace of $\mathbf{C}^2$, described by a unit vector ψ spanning this subspace (and determined only up to a phase).

Polarization filter or generalized $\triangleq$ orthogonal projection P onto a complex 0–1-valued random variable one-dimensional subspace.

(Also for left- or right-circular polarization there exist physical filters.)

Probability for a photon, described $\triangleq$ $\langle \psi, P\psi \rangle$. by ψ, to pass through a filter, described by P

The set of all states is conveniently parametrized by the unit vectors of the form

$$(\cos \alpha, e^{i\phi} \sin \alpha) \in \mathbf{C}^2 , \quad -\frac{\pi}{2} \le \alpha \le \frac{\pi}{2}, \quad 0 \le \phi \le \pi . \tag{22}$$

3.3 Finite-Dimensional Models

The mathematical model that is used by quantum mechanics is the straightforward generalization of the above description. In order to keep things simple, in the following we restrict ourselves to the quantum mechanics in finite dimension. This generalizes the probability theory of systems with only finitely many states. As in classical probability, the generalization to systems with a countably infinite number of states or a continuum of states is analytically more involved.

The model is as follows: *States* correspond to one-dimensional subspaces of $\mathbf{C}^n$, where the dimension n is determined by the model. Again, a state is described conveniently by some unit vector spanning this subspace.

0–1-Valued random variables or *events* are described by orthogonal projections onto linear subspaces of $\mathbf{C}^n$. Here also projections onto higher dimensional subspaces make sense.

The *probability* that a measurement of a random variable P on a system in a state ψ gives the value 1 is given by $\langle \psi, P\psi \rangle$.

Note that we do not assume that every orthogonal projection corresponds to a meaningful random variable. Specification of random variables is part of the description of the mathematical model for a given system. In a truly quantum mechanical situation, typically all projections are used. In contrast to this, classical probability is obtained by allowing only very few projections, as follows.

3.4 Finite Classical Models

A finite probability space is usually described by a finite set $\Omega = \{\omega_1, \ldots, \omega_n\}$ and a probability distribution $(p_1, \ldots, p_n)$, $0 \le p_i \le 1$, $\sum_i p_i = 1$, such that the probability for ω_i is p_i. A 0–1-valued random variable is a 0–1-valued function on Ω, i.e., a characteristic function χ_A of some subset $A \subseteq \Omega$. In order to describe such a system in our model, we think of $\mathbf{C}^n$ as the space of complex-valued functions on Ω and use the functions δ_i with $\delta_i(\omega_j) = \delta_{i,j}$ as basis. The states of the system, i.e., the points ω_i of Ω, are now represented by the unit vectors δ_i, $1 \le j \le n$. The random variable χ_A is identified with the orthogonal projection P_A onto the linear span of the vectors $\{\delta_i : \omega_i \in A\}$. In our basis χ_A becomes a diagonal matrix with a 1 at the ith place of the diagonal if $\omega_i \in A$, and a 0 otherwise. It is obvious that $\omega_i \in A$ if and only if $\chi_A(\omega_i) = 1$ if and only if $\langle \delta_i, P_A \delta_i \rangle = 1$.

Conversely, any set of pairwise commuting projections on $\mathbf{C}^n$ can be diagonalized simultaneously and thus have an interpretation as a set of classical 0–1-valued random variables. Therefore

Classical probability corresponds to sets of pairwise commuting projections.

3.5 Mixed States

In the above sketch of quantum probability an important point is still missing: How can we describe a situation where a photon has one polarization with some probability q and in another with probability $1-q$? Since states must play the role of probability distributions, this combination should be expressed as a single state of the photon.

In general, if P is any 0–1-valued random variable and $\psi_1, \ldots, \psi_k$ are arbitrary quantum states, each occurring with a probability p_i, $1 \le i \le k$, $\sum_i p_i = 1$, then the probability that a measurement of P gives 1 is clearly given by

$$\sum_i p_i \langle \psi_i, P\psi_i \rangle \ . \tag{23}$$

A more convenient description of mixed states is obtained as follows (compare chapter "Hilbert Space Methods for Quantum Mechanics"). For a unit vector $\psi \in \mathbb{C}^n$ denote by ρ_ψ the orthogonal projection onto the one-dimensional subspace generated by ψ. In the physics literature, ρ_ψ is often denoted by $|\psi\rangle\langle\psi|$. Let Tr denote the trace operation (see chapter "Hilbert Space Methods for Quantum Mechanics", Eq. (47)) on the $n \times n$ matrices, summing up the diagonal entries of such a matrix. Then one obtains

$$\langle \psi, P\psi \rangle \ = \ \mathrm{Tr}\,(\rho_\psi P) \ . \tag{24}$$

Hence

$$\sum_i p_i \langle \psi_i, P\psi_i \rangle \ = \ \mathrm{Tr}\left(\sum_i p_i \rho_{\psi_i} \cdot P \right) \ = \ \mathrm{Tr}(\rho P) \ , \tag{25}$$

where $\rho := \sum_i p_i \rho_{\psi_i}$.

Being a convex combination of one-dimensional projections, ρ is obviously a positive (i.e., self-adjoint positive semi-definite) $n \times n$ matrix with $\mathrm{Tr}\,(\rho) = 1$.

Conversely, from diagonalizing positive matrices it is clear that any such positive matrix ρ with $\mathrm{Tr}\,(\rho) = 1$ can be written as a convex combination of one-dimensional projections. The set of these matrices forms a closed (even compact) convex set, and its extreme points are precisely the one-dimensional projections which in turn correspond to pure states, represented also by unit vectors. Therefore it is this class of so-called *density matrices* which represents mixed states. Thus, a general mixed state is described by a density matrix ρ and the probability for an observation of P to yield the value 1 is given by $\mathrm{Tr}\,(\rho P)$.

Remark 2

1. A bounded closed set $\mathcal{U}$ of a real linear space $\mathcal{V}$ is convex if with two points $x_{1,2} \in \mathcal{U}$ it also contains the points $x_\lambda = \lambda x_1 + (1 - \lambda)x_2$, $0 \le \lambda \le 1$ of the connecting segment. Those points $x \in \mathcal{U}$ which can be points x_λ of a segment only if $\lambda = 0$ or $\lambda = 1$ are called *extremal*: any point $x \in \mathcal{U}$ can be written as a convex combination of the extremal points of $\mathcal{U}$. Such a representation is unique if and only if $\mathcal{U}$ is a simplex; an n-dimensional simplex is a closed convex set generated by convex combinations of $n + 1$ points $\{x_i\}_{i=1}^{n+1}$ such that the n connecting lines $x_i - x_{n+1}$ are linearly independent. Given a subset $\mathcal{Y} \subseteq \mathcal{V}$, its closed convex hull $\mathcal{U}$ is the smallest convex subset of $\mathcal{V}$ such that $\mathcal{Y} \subset \mathcal{U}$.

2. The decomposition of a density matrix ρ into a convex combination of one-dimensional projections is by no means unique. This point will be further elaborated in Proposition 6. So the compact convex set of density matrices is not a simplex at all. Indeed, on $\mathbf{C}^2$ it can be identified with a full ball in $\mathbb{R}^3$, by taking in $\mathbb{R}^3$ the convex hull of the sphere that was described above.

3. In classical probability the convex set of mixed states is the simplex of all probability distributions. In our picture, if we insist on decomposing a mixed state given by $\rho = \sum_i p_i P_{\delta_i}$ into a convex combination of pure states (within the convex hull of $\{P_{\delta_i} : 1 \leq i \leq n\}$ which is a simplex), then it becomes unique.

4. Physically, a state ρ is completely described by all of its values $\mathrm{Tr}\,(\rho\,P)$, where P runs through the random variables of the model. Thus, if we consider only subsets of projections, then two different density matrices can represent the same physical state of the system. As a drastic example, consider the classical system $\Omega = \{\omega_1, \ldots, \omega_n\}$ with equidistribution, i.e., $p_i(\omega_i) = \frac{1}{n}$, leading to the density matrix $\rho = \sum_i \frac{1}{n} P_{\delta_i} = \frac{1}{n} \cdot \mathbb{1}$. On the other hand, with the unit vector $\psi = (\frac{1}{\sqrt{n}}, \ldots, \frac{1}{\sqrt{n}}) \in \mathbf{C}^n$, we obtain for any subset $A \subseteq \Omega$:

$$\mathrm{Tr}\,(\rho\,P_A) = \frac{1}{n} \cdot |A| = \langle \psi, P_A \psi \rangle . \tag{26}$$

Therefore, on the random variables $\{P_A : A \subseteq \Omega\}$, the rank one density matrix P_ψ represents the same state as the density matrix $\frac{1}{n} \cdot \mathbb{1}$. Note, however, that P_ψ is not in the convex hull of $\{P_{\delta_i} : 1 \leq i \leq n\}$.

3.6 The Mathematical Model of Aspect's Experiment

As an illustration, we shall now explain the photon correlation in the Orsay experiment, given by the $\cos^2$ law. Note that here we cannot simply refer to the basic $\cos^2$ law of quantum probability, since the filters are acting on two different photons.

The polarization of a pair of photons is described by a unit vector in the tensor product $\mathbf{C}^2 \otimes \mathbf{C}^2 = \mathbf{C}^4$, where we use the basis

$$\begin{aligned}
(1, 0, 0, 0) &= e_1 \otimes e_1 =: e_{11}, \\
(0, 1, 0, 0) &= e_1 \otimes e_2 =: e_{12}, \\
(0, 0, 1, 0) &= e_2 \otimes e_1 =: e_{21}, \\
(0, 0, 0, 1) &= e_2 \otimes e_2 =: e_{22},
\end{aligned} \tag{27}$$

with $e_1 = (1, 0) \in \mathbf{C}^2$ and $e_2 = (0, 1) \in \mathbf{C}^2$. For example, in the pure state e_{12} the left-hand photon is vertically polarized and the right-hand photon horizontally. As it turns out, the state of the pair of photons as produced by the calcium atom is described by the state

$$\psi = \frac{1}{\sqrt{2}}(e_{12} - e_{21}).$$ (28)

Now, the filters $P(\alpha)$ on the left and $Q(\beta)$ on the right, introduced in Sect. 2.3, are represented by two-dimensional projection operators on $\mathbf{C}^4$, which are the "two-right amplification" and the "two-left-amplification" of the polarization matrix

$$\begin{pmatrix} \cos^2 \alpha & \cos \alpha \sin \alpha \\ \cos \alpha \sin \alpha & \sin^2 \alpha \end{pmatrix},$$ (29)

namely

$$\begin{aligned} P(\alpha) &= \begin{pmatrix} \cos^2 \alpha & \cos \alpha \sin \alpha \\ \cos \alpha \sin \alpha & \sin^2 \alpha \end{pmatrix} \otimes \begin{pmatrix} 1 & 0 \\ 0 & 1 \end{pmatrix} \\ &= \begin{pmatrix} \cos^2 \alpha & 0 & \cos \alpha \sin \alpha & 0 \\ 0 & \cos^2 \alpha & 0 & \cos \alpha \sin \alpha \\ \cos \alpha \sin \alpha & 0 & \sin^2 \alpha & 0 \\ 0 & \cos \alpha \sin \alpha & 0 & \sin^2 \alpha \end{pmatrix}, \end{aligned}$$ (30)

$$\begin{aligned} Q(\beta) &= \begin{pmatrix} 1 & 0 \\ 0 & 1 \end{pmatrix} \otimes \begin{pmatrix} \cos^2 \beta & \cos \beta \sin \beta \\ \cos \beta \sin \beta & \sin^2 \beta \end{pmatrix} \\ &= \begin{pmatrix} \cos^2 \beta & \cos \beta \sin \beta & 0 & 0 \\ \cos \beta \sin \beta & \sin^2 \beta & 0 & 0 \\ 0 & 0 & \cos^2 \beta & \cos \beta \sin \beta \\ 0 & 0 & \cos \beta \sin \beta & \sin^2 \beta \end{pmatrix}. \end{aligned}$$ (31)

We note that $P(\alpha)$ and $Q(\beta)$ are commuting projections for fixed α and β. It follows that $P(\alpha)Q(\beta)$ is again a projection, as well as the products

$$P(\alpha)(\mathbb{1} - Q(\beta)), \quad (\mathbb{1} - P(\alpha))Q(\beta), \quad (\mathbb{1} - P(\alpha))(\mathbb{1} - Q(\beta)).$$ (32)

So we obtain the description of a classical probability space with four states, to be interpreted as

$$\begin{aligned} &(\text{"left photon passes," "right photon passes"}), \\ &(\text{"left photon passes," "right photon is absorbed"}), \\ &(\text{"left photon is absorbed," "right photon passes"}), \\ &(\text{"left photon is absorbed," "right photon is absorbed"}). \end{aligned}$$ (33)

The probabilities of these four events are found by the actions on $\psi = \frac{1}{\sqrt{2}}(e_{12} - e_{21}) = \frac{1}{2}(0, 1, -1, 0)$ of the four projections. In particular, the probability that both photons pass is given by

$$\langle \psi, P(\alpha)Q(\beta)\psi \rangle = \frac{1}{2}(0, 1, -1, 0) \, M \begin{pmatrix} 0 \\ 1 \\ -1 \\ 0 \end{pmatrix}, \tag{34}$$

where, setting $C_a = \cos \alpha$, $C_b = \cos \beta$ and $S_a = \sin \alpha$, $S_b = \sin \beta$, the matrix 4×4 matrix M reads

$$M = \begin{pmatrix} C_a^2 C_b^2 & C_a^2 C_b S_b & C_a S_a C_b^2 & C_a S_a C_b S_b \\ C_a^2 C_b S_b & C_a^2 S_b^2 & C_a S_a C_b S_b & C_a S_a S_b^2 \\ C_a S_a C_b^2 & C_a S_a C_b S_b & S_a^2 C_b^2 & S_a^2 C_b S_b \\ C_a S_a C_b S_b & C_a S_a S_b^2 & S_a^2 C_b S_b & S_a^2 S_b^2 \end{pmatrix}. \tag{35}$$

It thus follows that

$$\langle \psi, P(\alpha)Q(\beta)\psi \rangle = \frac{1}{2}(\cos^2 \alpha \sin^2 \beta + \sin^2 \alpha \cos^2 \beta - 2 \cos \alpha \sin \alpha \cos \beta \sin \beta)$$
$$= \frac{1}{2}(\cos \alpha \sin \beta - \sin \alpha \cos \beta)^2 = \frac{1}{2} \sin^2(\alpha - \beta). \tag{36}$$

4 Quantum Probability

In classical probability a model – or *probability space* – is determined by giving a set Ω of outcomes ω, by specifying what subsets $S \subset \Omega$ are to be considered as *events*, and by associating a *probability* $P(S)$ with each of these events.

Requirements: The events must correspond to subsets from a σ-algebra that is a collection of sets that is closed with respect to all possible (infinite) unions and intersections of its subsets, which are called measurable; further, the probability measure P must be σ-additive, namely the probability of any union $S = \bigcup_j S_j$ of disjoint measurable subsets, $S_j \cap S_k = \emptyset$, must be the sum of the probabilities of the subsets, $P(S) = \sum_j P(S_j)$, and normalized, i.e., $P(\Omega) = 1$.

In quantum probability we must loosen this scheme somewhat. We must give up the set Ω of sample points: A point $\omega \in \Omega$ in a classical model decides about the occurrence or non-occurrence of all events simultaneously, and this we abandon. Following our polarization example of Sect. 2 we take as *events* certain *closed subspaces* of a *Hilbert space* or, equivalently, a set of *projections*. To all these projections we associate probabilities.

Requirements:

1. The set of $\mathcal{E}$ of all events of a quantum model must be the set of projections in some *-*algebra* $\mathcal{A}$ of operators on $\mathcal{H}$.
2. The probability function $P : \mathcal{E} \to [0, 1]$ must be σ-additive.

According to a theorem of Gleason [6], for $\dim(\mathcal{H}) \geq 3$ this implies that the probabilities are given by a *state* φ on $\mathcal{A}$:

$$\mathbf{P}(E) = \varphi(E) \qquad (E \in \mathcal{A} \text{ a projection}) . \tag{37}$$

In this section we shall work out the above notions in some detail.

4.1 *-Algebras of Operators and States

A *Hilbert space* is a complex linear space $\mathcal{H}$ with a sesquilinear function

$$\mathcal{H} \times \mathcal{H} \to \mathbf{C} : \quad (\psi, \chi) \mapsto \langle \psi, \chi \rangle , \tag{38}$$

the *inner* or *scalar* product. (For the defining properties of the inner product and the main facts about Hilbert spaces see chapter "Hilbert Space Methods for Quantum Mechanics".)

Let $\mathcal{H}$ be a finite-dimensional Hilbert space. By an *operator* on $\mathcal{H}$ we mean a linear map $A : \mathcal{H} \to \mathcal{H}$. Operators can be added and multiplied in the natural way. By the *adjoint* of an operator A we mean the unique operator $A^\dagger$ on $\mathcal{H}$ satisfying

$$\forall \psi, \vartheta \in \mathcal{H} : \quad \langle A^\dagger \psi, \vartheta \rangle = \langle \psi, A\vartheta \rangle . \tag{39}$$

The *norm* of an operator A is defined by

$$\| A \| := \sup \{ \| A\psi \| \mid \psi \in \mathcal{H}, \| \psi \| = 1 \} . \tag{40}$$

It has the property

$$\left\| A^\dagger A \right\| = \| A \|^2 . \tag{41}$$

Exercise 1 *Prove this!*

By a *(unital) *-algebra of operators on $\mathcal{H}$* we mean a subspace $\mathcal{A}$ of the space of all linear maps $A : \mathcal{H} \to \mathcal{H}$ such that $\mathbb{1} \in \mathcal{A}$ and

$$A, B \in \mathcal{A} \implies \lambda A, A + B, A \cdot B, A^\dagger \in \mathcal{A}. \tag{42}$$

By a *state* on $\mathcal{A}$ we mean a linear functional $\varphi : \mathcal{A} \to \mathbf{C}$ satisfying

1. $\forall A \in \mathcal{A} : \quad \varphi(A^\dagger A) \geq 0$ and
2. $\varphi(\mathbb{1}) = 1$.

We shall call a pair $(\mathcal{A}, \varphi)$ of the above kind a *quantum probability space*.

Examples

1. Let P_1, P_2, ..., P_k be mutually orthogonal projections on $\mathcal{H}$ with sum $\mathbb{1}$. Then
   their linear span

$$\mathcal{A} := \left\{ \sum_{j=1}^{k} \lambda_j P_j \,\middle|\, \lambda_1, \ldots, \lambda_k \in \mathbf{C} \right\} \tag{43}$$

forms a unital $*$-algebra of operators on $\mathcal{H}$. This is basically the classical model
of Sect. 2.4.: $\mathcal{A}$ is isomorphic to $\mathbf{C}(\Omega)$, the algebra of all complex functions
on the finite set $\Omega = \{1, \ldots, k\}$. If ψ is some vector in $\mathcal{H}$ of unit length, it
determines a state φ by

$$\varphi(A) := \langle \psi, A\psi \rangle . \tag{44}$$

The probabilities of this classical model are $p_j := \varphi(P_j) = \| P_j \psi \|^2$. Note that
there are many ψ's, and even more density matrices ρ (see Sect. 2.4) determining
the same state φ on $\mathcal{A}$.

2. Let $\mathcal{A}$ be the $*$-algebra M_n of all complex $n \times n$ matrices. Let $\varphi(A) := \mathrm{Tr}\,(\rho A)$
   with $\rho \geq 0$ and $\mathrm{Tr}\,(\rho) = 1$, as introduced in Sect. 2.4. The state φ is called a
   pure state if $\rho = |\psi\rangle\langle\psi|$ for some unit vector $\psi \in \mathcal{H}$.

   The qubit of Sect. 2.2 corresponds to the case $n = 2$.

   The most general way of representing M_n on a (finite-dimensional) Hilbert
   space is

$$\mathcal{H} = \mathbf{C}^m \otimes \mathbf{C}^n \quad (m \geq 1); \qquad \mathcal{A} = \left\{ \mathbb{1} \otimes A \,\middle|\, A \in M_n \right\}. \tag{45}$$

3. Let $k, n_1, \ldots, n_k, m_1, \ldots, m_k$ be natural numbers and let the Hilbert space $\mathcal{H}$ be
   given by

$$\mathcal{H} := \left(\mathbf{C}^{m_1} \otimes \mathbf{C}^{n_1} \right) \oplus \left(\mathbf{C}^{m_2} \otimes \mathbf{C}^{n_2} \right) \oplus \cdots \oplus \left(\mathbf{C}^{m_k} \otimes \mathbf{C}^{n_k} \right) . \tag{46}$$

Let $\mathcal{A}$ be the $*$-algebra given by

$$\mathcal{A} := \left\{ (\mathbb{1} \otimes A_1) \oplus \cdots \oplus (\mathbb{1} \otimes A_k) \,\middle|\, A_j \in M_{n_j} \text{ for } j = 1, \ldots, k \right\}. \tag{47}$$

Let $\psi = \psi_1 \oplus \cdots \oplus \psi_k$ be a unit vector in $\mathcal{H}$ and

$$\varphi(A) := \langle \psi, A\psi \rangle = \sum_{j=1}^{k} \langle \psi_j, A_j \psi_j \rangle . \tag{48}$$

If $m_j \geq n_j \forall j$ then every state on $\mathcal{A}$ is of the above form. Otherwise, density
matrices may be needed.

In finite dimension Example 1 is the only commutative possibility, Example 2 is the "purely quantum mechanical" situation, and Example 3 is the most general case.

Theorem 1 *Every Abelian, that is commutative, ∗-algebra of operators on a finite-dimensional Hilbert space is isomorphic to* $\mathbf{C}(\Omega)$ *for some finite* Ω.

This is the finite-dimensional version of Gel'fand's theorem [3] on commutative (C^*)-algebras.

Proof Since the operators in $\mathcal{A}$ all commute, there exists an orthonormal basis $e_1, \ldots, e_n$ in $\mathcal{H}$ on which they are all represented by diagonal matrices. Then the states $\omega_j : A \mapsto \langle e_j, Ae_j \rangle$ are multiplicative:

$$\omega_j(AB) = \langle e_j, ABe_j \rangle = \sum_{i=1}^{n} \langle e_j, Ae_i \rangle \langle e_i, Be_j \rangle = \langle e_j, Ae_j \rangle \langle e_j, Be_j \rangle$$

$$= \omega_j(A)\omega_j(B) \, . \tag{49}$$

These states need not all be different; let $\Omega := (\omega_{j_1}, \ldots, \omega_{j_k})$ be a maximal set of different ones. Then the map

$$\iota : \mathcal{A} \to \mathbf{C}(\Omega) : \iota(A)(\omega) := \omega(A) \tag{50}$$

is an isomorphism. The projections of Example 1 are found back as the operators $P_\omega := \iota^{-1}(\delta_\omega)$. $\qquad\square$

Exercise 2 *Check that the map ι defined above is indeed an isomorphism of ∗-algebras.*

Definition 1 By the commutant of a set $\mathcal{S}$ of operators on $\mathcal{H}$ we mean the ∗-algebra

$$\mathcal{S}' := \left\{ B : \mathcal{H} \to \mathcal{H} \text{ linear} \,\middle|\, \forall A \in \mathcal{S} : AB = BA \right\} . \tag{51}$$

The algebra generated by $\mathbb{1}$ and $\mathcal{S}$ we denote by $\text{alg}\,(\mathcal{S})$. The center of a ∗-algebra $\mathcal{A}$ is the (commutative) ∗-algebra $\mathcal{Z}$ given by

$$\mathcal{Z} := \mathcal{A} \cap \mathcal{A}' . \tag{52}$$

Exercise 3 *Find the center of $\mathcal{A}$ in each of the examples 1, 2, and 3.*

Theorem 2 (Double Commutant Theorem [3]) *Let $\mathcal{S}$ be a set of operators on a finite-dimensional Hilbert space $\mathcal{H}$, such that $X \in \mathcal{S} \Longrightarrow X^\dagger \in \mathcal{S}$. Then*

$$\text{alg}\,(\mathcal{S}) = \mathcal{S}'' . \tag{53}$$

Proof Clearly $S \subset S''$, and since S'' is a $*$-algebra, we have $\mathrm{alg}\,(S) \subset S''$. We shall now prove the converse inclusion. Let $B \in S''$ and let $\mathcal{A} := \mathrm{alg}\,(S)$. We must show that $B \in \mathcal{A}$.

Step 1: Choose $\psi \in \mathcal{H}$ and let P be the orthogonal projection onto $\mathcal{A}\psi$. Then for all $X \in S$ and $A \in \mathcal{A}$

$$XPA\psi = XA\psi \in \mathcal{A}\psi \quad \Longrightarrow \quad XPA\psi = PXA\psi. \tag{54}$$

So XP and PX coincide on the space $\mathcal{A}\psi$. But if $\vartheta \perp \mathcal{A}\psi$, then $P\vartheta = 0$ and for all $A \in \mathcal{A}$

$$\langle X\vartheta, A\psi \rangle = \langle \vartheta, X^\dagger A\psi \rangle = 0, \tag{55}$$

so $X\vartheta \perp \mathcal{A}\psi$ as well. Hence $PX\vartheta = 0 = XP\vartheta$, and the operators XP and PX also coincide on the orthogonal complement of $\mathcal{A}\psi$. We conclude that $XP = PX$, i.e., $P \in S'$. But then we also have $BP = PB$, since $B \in S''$. So

$$B\psi = BP\psi = PB\psi \in \mathcal{A}\psi, \tag{56}$$

and $B\psi$ is of the form $A\psi$ for some $A \in \mathcal{A}$.

Step 2: But this is not sufficient: We must show that $B\psi = A\psi$ for all ψ in a basis for $\mathcal{H}$.

So choose a basis $\psi_1, \ldots, \psi_n$ of $\mathcal{H}$. We define

$$\begin{aligned}
\widetilde{\mathcal{H}} &:= \mathcal{H} \oplus \mathcal{H} \oplus \cdots \oplus \mathcal{H} = \mathbf{C}^n \otimes \mathcal{H}, \\
\widetilde{\mathcal{A}} &:= \left\{ A \oplus A \oplus \cdots \oplus A \,\middle|\, A \in \mathcal{A} \right\} = \mathcal{A} \otimes \mathbb{1}, \\
\widetilde{\psi} &:= \psi_1 \oplus \psi_2 \oplus \cdots \oplus \psi_n.
\end{aligned} \tag{57}$$

Then $(\widetilde{\mathcal{A}})' = (\mathcal{A} \otimes \mathbb{1})' = \mathcal{A}' \otimes M_n$ and $(\widetilde{\mathcal{A}})'' = (\mathcal{A}' \otimes M_n)' = \mathcal{A}'' \otimes \mathbb{1}$. So $B \otimes \mathbb{1} \in (\widetilde{\mathcal{A}})''$. By step 1 we find an element $\widetilde{A}$ of $\widetilde{\mathcal{A}}$, such that

$$\widetilde{A}\widetilde{\psi} = (B \otimes \mathbb{1})\widetilde{\psi}. \tag{58}$$

But $\widetilde{A} \in \widetilde{\mathcal{A}}$ must be of the form $A \otimes \mathbb{1}$ with $A \in \mathcal{A}$, so

$$A\psi_1 \oplus \cdots \oplus A\psi_n = B\psi_1 \oplus \cdots \oplus B\psi_n. \tag{59}$$

This implies that $A = B$, hence $B \in \mathcal{A}$. $\qquad\square$

Exercise 4 *Find the algebra generated by* $\mathbb{1}$ *and the matrix*

$$\begin{pmatrix} 0 & 1 & 0 \\ 1 & 0 & 0 \\ 0 & 0 & 0 \end{pmatrix}. \tag{60}$$

We give the following proposition without proof. It characterizes the situation of Example 2.

Proposition 4 *If the center of $\mathcal{A}$ contains only multiples of $\mathbb{1}$, then $\mathcal{H}$ and $\mathcal{A}$ must be of the form*

$$\mathcal{H} = \mathbf{C}^m \otimes \mathbf{C}^n, \quad \text{with} \quad \mathcal{A} = \big\{ \mathbb{1} \otimes A \,\big|\, A \in M_n \big\}. \tag{61}$$

Proposition 5 *Let $\mathcal{H}$ be a finite-dimensional Hilbert space. Then every $*$-algebra of operators on $\mathcal{H}$ can be written in the form of Example 3.*

Proof The center $\mathcal{A} \cap \mathcal{A}'$ is an Abelian (commutative) $*$-algebra, so Theorem 1 applies, giving a set of projections P_j, $j = 1, \ldots, k$. Then it is not difficult to show that the unital $*$-algebras $P_j \mathcal{A} P_j$ on the Hilbert subspaces $P_j \mathcal{H}$ satisfy the condition of Proposition 4. The statement follows. $\qquad\square$

4.2 The Qubit

The simplest non-commutative $*$-algebra is M_2, the algebra of all 2×2 matrices with complex entries. And the simplest state on M_2 is $\frac{1}{2}\mathrm{Tr}$, the quantum analogue of a fair coin.

The events in this probability space are the orthogonal projections in M_2: the complex 2×2 matrices E satisfying

$$E^2 = E = E^\dagger. \tag{62}$$

Let us see what these projections look like. Since E is self-adjoint, it must have two real eigenvalues, and since $E^2 = E$ these must be both 0 and 1. So we have three possibilities:

- Both are 0, i.e., $E = 0$.
- One of them is 0 and the other is 1.
- Both are 1, i.e., $E = \mathbb{1}$.

In the second case, E is a one-dimensional projection satisfying

$$\mathrm{Tr}\, E = 0 + 1 = 1, \quad \det E = 0 \cdot 1 = 0. \tag{63}$$

As $E^\dagger = E$ and $\mathrm{Tr}\, E = 1$ we may write

$$E = E(x, y, z) = \frac{1}{2} \begin{pmatrix} 1+z & x-iy \\ x+iy & 1-z \end{pmatrix}. \tag{64}$$

Then $\det E = 0$ implies that

$$\tfrac{1}{4}((1-z^2) - (x^2 + y^2)) = 0 \quad \implies \quad x^2 + y^2 + z^2 = 1. \tag{65}$$

So the one-dimensional projections in M_2 are parametrized by the unit sphere S_2.
Notation: For $a = (a_1, a_2, a_3) \in \mathbb{R}^3$ let us write

$$\sigma(a) := \begin{pmatrix} a_3 & a_1 - ia_2 \\ a_1 + ia_2 & -a_3 \end{pmatrix} = a_1\sigma_1 + a_2\sigma_2 + a_3\sigma_3, \tag{66}$$

where $\sigma_1, \sigma_2,$ and σ_3 are the *Pauli matrices*

$$\sigma_1 := \begin{pmatrix} 0 & 1 \\ 1 & 0 \end{pmatrix}, \quad \sigma_2 := \begin{pmatrix} 0 & -i \\ i & 0 \end{pmatrix}, \quad \sigma_3 := \begin{pmatrix} 1 & 0 \\ 0 & -1 \end{pmatrix}. \tag{67}$$

We note that for all $a, b \in \mathbb{R}^3$ we have

$$\sigma(a)\sigma(b) = \langle a, b \rangle \cdot \mathbb{1} + i\sigma(a \times b). \tag{68}$$

We may now write (64) as

$$E(a) := \tfrac{1}{2}(\mathbb{1} + \sigma(a)) \quad (\| a \| = 1). \tag{69}$$

In the same way the possible states on M_2 can be calculated. We find that

$$\varphi(A) = \mathrm{Tr}\,(\rho A) \quad \text{where} \quad \rho = \rho(a) := \tfrac{1}{2}(\mathbb{1} + \sigma(a)), \quad \| a \| \le 1. \tag{70}$$

The probability of the event $E(a)$ in the state $\rho(b)$ is given by $\mathrm{Tr}\,(\rho(b)E(a)) = \tfrac{1}{2}(1 + \langle a, b \rangle)$. The events $E(a)$ and $E(b)$ are compatible if and only if $a = \pm b$. Moreover we have for all $a \in S_2$: $E(a) + E(-a) = \mathbb{1}$, $E(a)E(-a) = 0$.
Interpretation: The state of the qubit is given by a vector b in the three-dimensional unit ball. For every a on the unit sphere we can say with probability 1 that of the two events $E(a)$ and $E(-a)$ exactly one will occur, $E(a)$ having probability $\tfrac{1}{2}(1 + \langle a, b \rangle)$. So we have a classical coin toss (with probability for heads equal to $\tfrac{1}{2}(1 + \langle a, b \rangle)$) for every direction in $\mathbb{R}^3$. The coin tosses in different directions are incompatible (see Fig. 5).

The quantum coin toss is realized in nature: apart from photon polarization (see Sect. 3.2), the spin direction of a particle with total spin $\tfrac{1}{2}$ behaves in this way.

4.3 Photons

There is a second natural way to parametrize the one-dimensional projections in M_2, which is closer to the description of polarization of photons, as treated in Sect. 3.2.

The projection onto the one-dimensional subspace spanned by the unit vector $(\cos\alpha, e^{i\varphi}\sin\alpha)$ mentioned in (22) of that section is given by

$$F(\alpha, \varphi) = \begin{pmatrix} \cos^2\alpha & e^{-i\varphi}\cos\alpha\sin\alpha \\ e^{i\varphi}\cos\alpha\sin\alpha & \sin^2\alpha \end{pmatrix}. \tag{71}$$

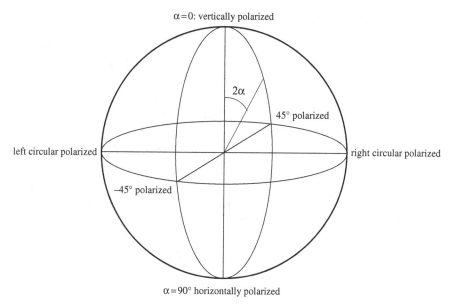

Fig. 5 Bloch sphere of the qubit

Equating this projection to $E(x, y, z)$ in (64) we obtain the relations $x = \sin 2\alpha \cos \varphi$, $y = \sin 2\alpha \sin \varphi$, and $z = \cos 2\alpha$; they define a mapping between the polarization states of a photon and the points of the unit sphere in $\mathbb{R}^3$, called the *Bloch sphere* in this context.

In particular, the projection $F(\alpha, 0)$ onto the line in $\mathbb{C}^2$ with real slope $\tan \alpha$ with $\alpha \in [-\pi/2, \pi/2)$ is given by

$$F(\alpha, 0) = \begin{pmatrix} \cos^2 \alpha & \cos \alpha \sin \alpha \\ \cos \alpha \sin \alpha & \sin^2 \alpha \end{pmatrix} = E(\sin 2\alpha, 0, \cos 2\alpha) . \tag{72}$$

Finally, any atomic or molecular system, only two energy levels of which are of importance in the experiment, can be described by some (M_2, φ).

Exercise 5 *Let $f : \mathbb{C} \cup \{\infty\} \to S_2$ be given by*

$$\begin{aligned}
f(0) &:= (0, 0, 1) ; \\
f(\infty) &:= (0, 0, -1) ; \\
f(re^{i\varphi}) &:= (\sin \vartheta \cos \varphi, \sin \vartheta \sin \varphi, \cos \vartheta) \\
&\quad \text{with} \quad \vartheta = 2 \arctan r, \quad r \in (0, \infty), \varphi \in [0, \pi) .
\end{aligned} \tag{73}$$

Show that $E(f(z))$ is the one-dimensional projection onto the line in $\mathbb{C}^2$ with slope $z \in \mathbb{C}$.

5 Operations on Probability Spaces

Our main objects of study will be *operations* on probability spaces. This means that we shall focus attention on the input–output aspect of probabilistic systems.

5.1 Operations on Classical Probability Spaces

It could be maintained that operations are already the core of *classical* probability. We start with a definition on the level of points.

Definition 2 By an operation from a finite classical probability space Ω to a finite classical probability space Ω' we mean an $\Omega \times \Omega'$ transition matrix, i.e. a matrix $(t_{\omega\omega'})$ of nonnegative numbers satisfying

$$\forall \omega \in \Omega : \quad \sum_{\omega' \in \Omega'} t_{\omega\omega'} = 1. \tag{74}$$

Example 1

1. Let τ be a bijection $\Omega \to \Omega'$. We may think of rearranging a deck of cards, ($\Omega = \Omega' = \{\text{cards}\}$), or the time evolution of a mechanical system ($\Omega = \Omega' =$ phase space), or the shift on sequences of letters, or just some relabeling of the outcomes of a statistical experiment. The associated matrix is

$$t_{\omega\omega'} := \begin{cases} 1 & \text{if} \quad \omega' = \tau(\omega), \\ 0 & \text{otherwise.} \end{cases} \tag{75}$$

2. Let $X : \Omega \to \Omega'$ be surjective. We think of X as an Ω'-valued random variable, where Ω' is usually some subset of $\mathbb{R}$ or $\mathbb{R}^n$ or so. The associated operation is that of "measuring X" or "forgetting everything about ω except the value of X." The associated matrix is again

$$t_{\omega\omega'} := \begin{cases} 1 & \text{if} \quad \omega' = X(\omega), \\ 0 & \text{otherwise.} \end{cases} \tag{76}$$

3. An inverse to the operation of Example 2 is given by

$$t_{\omega'\omega} := \begin{cases} \dfrac{\pi(\{\omega\})}{\pi(X^{-1}(\{\omega'\}))} & \text{if} \quad \omega' = X(\omega), \\ 0 & \text{otherwise.} \end{cases} \tag{77}$$

Here π is some probability distribution, which we assume to be everywhere nonzero. This operation describes the immersion of a system Ω' into the larger system Ω.

It can be shown that every transition matrix can be decomposed as a product of matrices of the types 3, 1, and 2. So every operation can be decomposed as an immersion, followed by a rearrangement and a restriction. Such a decomposition is called a *dilation* of the operation in question.

5.2 Quantum Operations

If $\mathcal{A}$ is a unital $*$-algebra describing a quantum system, then we denote by $\mathcal{A}^*$ the dual of $\mathcal{A}$ and by $\mathcal{A}^*_{+,1}$ the positive normalized functionals, i.e., the *states* on $\mathcal{A}$. By $M_n(\mathcal{A})$ we denote the unital $*$-algebra of all $n \times n$ matrices with entries in $\mathcal{A}$. Note that $M_n(\mathcal{A})$ is isomorphic to $M_n \otimes \mathcal{A}$.

Now suppose that we perform a physical operation which takes as input a state on the system $\mathcal{A}$ and yields as its output a state on the system $\mathcal{B}$. Which maps $f : \mathcal{A}^*_{+,1} \to \mathcal{B}^*_{+,1}$ can occur as descriptions of such an operation? We formulate three natural requirements:

1. f must be an affine map. This means that for all $\rho, \theta \in \mathcal{A}^*_{+,1}$ and all $\lambda \in [0, 1]$

$$\lambda f(\rho) + (1 - \lambda) f(\vartheta) = f\big(\lambda \rho + (1 - \lambda) \vartheta\big). \tag{78}$$

This request comes from the *stochastic equivalence principle* which states that a system which is in state ρ with probability λ and in state ϑ with probability $1 - \lambda$ cannot be distinguished from a system in the state $\lambda \rho + (1 - \lambda) \vartheta$. A map f satisfying this condition can be extended to a unique linear map $\mathcal{A}^* \to \mathcal{B}^*$, since every element of $\mathcal{A}^*$ can be written as a linear combination of (at most four) states on $\mathcal{A}$. So f must be the adjoint (or dual, see chapter "Bipartite Quantum Entanglement", Sect. 3) of some linear map $T : \mathcal{B} \to \mathcal{A}$. We shall henceforth write T^* instead of f.

2. Of course, $f = T^*$ must still map $\mathcal{A}^*_{+,1}$ to $\mathcal{B}^*_{+,1}$ for all $\rho \in \mathcal{A}^*$,

$$\begin{aligned}
\operatorname{Tr}(T^*(\rho)) &= \operatorname{Tr}(\rho), \\
T^*(\rho) &\geq 0 \quad \text{if} \quad \rho \geq 0.
\end{aligned} \tag{79}$$

3. It would seem at first sight that nothing more can be said a priori about T^*. However, it was realized in the early 1980s by Karl Kraus [7] (see chapter "Hilbert Space Methods for Quantum Mechanics", Sect. 2.4) that the positivity property has to be strengthened in quantum mechanics: If the system under consideration is in a combined state with some other system, then after performing the operation T^* on the former system, the whole combination must still be in some (positive) state. Surprisingly, this is not automatic in the quantum situation, where "entanglement," as treated in Sect. 2 (see chapter "Bipartite Quantum Entanglement" for more details on this point), can occur between the two systems. See Example 2. Therefore this stronger form of positivity must be added as a requirement: For all $n \in \mathbb{N}$

$$\operatorname{id}_n \otimes T^* \quad \text{maps states on} \quad M_n \otimes \mathcal{A} \quad \text{to states on} \quad M_n \otimes \mathcal{B}. \tag{80}$$

Requirement (3) is called *complete positivity* of the map T^* (or T for that matter). Summarizing we arrive at the following definition, which we shall formulate in the contravariant, "Heisenberg" picture.

Definition 3 A linear map $T : \mathcal{B} \to \mathcal{A}$ is called an operation (from $\mathcal{A}$ to $\mathcal{B}$!) if the following conditions hold:

1. $T(\mathbb{1}_{\mathcal{B}}) = \mathbb{1}_{\mathcal{A}}$.
2. T is completely positive, i.e., id$_n \otimes T$ is positive $M_n(\mathcal{B}) \to M_n(\mathcal{A})$ for all $n \in \mathbb{N}$.

Here $M_n(\mathcal{A})$ stands for the algebra of $n \times n$ matrices with entries in $\mathcal{A}$. This algebra is isomorphic to $M_n \otimes \mathcal{A}$.

Example 2 A map which is positive, but not completely positive:
Let $\mathcal{A} := M_2$ and let

$$T^* : \mathcal{A}^* \to \mathcal{A}^* : \begin{pmatrix} a & b \\ c & d \end{pmatrix} \mapsto \begin{pmatrix} a & c \\ b & d \end{pmatrix} \tag{81}$$

be the transposition map. Then T^* is linear, positive, and preserves the trace. However, T^* is not completely positive since

$$\text{id}_2 \otimes T^* : \frac{1}{2} \begin{pmatrix} 1 & 0 & 0 & 1 \\ 0 & 0 & 0 & 0 \\ 0 & 0 & 0 & 0 \\ 1 & 0 & 0 & 1 \end{pmatrix} \mapsto \frac{1}{2} \begin{pmatrix} 1 & 0 & 0 & 0 \\ 0 & 0 & 1 & 0 \\ 0 & 1 & 0 & 0 \\ 0 & 0 & 0 & 1 \end{pmatrix}. \tag{82}$$

The matrix on the left is a projection (on the vector $(e_0 \otimes e_0 + e_1 \otimes e_1)/\sqrt{2} \in \mathbb{C}^2 \otimes \mathbb{C}^2$; compare the entangled state of Sect. 2.5), whereas the matrix on the left has eigenvalues $\frac{1}{2}, \frac{1}{2}, \frac{1}{2}$, and $-\frac{1}{2}$, hence is not a valid density matrix. However, if $\mathcal{A}$ or $\mathcal{B}$ is Abelian, that is commutative, then any positive operator $T : \mathcal{A} \to \mathcal{B}$ is automatically completely positive.

5.3 Examples of Quantum Operations

- Let $U \in M_n$ be unitary. Then the automorphism $T : M_n \to M_n : A \mapsto U^\dagger A U$ is an operation (see Lemma 1).
- The $*$-homomorphism $j : M_k \to M_l \otimes M_k : A \mapsto \mathbb{1} \otimes A$ is an operation (see Lemma 1).
- Let φ be a state on M_k. Then the map $E : M_l \otimes M_k \to M_k : B \otimes A \mapsto \varphi(B)A$ is an operation.

The above examples are to be compared with those in Sect. 4.1. We shall prove their validity in two lemmas.

Lemma 1 *If $\mathcal{A} \subset M_k$ and $T : \mathcal{A} \to \mathcal{B} \subset M_l$ is a $*$-homomorphism, i.e., if for all A, $B \in \mathcal{A}$ we have $T(AB) = T(A)T(B)$ and $T(A^\dagger) = T(A)^\dagger$, then T is completely positive.*

Proof We must show that for all $n \in \mathbb{N}$ the map

$$\mathrm{id}_n \otimes T : \left(A_{ij}\right)_{i,j=1}^n \mapsto \left(T(A_{ij})\right)_{i,j=1}^n \tag{83}$$

is positive. Indeed, for all $\psi = (\psi_1, \ldots, \psi_n) \in (\mathbf{C}^l)^n$, putting $A = X^\dagger X$ with $X \in M_n(\mathcal{A})$

$$
\begin{aligned}
\langle \psi, (\mathrm{id}_n \otimes T)(X^\dagger X)\psi \rangle &= \sum_{i,i'=1}^l \langle \psi_i, T\left((X^\dagger X)_{ii'}\right)\psi_{i'} \rangle \\
&= \sum_{i,i'=1}^l \sum_{j=1}^n \langle \psi_i, T\left(X_{ji}^\dagger X_{ji'}\right)\psi_{i'} \rangle \\
&= \sum_{i,i'=1}^l \sum_{j=1}^n \langle \psi_i, T(X_{ji})^\dagger T(X_{ji'})\psi_{i'} \rangle \\
&= \sum_{j=1}^n \left\| \sum_{i=1}^l T(X_{ji})\psi_i \right\|^2 \geq 0 .
\end{aligned} \tag{84}
$$

$\square$

Lemma 2 *Let $\mathcal{A} \subset M_k$, $\mathcal{B} \subset M_l$ and let V be a linear map $\mathbf{C}^l \to \mathbf{C}^k$. Then*

$$T : \mathcal{A} \to \mathcal{B} : A \mapsto V^\dagger A V \tag{85}$$

is completely positive.

Proof If $(A_{ij})_{i,j=1}^n \in M_n(\mathcal{A})$ is positive, then for all $(\psi_1, \ldots, \psi_n) \in (\mathbf{C}^l)^n = \mathbf{C}^n \otimes \mathbf{C}^l$ we have

$$
\begin{aligned}
\langle \psi, (\mathrm{id}_n \otimes T)(A)\psi \rangle &= \sum_{i,j=1}^n \langle \psi_i, T(A_{ij})\psi_j \rangle = \sum_{i,j=1}^n \langle \psi_i, V^\dagger A_{ij} V\psi_j \rangle \\
&= \sum_{i,j=1}^n \langle V\psi_i, A_{ij} V\psi_j \rangle \geq 0 .
\end{aligned} \tag{86}
$$

$\square$

Lemma 2 covers the third case in Example 5.3 since φ can be decomposed into pure states as $\varphi = \sum_i \langle \psi, \cdot \psi \rangle$ and

$$\varphi(B)A = \sum_{i=1}^l \lambda_i \langle \psi_i, B\psi_i \rangle A = \sum_{i=1}^l \lambda_i V_i^\dagger (B \otimes A) V_i , \tag{87}$$

where $V_i : \mathbf{C}^k \to \mathbf{C}^l \otimes \mathbf{C}^k : \vartheta \mapsto \psi_i \otimes \vartheta$.

5.4 Unraveling Quantum Operations

The following important theorem, together with Proposition 5, characterizes all completely positive maps on finite-dimensional matrix algebras (the version of this result given by Kraus has been discussed in chapter "Quantum Probability and Quantum Information Theory", Sect. 2.4).

Theorem 3 (Stinespring 1955) *Let T be a linear map $M_k \to M_l$. Then T is completely positive if and only if there exist $m \in \mathbb{N}$ and operators $V_1, \ldots, V_m : \mathbf{C}^l \to \mathbf{C}^k$ such that for all $A \in M_k$*

$$T(A) = \sum_{i=1}^{m} V_i^\dagger A V_i .$$ (88)

We shall give a proof based on a physical argument (cf. [10]). The system is put in an entangled state with a second system, which for convenience we describe by the *opposite algebra* (see below). Then we act on the main system with our operation T, and by complete positivity we get a new state on the pair. Surprisingly, this state fully characterizes the operation T. By decomposing the state into vector states we shall obtain the unraveling we wanted.

Let us first introduce some notation. If $\mathcal{H}$ is a (finite-dimensional) Hilbert space, let $\mathcal{H}'$ denote its *dual*, the space of all linear functionals $\mathcal{H} \to \mathbf{C}$. The elements of $\mathcal{H}'$ are of the form $\overline{\vartheta} : \chi \mapsto \langle \vartheta, \chi \rangle$; in Dirac notation $\overline{\vartheta}$ is denoted as $\langle \vartheta |$. This dual $\mathcal{H}'$ is actually isomorphic to $\mathcal{H}$ itself, but it is convenient to maintain the distinction, as we shall see below. In particular, if $\mathcal{H} = \mathbf{C}^n$, then there is a natural action on $\mathcal{H}'$ of the algebra M_n^t, the *opposite algebra* of M_n, which has the multiplication reflected: $A^t B^t = (BA)^t$. The operator A^t acts on $\overline{\chi}$ as $A^t \overline{\chi} := \overline{\chi} \circ A$.

Now consider the tensor product $\mathcal{H}_{kl} := \mathbf{C}^k \otimes (\mathbf{C}^l)'$ of the Hilbert space $\mathbf{C}^k$ and the dual of $\mathbf{C}^l$. By identifying the vector $\psi \otimes \overline{\vartheta} \in \mathcal{H}_{kl}$ with the operator $|\psi\rangle\langle\vartheta|$: $\chi \mapsto \langle \vartheta, \chi \rangle \cdot \psi$, the Hilbert space $\mathcal{H}_{kl}$ can alternatively be viewed as the space of all operators $\mathbf{C}^l \to \mathbf{C}^k$. On this Hilbert space the algebra $M_k \otimes M_l^t$ acts naturally as follows:

$$A \otimes B^t : \psi \otimes \overline{\vartheta} \mapsto A\psi \otimes B^t\overline{\vartheta} \quad \left[\approx A|\psi\rangle\langle\vartheta|B \right].$$ (89)

The space $\mathcal{H}_{ll}$ has a rotation-invariant vector (the so-called fully entangled state on $M_l \otimes M_l^t$), given by

$$\Omega := \frac{1}{\sqrt{l}} \sum_{i=1}^{l} e_i \otimes \overline{e_i} \quad \left[\approx \frac{1}{\sqrt{l}} \sum_{i=1}^{l} |e_i\rangle\langle e_i| = \mathbb{1}_l/\sqrt{l} \right],$$ (90)

for *any* orthonormal basis $e_1, \ldots, e_l$ of $\mathbf{C}^l$. This vector has the property that

$$\langle \Omega, (A \otimes B')\Omega \rangle = \frac{1}{l} \sum_{i=1}^{l} \sum_{j=1}^{l} \langle e_i \otimes \overline{e_i}, (A \otimes B')e_j \otimes \overline{e_j} \rangle$$

$$= \frac{1}{l} \sum_{i=1}^{l} \sum_{j=1}^{l} \langle e_i, Ae_j \rangle \langle \overline{e_i}, B'\overline{e_j} \rangle$$

$$= \frac{1}{l} \sum_{i=1}^{l} \sum_{j=1}^{l} \langle e_i, Ae_j \rangle \langle e_j, Be_i \rangle = \frac{1}{l} \mathrm{Tr}\,(AB). \qquad (91)$$

Proof of Stinespring's Theorem The "if" part follows immediately from Lemma 2. For the "only if" part, assume that $T : M_k \to M_l$ is completely positive. Let $\mathcal{H}_{ll} := \mathbf{C}^l \otimes (\mathbf{C}^l)'$ as above and let ω denote the state

$$\omega(X) := \langle \Omega, X\Omega \rangle \qquad (92)$$

on $\mathcal{B}(\mathcal{H}_{ll}) \approx M_l \otimes M_l'$.

Since T is completely positive, the functional ω_T on $\mathcal{B}(\mathcal{H}_{kl}) \approx M_k \otimes M_l'$, given by

$$\omega_T(A \otimes B') := \omega(T(A) \otimes B') \qquad (93)$$

is also a state. Decompose ω_T into pure states given by vectors $v_1, v_2, \ldots, v_m \in \mathcal{H}_{kl}$:

$$\omega_T(X) = \sum_{i=1}^{m} \langle v_i, X v_i \rangle. \qquad (94)$$

Now, as noted above, $v_i \in \mathcal{H}_{kl}$ can be considered as an operator $V_i : \mathbf{C}^l \to \mathbf{C}^k$. We shall show that these operators satisfy requirement (88) of the theorem. Indeed, for all $\psi, \vartheta \in \mathbf{C}^l$

$$\sum_{i=1}^{m} \langle \psi, V_i^\dagger A V_i \vartheta \rangle = \sum_{i=1}^{m} \langle V_i \psi, A V_i \vartheta \rangle = \sum_{i=1}^{m} \langle v_i, \left(A \otimes (|\overline{\psi}\rangle \langle \overline{\vartheta}|) \right) v_i \rangle_{\mathcal{H}_{kl}}$$

$$= \omega_T \left(A \otimes (|\overline{\psi}\rangle \langle \overline{\vartheta}|) \right) = \omega \left(T(A) \otimes (|\overline{\psi}\rangle \langle \overline{\vartheta}|) \right) \qquad (95)$$

$$= \mathrm{Tr}\left(T(A)(|\vartheta\rangle \langle \psi|) \right) = \langle \psi, T(A)\vartheta \rangle.$$

$\square$

The second step is verified by substituting $V_i = \sum_j |\alpha_j^i\rangle \langle \beta_j^i|$ with $\alpha_j^i \in \mathbf{C}^k$, $\beta_j^i \in \mathbf{C}^l$ and realizing that $v_i = \sum_j \alpha_j^i \otimes \overline{\beta_j^i}$.

5.5 *Uniqueness of Unravelings*

Unraveling (88) is not unique.[1] If the matrices $V_1, \ldots, V_m$ are linearly independent, then they are determined by the completely positive map T up to a transformation of the form

$$V_i' := \sum_{j=1}^{m} u_{ij} V_j , \tag{96}$$

where u is a unitary $m \times m$ matrix of complex numbers. In this independent case the number m of terms in the unraveling takes its minimal value, which we shall call the *rank* of the operation T.

In general, any number m of terms, also larger than the rank, can occur in the unraveling of T. But in that case the operators V_i are not linearly independent. In fact, the space $\mathcal{D}$ of *dependencies*, given by

$$\mathcal{D} := \left\{ \lambda \in \mathbf{C}^m \mid \sum_{i=1}^{m} \lambda_i^* V_i = 0 \right\} , \tag{97}$$

has dimension $m - \text{rank}(T)$ and the matrix u of (96) is a partial isometry with initial space $\mathcal{D}^\perp$ and final space $(\mathcal{D}')^\perp$, where $\mathcal{D}'$ denotes the space of dependencies of the V_i'.

We shall now prove these statements in the context of the decomposition of states. From the proof of Theorem 3 it is clear that they carry over to operations.

Proposition 6 *Let φ be a state on $\mathcal{A} := M_k$ and let two decompositions of φ into pure states be given*

$$\varphi(A) = \sum_{i=1}^{m} \langle \psi_i, A\psi_i \rangle = \sum_{j=1}^{n} \langle \vartheta_j, A\vartheta_j \rangle . \tag{98}$$

Let $\mathcal{D} \subset \mathbf{C}^m$ and $\mathcal{D}' \subset \mathbf{C}^n$ denote the dependency spaces of $\psi = (\psi_1, \ldots, \psi_m)$ and $\vartheta = (\vartheta_1, \ldots, \vartheta_n)$, respectively. Then ψ and ϑ are connected by a transformation of the form

$$\vartheta_j = \sum_{i=1}^{m} u_{ji} \psi_i , \tag{99}$$

where the $n \times m$ matrix u describes a partial isometry $\mathbf{C}^m \to \mathbf{C}^n$ with initial space $\mathcal{D}^\perp$ and final space $(\mathcal{D}')^\perp$. In particular, if the m-tuple $(\psi_1, \ldots, \psi_m)$ and

---

[1] This section elaborates on a remark by Mark Fannes and can be skipped in a first reading.

the n-tuple $(\vartheta_1, \ldots, \vartheta_n)$ are both sequences of independent vectors, then $n = m$ and u is unitary.

Proof Consider ψ and ϑ as vectors in $\mathcal{H} := (\mathbf{C}^k)^m = \mathbf{C}^m \otimes \mathbf{C}^k$ and $\mathcal{H}' := (\mathbf{C}^k)^n = \mathbf{C}^n \otimes \mathbf{C}^k$, respectively. Then (98) can be written in the form

$$\varphi(A) = \langle \psi, (\mathbb{1}_m \otimes A)\psi \rangle = \langle \vartheta, (\mathbb{1}_n \otimes A)\vartheta \rangle . \tag{100}$$

Let $\Lambda \subset \mathcal{H}$ and $\Lambda' \subset \mathcal{H}'$ be the subspaces consisting of the vectors $(\mathbb{1}_m \otimes A)\psi$ and $(\mathbb{1}_n \otimes A)\vartheta$, respectively, where A runs through the matrix algebra $\mathcal{A} = M_k$. Let $U : \Lambda \to \Lambda'$ be given by

$$U(\mathbb{1}_m \otimes A)\psi := (\mathbb{1}_n \otimes A)\vartheta . \tag{101}$$

Then U is well defined, isometric, and onto since

$$\begin{aligned} \| (\mathbb{1}_n \otimes A)\vartheta \|^2 &= \langle (\mathbb{1}_n \otimes A)\vartheta, (\mathbb{1}_n \otimes A)\vartheta \rangle = \langle \vartheta, (\mathbb{1}_n \otimes A^\dagger A)\vartheta \rangle \\ &= \varphi(A^\dagger A) = \| (\mathbb{1}_m \otimes A)\psi \|^2 . \end{aligned} \tag{102}$$

We extend U to a map $\mathcal{H} \to \mathcal{H}'$ by putting $U\chi = 0$ for all $\chi \in \mathcal{H}$ which are orthogonal to Λ.

Next, let us show that U is actually of the form $u \otimes \mathbb{1}_k$ for some partial isometry $u : \mathbf{C}^m \to \mathbf{C}^n$. This is equivalent to the statement that for all $A \in M_k$:

$$U(\mathbb{1}_m \otimes A) = (\mathbb{1}_n \otimes A)U , \tag{103}$$

which is true since $(\mathbb{1}_m \otimes A)$ leaves $\Lambda^\perp$ invariant, so that both sides vanish on $\Lambda^\perp$. And for $\chi \in \Lambda$, i.e., for $\chi = (\mathbb{1}_m \otimes X)\psi$ with $X \in M_k$, we have

$$\begin{aligned} U(\mathbb{1}_m \otimes A)\chi &= U(\mathbb{1}_m \otimes A)(\mathbb{1}_m \otimes X)\psi = U(\mathbb{1}_m \otimes AX)\psi = (\mathbb{1}_n \otimes AX)\vartheta \\ &= (\mathbb{1}_n \otimes A)(\mathbb{1}_n \otimes X)\vartheta = (\mathbb{1}_n \otimes A)U(\mathbb{1}_m \otimes X)\psi = (\mathbb{1}_n \otimes A)U\chi . \end{aligned} \tag{104}$$

It remains to be shown that

$$\Lambda^\perp = \mathcal{D} \otimes \mathbf{C}^k \tag{105}$$

(and analogously $(\Lambda')^\perp = \mathcal{D}' \otimes \mathbf{C}^k$). Clearly, for all $\lambda \in \mathbf{C}^m$ and $\mu \in \mathbf{C}^k$,

$$\langle \lambda \otimes \mu, (\mathbb{1} \otimes A)\psi \rangle = \sum_{i=1}^m \lambda_i^* \langle \mu, A\psi_i \rangle = \left\langle A^\dagger \mu, \left(\sum_{i=1}^m \lambda_i^* \psi_i \right) \right\rangle . \tag{106}$$

It follows that for $\lambda \in \mathcal{D}$ the vector $\lambda \otimes \mu$ is orthogonal to Λ, so we have $\mathcal{D} \otimes \mathbf{C}^k \subset \Lambda^\perp$. To prove the converse inclusion, we first note that the orthogonal projection onto Λ is $U^\dagger U = u^\dagger u \otimes \mathbb{1}_k$, hence $\Lambda = \mathcal{E} \otimes \mathbf{C}^k$ for some subspace $\mathcal{E}$ of $\mathbf{C}^m$. We must show that $\mathcal{E}^\perp \subset \mathcal{D}$. So suppose that $\lambda \perp \mathcal{E}$, so that $\lambda \otimes \mu \perp \Lambda$ for all $\mu \in \mathbf{C}^k$. Putting $A = \mathbb{1}$ in (106) we find that the left-hand side, and hence the right-hand side, is 0 for all μ, so $\sum_{i=1}^m \lambda_i^* \psi_i = 0$ and $\lambda \in \mathcal{D}$. $\square$

5.6 Properties of Quantum Operations

When A and B are operators on a Hilbert space, we mean by $A \geq B$ that the difference $A - B$ is a positive operator. The following is an extremely useful inequality for operations.

Proposition 7 (Cauchy–Schwartz for operations) *Let $\mathcal{A}$ and $\mathcal{B}$ be *-algebras of operators on Hilbert spaces $\mathcal{H}$ and $\mathcal{K}$ and let $T : \mathcal{A} \to \mathcal{B}$ be an operation. Then we have for all $A \in \mathcal{A}$*

$$T(A^\dagger A) \geq T(A)^\dagger T(A). \tag{107}$$

Proof The operator $X \in M_2 \otimes \mathcal{A}$ given by

$$X := \begin{pmatrix} A^\dagger A & -A^\dagger \\ -A & \mathbb{1} \end{pmatrix} = \begin{pmatrix} A & -\mathbb{1} \\ 0 & 0 \end{pmatrix}^\dagger \begin{pmatrix} A & -\mathbb{1} \\ 0 & 0 \end{pmatrix} \tag{108}$$

is positive. Since T is completely positive and $T(\mathbb{1}) = \mathbb{1}$, it follows that also

$$(\mathrm{id} \otimes T)(X) = \begin{pmatrix} T(A^\dagger A) & -T(A)^\dagger \\ -T(A) & \mathbb{1} \end{pmatrix} \tag{109}$$

is a positive operator. Putting $\xi := \psi \oplus T(A)\psi$ we find that

$$\langle \xi, (\mathrm{id} \otimes T) X \xi \rangle = \langle \psi, \big(T(A^\dagger A) - T(A)^\dagger T(A) \big) \psi \rangle \tag{110}$$

is positive for all $\psi \in \mathcal{H}$. $\square$

Theorem 4 (Multiplication theorem) *If $T : \mathcal{A} \to \mathcal{B}$ is an operation and $T(A^\dagger A) = T(A)^\dagger T(A)$ for some $A \in \mathcal{A}$, then $T(A^\dagger B) = T(A)^\dagger T(B)$ and $T(B^\dagger A) = T(B)^\dagger T(A)$ for all $B \in \mathcal{A}$.*

Proof Take any $B \in \mathcal{A}$ and $\lambda \in \mathbb{R}$. Then

$$T\big((A^\dagger + \lambda B^\dagger)(A + \lambda B)\big) = T(A)^\dagger T(A) + \lambda T(A^\dagger B + B^\dagger A) + \lambda^2 T(B^\dagger B), \tag{111}$$

while by Cauchy–Schwartz

$$\begin{aligned} T\big((A^\dagger + \lambda B^\dagger)(A + \lambda B)\big) \\ \geq T(A)^\dagger T(A) + \lambda(T(A)^\dagger T(B) + T(B)^\dagger T(A)) + \lambda^2 T(B)^\dagger T(B). \end{aligned} \tag{112}$$

This inequality holds for all $\lambda \in \mathbb{R}$ which implies

$$T(A^\dagger B + B^\dagger A) \ge T(A)^\dagger T(B) + T(B)^\dagger T(A). \tag{113}$$

Replacing A by iA and B by $-iB$ shows that the opposite inequality also holds, so we have equality. Finally replacing only B by iB shows that $T(A^\dagger B) = T(A)^\dagger T(B)$ and $T(B^\dagger A) = T(B)^\dagger T(A)$. $\qquad\square$

In particular, if a Cauchy–Schwartz *equality* holds for an operation T then T is a *-homomorphism.

Theorem 5 (Embedding theorem) *Let $(\mathcal{A}, \varphi)$ and $(\mathcal{B}, \psi)$ be non-degenerate quantum probability spaces and let $j : \mathcal{A} \to \mathcal{B}$, $E : \mathcal{B} \to \mathcal{A}$ be operations which preserve the states. If*

$$E \circ j = \mathrm{id}_{\mathcal{A}}, \tag{114}$$

*then j is an injective *-homomorphism and $P := j \circ E$ is a conditional expectation, i.e.,*

$$P(C_1 \, B \, C_2) = C_1 \, P(B) \, C_2 \tag{115}$$

for all $C_1, C_2 \in j(\mathcal{A})$ and all $B \in \mathcal{B}$.

Following the language used in Sect. 4.1 we shall call j a *random variable* and P the *conditional expectation with respect to ψ, given j*. Compare the following proof with that of Theorem 4.

Proof For any $A \in \mathcal{A}$ we have by Cauchy–Schwartz

$$A^\dagger A = E \circ j(A^\dagger A) \ge E\big(j(A)\big)^\dagger j(A)\big) \ge \big(E \circ j(A)\big)^\dagger \big(E \circ j(A)\big) = A^\dagger A, \tag{116}$$

so we have equalities here. In particular

$$\psi\big(j(A^\dagger A) - j(A)^\dagger j(A)\big) = \varphi \circ E\big(j(A^\dagger A) - j(A)^\dagger j(A)\big) = 0. \tag{117}$$

As $(\mathcal{B}, \psi)$ is non-degenerate, $j(A^\dagger A) = j(A)^\dagger j(A)$, i.e., j is a *-homomorphism. j is injective since it has the left-inverse E.

But also from (116) we have $E\big(j(A)^\dagger j(A)\big) = E \circ j(A)^\dagger E \circ j(A)$. The Multiplication Theorem 4 then implies that for all $B \in \mathcal{B}$ and $A_1 \in \mathcal{A}$,

$$E(j(A_1)^\dagger B) = E \circ j(A_1)^\dagger E(B) = A_1^\dagger E(B), \tag{118}$$

and similarly, with $A_2 \in \mathcal{A}$

$$E\big(j(A_1)^\dagger B j(A_2)\big) = E\big(j(A_1)^\dagger B\big) E \circ j(A_2) = A_1^\dagger E(B) A_2. \tag{119}$$

Applying j to both sides we find (115). $\qquad\square$

6 Quantum Impossibilities

The result of any physical operation applied on a probabilistic system (quantum or not) is described by a completely positive identity preserving map from the state space of that system to the state space of the resulting system. This imposes strong restrictions on what can be done. Some of these are well-known quantum principles, such as the Heisenberg principle ("no measurement without disturbance"), some are surprising and relatively recent discoveries ("no cloning"), but all of them obtain quite neat formulations in the language of quantum probability.

6.1 No-Cloning

In its original formulation [5, 11] the "No-Cloning Theorem" dealt with the reproduction of nonorthogonal vector states. Here we give an algebraic version, which distinguishes clearly between the classical and the quantum cases.

"*Cloning*", or – more mundanely – *copying* a stochastic object is an operation which takes as input an object in some state ρ and yields as its output a pair of objects with identical state spaces, such that, if we throw away one of them, we are left with a single object in the state ρ (cf. Fig. 6, which is actually not complete: the same equality should hold with the other output line blocked).

In a formula, for all $\rho \in \mathcal{A}^*_{+,1}$

$$(\mathrm{Tr} \otimes \mathrm{id}) \circ C^*(\rho) = (\mathrm{id} \otimes \mathrm{Tr}) \circ C^*(\rho) = \rho. \tag{120}$$

Reformulated in the Heisenberg picture: We call an operation $C : \mathcal{A} \otimes \mathcal{A} \to \mathcal{A}$ a *copying operation* or *copier* if for all $A \in \mathcal{A}$:

$$C(\mathbb{1} \otimes A) = C(A \otimes \mathbb{1}) = A. \tag{121}$$

As is well known, copying presents no problem in classical physics or classical probability. Here is an example of a classical copying operation. For simplicity, let us think of the operation of copying n bits. Let Ω denote the space $\{0, 1\}^n$ of all strings of n bits and let $\exp^{(1)}$ be the "copying" map $\Omega \to \Omega \times \Omega : \omega \mapsto (\omega, \omega)$. This map induces an operation

$$C : \quad \mathbf{C}(\Omega) \times \mathbf{C}(\Omega) \to \mathbf{C}(\Omega) : \quad Cf(\omega) := f \circ \exp^{(1)}(\omega) = f(\omega, \omega). \tag{122}$$

Clearly, for all $f \in \mathbf{C}(\Omega)$:

$$C(\mathbb{1} \otimes f)(\omega) = (\mathbb{1} \otimes f)(\omega, \omega) = f(\omega), \tag{123}$$

Fig. 6 Definition of a copier

and the same holds for $C(f \otimes \mathbb{1})$, so (121) is satisfied. In the Schrödinger picture our operation looks as follows: for any probability distribution π on Ω,

$$(C^*\pi)(v, \omega) = \delta_{v\omega}\pi(\omega), \tag{124}$$

and we see that (120) is satisfied:

$$(\mathrm{Tr} \otimes \mathrm{id}) \circ C^*(\pi)(\omega) = \sum_{v\in\Omega} \delta_{v\omega}\pi(\omega) = \pi(\omega). \tag{125}$$

The following theorem says that this construction is only possible in the Abelian (i.e., commutative) case.

Theorem 6 ("No-cloning") *Let $\mathcal{A}$ be a ∗-algebra of operators on a (finite-dimensional) Hilbert space. Then $\mathcal{A}$ admits a copying operation if and only if $\mathcal{A}$ is Abelian.*

Proof If $\mathcal{A}$ is Abelian, by Gel'fands Theorem 1, $\mathcal{A}$ is isomorphic to $\mathbf{C}(\Omega)$ for some finite set Ω, and the above construction of a copier applies. Conversely, suppose that $C : \mathcal{A} \otimes \mathcal{A} \to \mathcal{A}$ is a copying operation. Then (121) implies that for all $A \in \mathcal{A}$

$$C\big((\mathbb{1} \otimes A)^\dagger(\mathbb{1} \otimes A)\big) = C(\mathbb{1} \otimes A^\dagger A) = A^\dagger A = C(\mathbb{1} \otimes A)^\dagger C(\mathbb{1} \otimes A). \tag{126}$$

Then it follows from the Multiplication Theorem 4 that for all $A, B \in \mathcal{A}$

$$\begin{aligned}
A B &= C(A \otimes \mathbb{1})C(\mathbb{1} \otimes B) = C\big((A \otimes \mathbb{1})(\mathbb{1} \otimes B)\big) \\
&= C\big((\mathbb{1} \otimes B)(A \otimes \mathbb{1})\big) = C(\mathbb{1} \otimes B)C(A \otimes \mathbb{1}) = B A.
\end{aligned} \tag{127}$$

$\square$

6.2 No Classical Coding

Closely related to the above is the rule that "quantum information cannot be classically coded": It is not possible to operate on a quantum system, extracting some information from it, and then from this information reconstruct the quantum system in its original state:

$$\rho \in \mathcal{A}^* \overset{C^*}{\longmapsto} \pi \in \mathcal{B}^* \overset{D^*}{\longmapsto} \rho \in \mathcal{A}^*. \tag{128}$$

We formulate this theorem in the contravariant ("Heisenberg") picture:

Theorem 7 *Let $\mathcal{A}$ and $\mathcal{B}$ be ∗-algebras and let $C : \mathcal{B} \to \mathcal{A}$ and $D : \mathcal{A} \to \mathcal{B}$ be operations, ("coding" and "decoding"), such that $C \circ D = \mathrm{id}_{\mathcal{A}}$. Then if $\mathcal{B}$ is Abelian, so is $\mathcal{A}$.*

Proof We have for all $A \in \mathcal{A}$

$$A^\dagger A = C \circ D(A^\dagger A) \geq C\left(D(A)^\dagger D(A)\right) \geq A^\dagger A \qquad (129)$$

and

$$AA^\dagger = C \circ D(AA^\dagger) \geq C\left(D(A)D(A)^\dagger\right) \geq AA^\dagger, \qquad (130)$$

whence equality holds everywhere. If $\mathcal{B}$ is Abelian, then $D(A)^\dagger D(A) = D(A)D(A)^\dagger$ and $A^\dagger A = AA^\dagger$. □

Exercise 6 *Prove that, if $A^\dagger A = AA^\dagger$ for all $A \in \mathcal{A}$, then $\mathcal{A}$ is Abelian.*

6.3 The Heisenberg Principle

The *Heisenberg principle* states – roughly speaking – that no information on a quantum system can be obtained without changing its state. In this form, the statement is not so interesting: If we realize that the *state* of the system expresses the expectations of its observables, given the information we have on it, it is no wonder that this state changes once we gain information!

A more precise formulation is the following:

> *If we extract information from a system whose algebra $\mathcal{A}$ is a factor (i.e., $\mathcal{A} \cap \mathcal{A}' = \mathbf{C}\mathbb{1}$), and if we throw away (disregard) this information, then still it cannot be avoided that some initial states are altered.*

Let us work toward a mathematical formulation: A *measurement* is an operation performed on a physical system which results in the extraction of information from that system, while possibly changing its state. So a measurement is an operation

$$M^* : \mathcal{A}^* \to \mathcal{A}^* \otimes \mathcal{B}^*, \qquad (131)$$

where $\mathcal{A}$ describes the physical system and $\mathcal{B}$ the output part of a measurement apparatus which we couple to it. $\mathcal{A}^*$ consists of states and $\mathcal{B}^*$ of probability distributions on the outcomes. So $\mathcal{B}$ will be commutative, but we do not need this property here. Now suppose that no initial state is altered by the measurement:

$$(\mathrm{id} \otimes \mathrm{Tr})M^*(\rho) = \rho \qquad \forall \rho \in \mathcal{A}^*. \qquad (132)$$

Suppose also that $\mathcal{A}$ is a factor. We claim that no information can be obtained on ρ:

$$(\mathrm{Tr} \otimes \mathrm{id})M^*(\rho) = \vartheta, \qquad (133)$$

where ϑ does not depend on ρ. Figure 7 symbolically expresses this fact.
We again formulate and prove the theorem in the contravariant picture:

Fig. 7 The Heisenberg principle

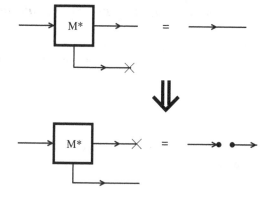

Theorem 8 (Heisenberg's principle) *Let M be an operation $\mathcal{A} \otimes \mathcal{B} \to \mathcal{A}$ such that for all $A \in \mathcal{A}$,*

$$M(A \otimes \mathbb{1}) = A, \tag{134}$$

then

$$M(\mathbb{1} \otimes B) \in \mathcal{A} \cap \mathcal{A}'. \tag{135}$$

In particular, if $\mathcal{A}$ is a factor, then for some fixed state ϑ on $\mathcal{B}$

$$M(\mathbb{1} \otimes B) = \vartheta(B) \cdot \mathbb{1}_{\mathcal{A}}. \tag{136}$$

We note that (136) implies (133), since for all ρ on $\mathcal{A}$ and all $B \in \mathcal{B}$

$$\big((\mathrm{Tr} \otimes \mathrm{id})M^*\rho\big)(B) = \rho\big(M(\mathbb{1} \otimes B)\big) = \rho\big(\vartheta(B)\mathbb{1}_{\mathcal{A}}\big) = \vartheta(B). \tag{137}$$

Proof As in the proof of the "no cloning" theorem we have by the multiplication theorem for all $A \in \mathcal{A}$, $B \in \mathcal{B}$

$$M(\mathbb{1} \otimes B) \cdot A = M(\mathbb{1} \otimes B)M(A \otimes \mathbb{1}) = M(A \otimes B). \tag{138}$$

But also,

$$A \cdot M(\mathbb{1} \otimes B) = M(A \otimes \mathbb{1})M(\mathbb{1} \otimes B) = M(A \otimes B). \tag{139}$$

So $M(\mathbb{1} \otimes B)$ lies in the center of $\mathcal{A}$. If $\mathcal{A}$ is a factor, then $B \mapsto M(\mathbb{1} \otimes B)$ is an operation from $\mathcal{B}$ to $\mathbf{C} \cdot \mathbb{1}_{\mathcal{A}}$, i.e., a state on $\mathcal{B}$ times $\mathbb{1}_{\mathcal{A}}$. $\qquad\square$

6.4 Random Variables and von Neumann Measurements

Following the suggestion made in Sect. 4.2 (in particular case 2), we define a *random variable* to be a *-homomorphism from one algebra $\mathcal{B}$ to a (larger) algebra $\mathcal{A}$:

$$\mathcal{A} \xleftarrow{\ j\ } \mathcal{B}. \tag{140}$$

In the covariant ("Schrödinger") picture this describes the operation j^* of *restriction to* the subsystem $\mathcal{B}$:

$$\mathcal{A}^* \xrightarrow{\ j^*\ } \mathcal{B}^*. \tag{141}$$

An important case is when $\mathcal{B} = \mathbf{C}(\Omega)$ for some finite set Ω; then j is to be viewed as an Ω-*valued random variable*. Let $\Omega = \{x_1, \ldots, x_n\}$. Then $j(1_{\{x_i\}})$ is a projection, P_i say, in $\mathcal{A}$, with the properties that

$$\sum_{i=1}^{n} P_i = \sum_{i=1}^{n} j(1_{\{x_i\}}) = j(1_{\mathcal{B}}) = 1_{\mathcal{A}} \tag{142}$$

and for $i \neq j$,

$$P_i P_k = j(1_{\{x_i\}}) j(1_{\{x_k\}}) = j(1_{\{x_i\}} \cdot 1_{\{x_k\}}) = 0. \tag{143}$$

We interpret P_i as the event "the random variable described by j takes the value x_i." Note that j can be written as

$$j(f) = j\left(\sum_{i=1}^{n} f(x_i) 1_{\{x_i\}}\right) = \sum_{i=1}^{n} f(x_i) P_i. \tag{144}$$

In particular, if $\Omega \subset \mathbb{R}$, then j defines a Hermitian operator

$$j(\mathrm{id}) = \sum_{i=1}^{n} x_i P_i =: X, \tag{145}$$

which completely determines j.

Proposition 8 *Let $\mathcal{A}$ be a finite-dimensional *-algebra with unit. Then there is a one-to-one correspondence between injective *-homomorphisms $j : \mathbf{C}(\Omega) \to \mathcal{A}$ for some finite $\Omega \subset \mathbb{R}$ and self-adjoint operators $X \in \mathcal{A}$, given by*

$$j(\mathrm{id}) = X. \tag{146}$$

Proof If j is a *-homomorphism $\mathbf{C}(\{x_1, \ldots, x_n\}) \to \mathcal{A}$ with $x_1, \ldots, x_n$ real, then

$$X := j(\mathrm{id}) = \sum_{i=1}^{n} x_i j(1_{\{x_i\}}) =: \sum_{i=1}^{n} x_i P_i \tag{147}$$

is a Hermitian element of $\mathcal{A}$. Conversely, if $X \in \mathcal{A}$ is Hermitian, then let $x_1, \ldots, x_n$ be its eigenvalues. Let $p : \mathbf{C} \to \mathbf{C}$ denote the polynomial

$$p(x) := (x - x_1) \cdots (x - x_n) \tag{148}$$

and let, for $i = 1, \ldots, n$, the (Lagrange interpolation) polynomial p_i be given by

$$p_i(x) := \frac{p(x)}{(x - x_i) p(x_i)} \, . \tag{149}$$

Then $p_i(x_k) = \delta_{ik} p_k$, so we have on the spectrum $\mathrm{sp}(X) = \{x_1, \ldots, x_n\}$ of X

$$\sum_{i=1}^{n} p_i = 1 \quad \text{and} \quad p_i \cdot p_k = \delta_{ik} p_k \, . \tag{150}$$

It follows that the projections $P_i := p_i(X)$, with $i = 1, \ldots, n$, lie in the algebra $\mathcal{A}$ and satisfy

$$\sum_{i=1}^{n} P_i = \mathbb{1} \quad \text{and} \quad P_i P_k = \delta_{ik} P_k \, . \tag{151}$$

Hence, if we define

$$j(f) := \sum_{i=1}^{n} f(x_i) P_i \, , \tag{152}$$

then j is a *-homomorphism with the property that $j(\mathrm{id}) = X$. Clearly, different X's correspond to different j's. $\qquad \square$

6.5 The Joint Measurement Apparatus

Let X and Y be self-adjoint elements of the *-algebra $\mathcal{A}$. We consider X and Y as random variables taking values in the spectra $\mathrm{sp}(X)$ and $\mathrm{sp}(Y)$.

By a *joint measurement* M^* of these random variables we mean an operation that takes a state ρ on $\mathcal{A}$ as input and yields a probability distribution π on $\mathrm{sp}(X) \times \mathrm{sp}(Y)$ as output, in such a way that for all functions f on $\mathrm{sp}(X)$ and g on $\mathrm{sp}(Y)$

$$\rho(f(X)) = \sum_{x \in sp(X)} \sum_{y \in sp(Y)} \pi(x, y) f(x), \tag{153}$$

$$\rho(g(Y)) = \sum_{x \in sp(X)} \sum_{y \in sp(Y)} \pi(x, y) g(y). \tag{154}$$

A contravariant formulation of these requirements is

$$M(f \otimes \mathbb{1}) = f(X), \tag{155}$$
$$M(\mathbb{1} \otimes g) = g(Y). \tag{156}$$

Theorem 9 *If two random variables X and Y allow a joint measurement operation, then they commute.*

Proof Let us denote by x the identity function on $sp(X)$ and by y on $sp(Y)$. We apply the multiplication theorem on the measurement operation M, which is supposed to exist. Since

$$M\big((x \otimes \mathbb{1})^\dagger (x \otimes \mathbb{1})\big) = M(x^2 \otimes \mathbb{1}) = X^2 = M(x \otimes \mathbb{1})^\dagger M(x \otimes \mathbb{1}), \tag{157}$$

we have

$$M\big((x \otimes \mathbb{1})^\dagger (\mathbb{1} \otimes y)\big) = M(x \otimes \mathbb{1})^\dagger M(\mathbb{1} \otimes y) = XY \tag{158}$$

and

$$M\big((\mathbb{1} \otimes y)^\dagger (x \otimes \mathbb{1})\big) = M(\mathbb{1} \otimes y)^\dagger M(x \otimes \mathbb{1}) = YX. \tag{159}$$
As $(x \otimes \mathbb{1})^\dagger (\mathbb{1} \otimes y) = x \otimes y = (\mathbb{1} \otimes y)^\dagger (x \otimes \mathbb{1})$, we have $XY = YX$. \tag{160}

$\square$

7 Quantum Novelties

In the previous section we saw certain strange limitations that quantum operations are subject to. Let us now look at the other side of the coin: some surprising possibilities. We leave treatment of the really sensational features to other contributions in this volume, such as very fast computation and secure cryptography. Here we shall treat "teleportation" of quantum states and "dense coding."

7.1 Teleportation of Quantum States

Suppose that Alice wishes to send to Bob the quantum state ρ of a qubit over a (classical) telephone line.

Fig. 8 Teleportation based
on shared entanglement

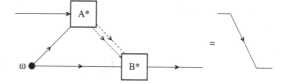

In Sect. 6.2 we have seen that, without any further tools, this is impossible. If Alice performed measurements on the qubit and told the results to Bob over the telephone, these would not enable Bob to reconstruct the state ρ. However, suppose that Alice and Bob have been together in the past, and that at that time they have created an entangled pair of qubits, as introduced in Sect. 2.3, each taking one qubit with them. It was discovered in 1993 by Bennett, Wootters, Peres, and others that by making use of this shared entanglement, Alice is indeed able to transfer her qubit to Bob. Of course, she cannot avoid destroying the original state ρ in the process; otherwise Alice and Bob would have copied the state ρ, which is impossible by Theorem 6.1 ("no cloning"). It is for this reason that the procedure is called "teleportation."

We illustrate the procedure in Fig. 8.

Here ω is the fully entangled state $X \mapsto \langle \Omega, X\Omega \rangle$ on $M_2 \otimes M_2$ (see the proof of Theorem 3, Stinespring's theorem).

The procedure runs as follows. Alice possesses two qubits, one from the entangled pair and one which she wishes to send to Bob. She performs a von Neumann measurement on these two qubits along the four *Bell projections*

$$Q_{00} := \frac{1}{2}\begin{pmatrix} 1 & 0 & 0 & 1 \\ 0 & 0 & 0 & 0 \\ 0 & 0 & 0 & 0 \\ 1 & 0 & 0 & 1 \end{pmatrix}, \quad Q_{01} := \frac{1}{2}\begin{pmatrix} 1 & 0 & 0 & -1 \\ 0 & 0 & 0 & 0 \\ 0 & 0 & 0 & 0 \\ -1 & 0 & 0 & 1 \end{pmatrix}, \quad (161)$$

$$Q_{10} := \frac{1}{2}\begin{pmatrix} 0 & 0 & 0 & 0 \\ 0 & 1 & 1 & 0 \\ 0 & 1 & 1 & 0 \\ 0 & 0 & 0 & 0 \end{pmatrix}, \quad Q_{11} := \frac{1}{2}\begin{pmatrix} 0 & 0 & 0 & 0 \\ 0 & 1 & -1 & 0 \\ 0 & -1 & 1 & 0 \\ 0 & 0 & 0 & 0 \end{pmatrix}. \quad (162)$$

The operation performed by Alice has the contravariant description:

$$A : \mathbf{C}_2 \otimes \mathbf{C}_2 \to M_2 \otimes M_2 : \quad A(e_i \otimes e_j) := Q_{ij}. \quad (163)$$

The two bits Alice obtains in this way – (i, j) say – she sends to Bob over the telephone. He then takes his own qubit from the entangled pair, and if $j = 1$ performs the "phase flip" operation

$$Z : \begin{pmatrix} \rho_{00} & \rho_{01} \\ \rho_{10} & \rho_{11} \end{pmatrix} \mapsto \begin{pmatrix} \rho_{00} & -\rho_{01} \\ -\rho_{10} & \rho_{11} \end{pmatrix} = \begin{pmatrix} 1 & 0 \\ 0 & -1 \end{pmatrix}\begin{pmatrix} \rho_{00} & \rho_{01} \\ \rho_{10} & \rho_{11} \end{pmatrix}\begin{pmatrix} 1 & 0 \\ 0 & -1 \end{pmatrix} \quad (164)$$

and if $j = 0$ he does nothing. Then, if $i = 1$ he performs the "quantum NOT" operation

$$X : \begin{pmatrix} \rho_{00} & \rho_{01} \\ \rho_{10} & \rho_{11} \end{pmatrix} \mapsto \begin{pmatrix} \rho_{11} & \rho_{10} \\ \rho_{01} & \rho_{00} \end{pmatrix} = \begin{pmatrix} 0 & 1 \\ 1 & 0 \end{pmatrix} \begin{pmatrix} \rho_{00} & \rho_{01} \\ \rho_{10} & \rho_{11} \end{pmatrix} \begin{pmatrix} 0 & 1 \\ 1 & 0 \end{pmatrix} \qquad (165)$$

and if $i = 0$ he does nothing. In the Heisenberg picture, the result of Bob's actions is the operation

$$B : M_2 \rightarrow \mathbf{C}_2 \otimes \mathbf{C}_2 \otimes M_2 : \quad M \mapsto M \oplus \sigma_3 M \sigma_3 \oplus \sigma_1 M \sigma_1 \oplus \sigma_2 M \sigma_2, \quad (166)$$

where $\sigma_1 := \begin{pmatrix} 0 & 1 \\ 1 & 0 \end{pmatrix}$, $\sigma_2 := \begin{pmatrix} 0 & -i \\ i & 0 \end{pmatrix}$, and $\sigma_3 := \begin{pmatrix} 1 & 0 \\ 0 & -1 \end{pmatrix}$ are Pauli's spin matrices. Bob ends up with a qubit in exactly the same state as Alice wanted to send. We formulate this result in the Heisenberg picture.

Proposition 9 *The state ω and the operations A and B described above satisfy*

$$(\mathrm{id}_{M_2} \otimes \omega) \circ (A \otimes \mathrm{id}_{M_2}) \circ B = \mathrm{id}_{M_2}. \qquad (167)$$

Proof We just calculate for $M \in M_2$:

$$M \xoverset{B}{\mapsto} M \oplus \sigma_3 M \sigma_3 \oplus \sigma_1 M \sigma_1 \oplus \sigma_2 M \sigma_2$$

$$\xoverset{A \otimes \mathrm{id}}{\mapsto} (Q_{00} \otimes M) + (Q_{01} \otimes \sigma_3 M \sigma_3) + (Q_{10} \otimes \sigma_1 M \sigma_1) + (Q_{11} \otimes \sigma_2 M \sigma_2)$$

$$= \frac{1}{2} \begin{pmatrix} M + \sigma_3 M \sigma_3 & 0 & 0 & M - \sigma_3 M \sigma_3 \\ 0 & \sigma_1 M \sigma_1 + \sigma_2 M \sigma_2 & \sigma_1 M \sigma_1 - \sigma_2 M \sigma_2 & 0 \\ 0 & \sigma_1 M \sigma_1 - \sigma_2 M \sigma_2 & \sigma_1 M \sigma_1 + \sigma_2 M \sigma_2 & 0 \\ M - \sigma_3 M \sigma_3 & 0 & 0 & M + \sigma_3 M \sigma_3 \end{pmatrix}$$

$$= \begin{pmatrix} m_{00} & 0 & 0 & 0 & | & 0 & 0 & 0 & m_{01} \\ 0 & m_{11} & 0 & 0 & | & 0 & 0 & m_{10} & 0 \\ 0 & 0 & m_{11} & 0 & | & 0 & m_{01} & 0 & 0 \\ 0 & 0 & 0 & m_{00} & | & m_{01} & 0 & 0 & 0 \\ - & - & - & - & | & - & - & - & - \\ 0 & 0 & 0 & m_{10} & | & m_{11} & 0 & 0 & 0 \\ 0 & 0 & m_{01} & 0 & | & 0 & m_{00} & 0 & 0 \\ 0 & m_{01} & 0 & 0 & | & 0 & 0 & m_{00} & 0 \\ m_{10} & 0 & 0 & 0 & | & 0 & 0 & 0 & m_{11} \end{pmatrix} \xoverset{\mathrm{id} \otimes \omega}{\mapsto} \begin{pmatrix} m_{00} & m_{01} \\ m_{10} & m_{11} \end{pmatrix} = M.$$

$$(168)$$

$\square$

Teleportation has been carried out successfully in the lab by Zeilinger et al. in Vienna in 1997 using polarized photons, and by other experimenters using different techniques later.

For the sake of such experiments explicit operations have been developed that form the "building blocks" of the diversity of quantum operations needed. For example the operation performed by Alice to prepare the teleportation of a qubit can be decomposed into an interaction and a measurement. Let j be the ordinary measurement operation of a qubit:

$$j : \mathbf{C}_2 \to M_2 : \qquad (f_0, f_1) \mapsto \begin{pmatrix} f_0 & 0 \\ 0 & f_1 \end{pmatrix}. \qquad (169)$$

Let H denote the *Hadamard gate*, which acts on states or observables by multiplication on the left and on the right by the *Hadamard matrix* $\frac{1}{\sqrt{2}} \begin{pmatrix} 1 & 1 \\ 1 & -1 \end{pmatrix}$, and let C denote the CNOT gate (compare chapter "Bipartite Quantum Entanglement", Sect. 2) with matrix $\begin{pmatrix} 1 & 0 & 0 & 0 \\ 0 & 0 & 0 & 1 \\ 0 & 0 & 1 & 0 \\ 0 & 1 & 0 & 0 \end{pmatrix}$. The operation C performs a NOT operation on the first qubit provided that the second is a 1, which is as shown in Fig. 9.

Check that, using the above building blocks, the procedure of quantum teleportation can be charted as in Fig. 10.

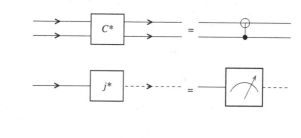

Fig. 9 Conventional signs used for the C and j operations

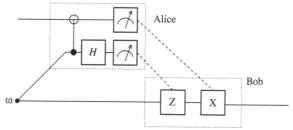

Fig. 10 More detailed scheme of teleportation

7.2 Dense Coding

We have seen that Alice can "teleport" a qubit using two classical bits, given a pre-entangled qubit pair. A kind of converse is also possible: Bob can communicate two classical bits to Alice by sending her a single qubit, again given a shared pre-

Fig. 11 Superdense coding: two bits in a single photon

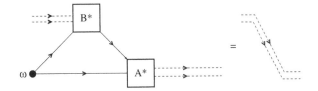

entangled qubit pair (Fig. 11). (We have interchanged the roles of Alice and Bob here because it turns out that in that case they can continue using exactly the same equipment as they used for teleportation!)

Proposition 10 *Taking ω, A, and B as in Proposition 9, we have*

$$(\mathrm{id}_{\,C_2 \otimes C_2} \otimes \omega) \circ (B \otimes \mathrm{id}_{\,M_2}) \circ A = \mathrm{id}_{\,C_2 \otimes C_2} . \tag{170}$$

We leave the proof as an exercise.

References

1. Aspect, A., Dalibard, J., Roger, G.: Phys. Rev. Lett. **49**, 1804 (1982)
2. Bell, J.S.: Physics **1**, 195 (1964)
3. Bratteli, O., Robinson, D.W.: Operator Algebras and Quantum Statistical Mechanics I. Springer, New York (1981)
4. Bohm, D.: Phys. Rev. **85**, 189 (1952)
5. Diecks, D.: Phys. Lett. A **92**, 271 (1982)
6. Dvurecenskij, A.: Gleason's theorem and its applications. In: Mathematics and Its Applications, vol. 60, p. 348. Kluwer, Dordrecht (1992)
7. Kraus, K.: *States, Effects and Operations*. Lect. Notes Phys. **190**. Springer, Berlin (1983)
8. Kümmerer, B., Maassen, H.: Elements of Quantum Probability. In: Hudson, R.L., Lindsay, J.M. (eds.) Quantum Probability Communications X, pp. 73–100. World Scientific, Singapore (1998)
9. Nelson, E.: Dynamical Theories of Brownian Motion. Princeton University Press, Princeton (1967)
10. Nielsen, M.A., Chuang, I.L.: Quantum Computation and Quantum Information. Cambridge University Press, Cambridge (2000)
11. Wootters, W.K., Zurek, W.H.: Nature **299**, 802 (1982)

Bipartite Quantum Entanglement

F. Benatti

1 Introduction

It was rather soon realized that quantum mechanics permitted the existence of states exhibiting statistical correlations that are unexplainable by classical probability theory (see chapter "Classical Information Theory") and conflict with a natural notion of locality. These properties led Einstein, Podolski, and Rosen to argue that quantum mechanics could not be a *complete description of reality* (see chapter "Quantum Probability and Quantum Information Theory", Sect. 2 and chapter "Photonic Realization of Quantum Information Protocols", Sect. 2.1 for more details) [14]. Schrödinger referred to the peculiar features of such quantum states as *verschränkung*, that is *entanglement* [31].

Quantum entanglement remained for years more an epistemological riddle than anything physical until Bell's inequalities [3] turned it into an experimentally accessible phenomenon; a real breakthrough occurred when quantum entanglement was recognized to provide unexplored possibilities from an informational viewpoint, that is, as a resource to be used in quantum cryptography (chapter "Quantum Cryptography"), information transmission (chapter "Quantum Entropy and Information"), and quantum computation (chapter "Physical Realization of Quantum Information"). The sudden change of perspective prompted a huge amount of studies that have been focussing on entanglement detection, quantification, and generation both theoretically and experimentally (see chapters "Photonic Realization of Quantum Information Protocols" and "Physical Realization of Quantum Information"), thus giving birth to a whole new theory [25, 13, 20].

This chapter will mainly be dealing with (bipartite) entanglement between pairs of finite-level systems; in particular, it will focus upon tools for entanglement detection as the von Neumann entropy and the quantum relative entropy [27] (chapter "Hilbert Space Methods for Quantum Mechanics") and positive and completely positive maps [13].

F. Benatti (✉)
Dipartimento di Fisica Teorica, Università di Trieste, Trieste, Italy, benatti@ts.infn.it

Benatti, F.: *Bipartite Quantum Entanglement*. Lect. Notes Phys. **808**, 109–149 (2010)
DOI 10.1007/978-3-642-11914-9_4   © Springer-Verlag Berlin Heidelberg 2010

A brief introductory comment on these two classes of linear maps is due at this point. Completely positive maps have been discussed in detail in chapters "Hilbert Space Methods for Quantum Mechanics" and "Quantum Probability and Quantum Information Theory" under the captions *state transformations*, respectively, *quantum operations*: Mathematically, completely positive maps describe how quantum states (or in the dual picture, quantum observables) transform under any possible physical processes such as the reversible time evolution that a system undergoes because of its Hamiltonian dynamics or the irreversible wave-packet reduction mechanism resulting from a measurement process, or the *reduced dynamics* of open quantum systems, that is, of subsystems immersed in their environment and weakly interacting with it.

Whatever the actual transformation, physical states change into physical states; since quantum states (of finite quantum systems) are generically described by density matrices, their corresponding changes in time must then be described by maps sending density matrices into density matrices. As seen in chapter "Hilbert Space Methods for Quantum Mechanics", density matrices are positive operators of trace 1, namely, their eigenvalues are probabilities and the whole statistical interpretation of quantum mechanics is based on this fact. Consequently, for the consistence of the theory, physical processes must be described by maps that preserve the positivity (of the spectrum) of the density matrices on which they act: linear maps with this property are called *positive*.

The standard quantum time evolution generated by the Schrödinger equation and the description of the effects of a measurement process by means of *POVMs* are instances of positive maps. As they are directly written in the so-called Kraus form (see chapter "Hilbert Space Methods for Quantum Mechanics", Sect. 2.4 and chapter "Quantum Probability and Quantum Information Theory", Sect. 2.4), these maps enjoy a stronger kind of positivity, called *complete positivity*. We briefly remind the reader what it amounts to. Suppose the states ρ_S of a system S undergoing a certain physical process transform according to a linear map $\rho \mapsto \mathbb{E}[\rho]$; if $\mathbb{E}$ is positive, starting with a state ρ, $\mathbb{E}[\rho]$ also is a positive operator of trace 1, namely, a new state of the system S. However, one may always imagine to couple S to another, completely insert, finite-level system of arbitrary dimensionality, that is, to an *ancilla* A; then, the physical process affecting S only, but now a subsystem of the compound system $S + A$, would be mathematically described by the linear positive map $\mathbb{E} \otimes \mathrm{id}$, where the identity operation id means that nothing happens to the (states of the) system A. Interestingly, despite $\mathbb{E}$ being positive, $\mathbb{E} \otimes \mathrm{id}$ is not necessarily such: matrix transposition T on 2×2 matrices has already been offered (see chapter "Quantum Probability and Quantum Information Theory", Sect. 5.2) as a prototypical instance of a positive, but non-completely positive map as $\mathrm{T} \otimes \mathrm{id}_2$ fails to be positive on the states of two two-level systems (qubits). In order to guarantee the positivity of $\mathbb{E} \otimes \mathrm{id}$ on the states of the compound system $S + A$ with A any n-level system, $\mathbb{E}$ must be not only positive but also completely positive, a property fully characterized by the Kraus representation (see chapter "Hilbert Space Methods for Quantum Mechanics", Sect. 2.4 and chapter "Quantum Probability and Quantum Information Theory", Sect. 5).

2 Bipartite Entanglement

In the following, we shall deal with finite-level systems S: Their observables are represented by Hermitian $d \times d$ matrices $X = X^\dagger \in M_d(\mathbf{C})$ acting on the Hilbert space $\mathbf{C}^d$ and their pure and mixed states by density matrices ρ, namely, by positive $d \times d$ matrices, $\rho \geq 0$, of trace 1, $\mathrm{Tr}(\rho) = 1$, forming a *convex set*, the *space of states* of S, $\mathcal{S}(S)$. Its extremal points are the pure states and it is closed with respect to the trace norm (see Exercise 1.1 and the discussion in chapter "Quantum Probability and Quantum Information Theory", Sect. 3.5).

In particular, we shall focus upon bipartite composite systems $S = S_1 + S_2$, consisting of two $d_{1,2}$-level parties $S_{1,2}$ usually identified with two distant, spatially well-separated quantum systems. The operators of S are matrices from the matrix-algebra $M_{d_1 d_2}(\mathbf{C}) = M_{d_1}(\mathbf{C}) \otimes M_{d_2}(\mathbf{C})$ which is the tensor product of the two *local* subalgebras $M_{d_1}(\mathbf{C}) \otimes \mathbb{1}_2$ and $\mathbb{1}_1 \otimes M_{d_2}(\mathbf{C})$, whence the term *non-local*.

Definition 1 A state $\rho \in \mathcal{S}(S_1 + S_2)$ is called separable if and only if it can be written as a linear convex combination of tensor products of density matrices

$$\rho = \sum_{ij} \lambda_{ij}\, \rho_1^i \otimes \rho_2^j\,, \quad \lambda_{ij} \geq 0\,, \quad \sum_{ij} \lambda_{ij} = 1\,, \tag{1}$$

where $\rho_1^i \in \mathcal{S}(S_1)$ and $\rho_2^j \in \mathcal{S}(S_2)$. States $\rho \in \mathcal{S}(S_1 + S_2)$ which cannot be written as in (1) are called entangled or non-separable.

Remark 1 Generic, separable mixed states can always be written as linear convex combinations of tensor products of pure states (just use the spectral decompositions of the density matrices appearing in (1)): They carry statistical correlations, but these are of classical nature and are related to how the pure states are mixed together. Indeed, a density matrix of the form

$$\rho = \sum_{i,j} \lambda_{ij} |\Psi_1^i\rangle\langle\Psi_1^i| \otimes |\Psi_2^j\rangle\langle\Psi_2^j|\,, \quad \lambda_{ij} \geq 0\,, \quad \sum_{ij} \lambda_{ij} = 1, \tag{2}$$

describes a statiscal ensemble that can always be thought of being assembled by collecting systems from a source which, in N uses, emits N_{ij} quantum systems described by state vectors $|\Psi_1^i\rangle \otimes |\Psi_2^j\rangle$ in such a way that $N_{ij}/N \to \lambda_{ij}$ with increasing N. The weights thus reflect the statistics of the source, viewed as a classical stochastic variable (see chapters "Classical Information Theory" and "Quantum Entropy and Information").

Notice that the local character of separable states cannot be modified by local actions of the form $\mathbb{E}_1 \otimes \mathbb{E}_2$ where $\mathbb{E}_{1,2}$ are completely positive, trace-preserving maps acting on the spaces of states $\mathcal{S}(S_{1,2})$. This means that in order to change the local character of a separable state into a non-local one, it is necessary to operate a non-local action (see Example 1.3).

Pure separable states correspond to projections onto state vectors of the form $|\Psi\rangle = |\Psi_1\rangle \otimes |\Psi_2\rangle$ for some $\Psi_{1,2} \in \mathbf{C}^{d_{1,2}}$, otherwise they are entangled. A generic state vector reads

$$|\Psi\rangle = \sum_{i,j=1}^{d_{1,2}} c_{ij} |\psi_1^i\rangle \otimes |\psi_2^j\rangle \tag{3}$$

with respect to two orthonormal bases $\{|\psi_1^i\rangle\}_{i=1}^{d_1}$ in $\mathbf{C}^{d_1}$, respectively, $\{|\psi_2^i\rangle\}_{i=1}^{d_2}$ in $\mathbf{C}^{d_2}$. At first sight, it is not at all obvious whether $|\Psi\rangle$ can or cannot be written in the separable tensor product form; in this case, the so-called *Schmidt decomposition* easily settles the issue [13].

Lemma 1 *Suppose $d_1 \leq d_2$; given $|\Psi\rangle \in \mathbf{C}^{d_1} \otimes \mathbf{C}^{d_2}$, one can find orthonormal bases $\{|\phi_1^i\rangle\}_{i=1}^{d_1}$, respectively, $\{|\phi_2^i\rangle\}_{i=1}^{d_2}$ yielding a diagonal decomposition*

$$|\Psi\rangle = \sum_{i=1}^{d_1} \sqrt{\lambda_i} |\phi_1^i\rangle \otimes |\phi_2^i\rangle \tag{4}$$

with non-negative Schmidt coefficients $\sqrt{\lambda_i}$.

Proof Consider the marginal density matrix (see chapter "Hilbert Space Methods for Quantum Mechanics", Sect. 2.3) $\rho_1 = \mathrm{Tr}_2(|\Psi\rangle\langle\Psi|)$ obtained by tracing with respect to any orthonormal basis in $\mathbf{C}^{d_2}$. Using as orthonormal basis $\{|\psi_1^i\rangle\}_{i=1}^{d_1}$ in (3) the eigenvectors $|\phi_1^i\rangle$ of ρ_1, write $|\Psi\rangle = \sum_{i=1}^{d_1} |\phi_1^i\rangle \otimes |\chi_2^i\rangle$, with the new vectors $|\chi_2^i\rangle = \sum_{j=1}^{d_2} c_{ij} |\psi_2^j\rangle$ neither normalized nor orthogonal, in general. Then, $|\Psi\rangle\langle\Psi| = \sum_{i,j=1}^{d_1} |\phi_1^i\rangle\langle\phi_1^j| \otimes |\chi_2^i\rangle\langle\chi_2^j|$ yields $\rho_1 = \sum_{i=1}^{d_1} \lambda_i |\phi_1^i\rangle\langle\phi_1^i| = \sum_{i,j=1}^{d_1} |\phi_1^i\rangle\langle\phi_1^j| \langle\chi_2^j|\chi_2^i\rangle$, whence $\langle\chi_2^j|\chi_2^i\rangle = \delta_{ij}\lambda_i$. The result follows by setting $|\chi_2^i\rangle = \sqrt{\lambda_i}|\phi_2^i\rangle$. $\qquad\square$

Being determined by the spectrum of the marginal density matrices (see Exercise 1.2), the number of non-zero Schmidt coefficients of a bipartite state vector is a representation-independent quantity. One thus has

Corollary 1 *A state $|\Psi\rangle \in \mathbf{C}^{d_1} \otimes \mathbf{C}^{d_2}$ is separable if and only if its marginal states are pure.*

Proof If $|\Psi\rangle$ is separable, $|\Psi\rangle\langle\Psi| = |\Psi_1\rangle\langle\Psi_1| \otimes |\Psi_2\rangle\langle\Psi_2|$ and $\rho_1 = |\Psi_1\rangle\langle\Psi_1|$. Vice versa, if ρ_1 is not a projection, it has more than one non-zero eigenvalue and, by the Schmidt decomposition, $|\Psi\rangle$ is not separable. $\qquad\square$

The Schmidt decomposition provides a useful representation of the pure states of a bipartite system; let us now consider a related tool, the so-called *purification*. It allows one to associate the mixed state of any quantum system with a pure state of a larger composite system. It is a simple consequence of the Schmidt decomposition.[1]

Given a state ρ_S of a quantum system S, it is possible to introduce another system, which we denote by A, and define a state vector $|\Psi_{SA}\rangle$ for the joint system, such that

$$\rho_S = \text{Tr}_A\left(|\Psi_{SA}\rangle\langle\Psi_{SA}|\right). \tag{5}$$

Just take the spectral decomposition $\rho_S = \sum_{i=1}^{d} r_S^i |r_S^i\rangle\langle r_S^i|$; let A be a copy of S, that is, another d-level system and set

$$\mathbf{C}^d \otimes \mathbf{C}^d \ni |\Psi_{SA}\rangle = \sum_{i=1}^{d} \sqrt{r_S^i} |r_S^i\rangle \otimes |r_S^i\rangle. \tag{6}$$

The pure state $|\Psi_{SA}\rangle$ is referred to as the *purification* of ρ_S. It reduces to ρ_S when we look at the system S alone (i.e., when we take a partial trace over the system A). This is a purely mathematical technique which allows us to associate pure states with mixed states. The jargon refers to A as an *ancilla*; it is a fictitious system that one can always append to a given one and has no direct physical significance, but is technically often and fruitfully employed (see chapters "Quantum Entropy and Information", "Quantum Cryptography", and "Quantum Algorithms").

Exercise 1

1. *Show that the space of states of a finite-level system is convex and closed with respect to the trace norm*

$$\|X_1 - X_2\|_1 = \text{Tr}\sqrt{(X_1 - X_2)^\dagger(X_1 - X_2)}, \qquad X_{1,2} \in M_d(\mathbf{C}). \tag{7}$$

   Show that pure state projections are the extremal elements of such a convex set, namely, those that cannot be further decomposed into convex combinations of other elements of the set.
2. *Using the Schmidt decomposition, show that the two marginal density matrices $\rho_{1,2}$ associated with a pure state $|\Psi\rangle \in \mathbf{C}^{d_1} \otimes \mathbf{C}^{d_2}$ have the same eigenvalues with the same multiplicities, apart, possibly, for the 0 eigenvalue. Identify the eigenvalue and relate the degeneracy of the null eigenvalue to the possible difference $d_1 \neq d_2$.*
3. *Well-known instances of entangled pure states are the Bell states [25] of two two-level systems (two qubits); they form an orthonormal basis in $\mathbf{C}^4$ (see chapters "Hilbert Space Methods for Quantum Mechanics" and "Quantum Probability and Quantum Information Theory"):*

---

[1] Purification is an instance of the GNS-representation (see [1, 10] and chapter "Quantum Probability and Quantum Information Theory").

$$|\Psi_{00}\rangle = \frac{|00\rangle + |11\rangle}{\sqrt{2}}, \quad |\Psi_{01}\rangle = \frac{|01\rangle + |10\rangle}{\sqrt{2}}, \tag{8}$$

$$|\Psi_{10}\rangle = \frac{|00\rangle - |11\rangle}{\sqrt{2}}, \quad |\Psi_{11}\rangle = \frac{|01\rangle - |10\rangle}{\sqrt{2}}. \tag{9}$$

Let $\mathbf{C}^2 \ni |x\rangle$, $x = 0, 1$, denote the basis vectors $\begin{pmatrix} 1 \\ 0 \end{pmatrix}$, respectively $\begin{pmatrix} 0 \\ 1 \end{pmatrix}$; these correspond to qubits in the so-called computational basis. Consider the following steps: (1) start from an initial state $|x, y\rangle$ of two qubits, (2) operate a Hadamard unitary rotation

$$\frac{1}{\sqrt{2}} \begin{pmatrix} 1 & 1 \\ 1 & -1 \end{pmatrix} \otimes \mathbb{1} \tag{10}$$

on the first qubit, and (3) a so-called Control-NOT unitary operation,

$$U_{CNOT} := |0\rangle\langle 0| \otimes \mathbb{1} + |1\rangle\langle 1| \otimes \sigma_1 = \begin{pmatrix} \mathbb{1} & 0 \\ 0 & \sigma_1 \end{pmatrix}, \tag{11}$$

on both qubits. Show that as a result of the action of the two unitary operations, one gets

$$|\Psi_{xy}\rangle := \frac{1}{\sqrt{2}} \sum_{i=0}^{1} (-1)^{ix} |i, y \oplus i\rangle = \frac{|0, y\rangle + (-1)^x |1, y \oplus 1\rangle}{\sqrt{2}} \tag{12}$$

(where $\oplus$ denotes binary summation) so that, by varying $x, y \in \{0, 1\}$, one obtains the Bell basis.

2.1 Entanglement and Entropy

As seen in Remark 1, the correlations carried by separable mixed states (1) are embodied in the weights λ_{ij} and are thus of classical nature. Instead, entangled states carry correlations that are purely quantum mechanical; they indeed conflict with what one expects from the classical behavior of the entropy as a measure of uncertainty.

With reference to Sect. 1.2 in chapter "Classical Information Theory", we shall denote by $h(X)$ the Shannon entropy of a classical stochastic variable X, by $h(X, Y)$ the joint entropy of two stochastic variables X and Y, and by $h(X|Y)$ the conditional entropy of X with respect to Y. Let us consider the case when Y refers to a whole classical system and X to a subsystem. Then, complete knowledge of Y would correspond to complete knowledge of X and thus to no uncertainty about the latter, $h(X|Y) = 0$ or, from (13) in chapter "Classical Information Theory", $h(X, Y) = h(Y)$. Consequently,

$$0 \leq h(Y|X) = h(X, Y) - h(X) = h(Y) - h(X) \Longrightarrow h(X) \leq h(Y). \tag{13}$$

This means that the Shannon entropy of any global state of classical systems cannot be smaller than the entropy of the reduced states of any of its constituent parts.

Consider instead a bipartite system consisting of two d-level systems in the *totally symmetric state*

$$|\Psi_+^{(d)}\rangle = \frac{1}{\sqrt{d}} \sum_{i=1}^{d} |i, i\rangle , \tag{14}$$

where $\{|i\rangle$ is any fixed orthonormal basis in $\mathbf{C}^d$ (this generalizes the two-qubit Bell state $|\Psi_{00}\rangle$ in (8)). By the partial trace of the corresponding projector (which we shall refer to as totally symmetric projector)

$$P_+^{(d)} := \frac{1}{d} \sum_{i,j=1}^{d} |i\rangle\langle j| \otimes |i\rangle\langle j| , \tag{15}$$

one obtains the marginal density matrices

$$\rho_{1,2} = \mathrm{Tr}\,(P_+^{(d)}) = \frac{1}{d} \sum_{i=1}^{d} |i\rangle\langle i| = \frac{\mathbb{1}}{d} , \tag{16}$$

namely the totally mixed state for both parties. As discussed in chapter "Hilbert Space Methods for Quantum Mechanics", Sect. 3.1, what replaces the Shannon entropy $h(X)$ of a stochastic variable is the von Neumann entropy $S(\rho)$ of a density matrix ρ, which amounts to the Shannon entropy of the spectrum of ρ:

$$S(\rho) = -\mathrm{Tr}\,\rho \log \rho = -\sum_i r_i \log r_i , \tag{17}$$

where the spectralization $\rho = \sum_i r_i |r_i\rangle\langle r_i|$, $r_i \geq 0$, $\sum_i r_i = 1$, $\langle r_i | r_j \rangle = \delta_{ij}$, has been used.

Therefore, the bipartite state $P_+^{(d)} := |\Psi_+^{(d)}\rangle\langle\Psi_+^{(d)}|$ of $S = S_1 + S_2$ has minimal entropy, while that of its constituent parties is maximal:

$$0 = S(P_+^{(d)}) \leq S(\rho_{1,2}) = \log d . \tag{18}$$

In other terms, taking the von Neumann entropy of a quantum state as a measure of its information content, the entangled, non-local, pure state $P_+^{(d)}$ of the whole quantum system $S = S_1 + S_2$ is fully determined, whereas the states of the local subsystems $S_{1,2}$ are totally random. This fact is true of any pure bipartite-entangled states.

Corollary 2 *A state vector $|\Psi\rangle \in \mathbf{C}^{d_1} \otimes \mathbf{C}^{d_2}$ is entangled if and only if its marginal density matrices have non-zero entropy.*

Proof By Corollary 1, $|\Psi\rangle$ is separable if and only if $\rho_{1,2} = \mathrm{Tr}_{2,1}(|\Psi\rangle\langle\Psi|)$ are projections which is true if and only if $S(\rho_{1,2}) = 0$.   □

As we shall see, the case of mixed entangled states is not as easy to deal with as for pure entangled states: for instance, there are entangled mixed states that have more entropy than their marginal states (see Example 1.3). However, the classical behavior, namely that the entropy of a whole system cannot be smaller than that of any of its parts, is true of all separable quantum states; in order to prove this fact, we need to inspect in more detail the notion of von Neumann entropy and of quantum relative entropy introduced in chapter "Hilbert Space Methods for Quantum Mechanics", Sect. 3.2.

Exercise 2

1. *Entangled pure states such that the entropy of their marginal density matrices is maximal are called maximally entangled (see Exercise 3.3). Show that, if $\cos\vartheta \neq \pm\sin\vartheta$, then the two-qubit state*

$$|\vartheta\rangle = \cos\vartheta \,|01\rangle + \sin\vartheta \,|10\rangle \tag{19}$$

   is not maximally entangled.
2. *Consider the symmetric projector $P_+^{(d)}$ and show that it remains maximally entangled under any local unitary transformation*

$$P_+^{(d)} \longmapsto \sum_{i=1}^{d} U_1 |i\rangle\langle i| U_1^\dagger \otimes U_2 |i\rangle\langle i| U_2^\dagger , \tag{20}$$

   where $U_{1,2} U_{1,2}^\dagger = U_{1,2}^\dagger U_{1,2} = 1$.
3. *Show that the totally symmetric state vector (14) is such that*

$$A \otimes B \,|\Psi_+^{(d)}\rangle = \mathbb{1}_d \otimes B A^{\mathrm{T}} \,|\Psi_+^{(d)}\rangle = A B^{\mathrm{T}} \otimes \mathbb{1}_d \,|\Psi_+^{(d)}\rangle , \tag{21}$$

   for all $A, B \in M_d(\mathbf{C})$, where T denotes transposition with respect to the orthonormal basis $\{|i\rangle\}_{i=1}^{d}$ and that

$$\langle\Psi_+^{(d)}| A \otimes B \,|\Psi_+^{(d)}\rangle = \frac{1}{d}\,\mathrm{Tr}\left(A B^{\mathrm{T}}\right) = \frac{1}{d}\,\mathrm{Tr}\left(A^{\mathrm{T}} B\right) . \tag{22}$$

2.2 von Neumann Entropy and Quantum Relative Entropy

As in the classical case, the concept of entropy in the quantum case gives rise to related notions, e.g., quantum analogues of the relative entropy, joint entropy, conditional entropy, and mutual information (see chapter "Quantum Entropy and Information", Sect. 4). Many important properties of the von Neumann entropy follow from those of the quantum relative entropy of two density matrices $\rho_{1,2}$ (see chapter

"Hilbert Space Methods for Quantum Mechanics", Sect. 3.1). We remind the reader of its definition and of some of its salient properties

$$S(\rho_1||\rho_2) = \text{Tr}\left(\rho_1(\log \rho_1 - \log \rho_2)\right). \tag{23}$$

It is the quantum analogue of the classical relative entropy $D(\underline{p}||\underline{q})$ between two probability distributions $\underline{p} = \{p_i\}$ and $\underline{q} = \{q_i\}$, defined as

$$D(\underline{p}||\underline{q}) = \sum_i p_i \log \frac{p_i}{q_i}. \tag{24}$$

Note that $S(\rho_1||\rho_2) < \infty$ if and only if $\text{supp}\,\rho_1 \subseteq \text{supp}\,\rho_2$, where $\text{supp}\,\rho$ denotes the support of the operator ρ, i.e., the subspace spanned by eigenvectors of ρ with non-zero eigenvalues.

Some of the many important properties of the quantum relative entropy are listed below:

1. *Non-negativity*

$$S(\rho_1||\rho_2) \geq 0, \tag{25}$$

with equality if and only if $\rho_1 = \rho_2$ (see Exercise 3.2).

2. *Joint convexity*: For $p_i \geq 0$, $\sum_i p_i = 1$ and density matrices $\rho_{1,2}^i$,

$$S\left(\sum_i p_i \rho_1^i || \sum_i p_i \rho_2^i\right) \leq \sum_i p_i S(\rho_1^i||\rho_2^i). \tag{26}$$

The above inequality implies that $S(\rho_1||\rho_2)$ is convex in each of its arguments.

3. *Monotonicity under CPT maps*: A quantum operation (CPT map) can never increase the relative entropy. For a completely positive, trace-preserving map $\mathbb{E}$ (see [33] and chapter "Hilbert Space Methods for Quantum Mechanics", Sect. 3.2),

$$S\left(\mathbb{E}[\rho_1]||\mathbb{E}[\rho_2]\right) \leq S(\rho_1||\rho_2). \tag{27}$$

Remark 2 Property 1 suggests that the relative entropy can be used to measure the distance between two density matrices [34]. Note, however, that the relative entropy is not technically a metric, since it is not symmetric and does not satisfy the triangle inequality.

One important class of inequalities relates the entropies of subsystems to that of a composite system. Let $\mathcal{H}_{AB} = \mathcal{H}_A \otimes \mathcal{H}_B$ and $\mathcal{H}_{ABC} = \mathcal{H}_A \otimes \mathcal{H}_B \otimes \mathcal{H}_C$ be the Hilbert spaces of a bipartite system and a tripartite system; $\rho_{AB} = \text{Tr}\,_C\rho_{ABC}$, $\rho_{BC} = \text{Tr}\,_A\rho_{ABC}$ and $\rho_B = \text{Tr}\,_{AC}\rho_{ABC}$ will denote the marginal density matrices of two and single parties which follow from partial tracing over the one or two Hilbert spaces [27].

1. *Concavity of the entropy difference $S(\rho_{AB}) - S(\rho_A)$:*

$$S\left(\sum_i p_i \rho^i_{AB}\right) - S\left(\sum_i p_i \rho^i_A\right) \geq \sum_i p_i \left(S(\rho^i_{AB}) - S(\rho^i_A)\right), \quad (28)$$

where $p_i \geq 0$ and $\sum_i p_i = 1$ are weights and ρ^i_{AB}, ρ^i_A are two- and single-party density matrices. A similar inequality holds exchanging A and B.

2. *Strong subadditivity (SSA):*   For any state ρ_{ABC} of a tripartite system

$$S(\rho_{ABC}) + S(\rho_B) \leq S(\rho_{AB}) + S(\rho_{BC}). \quad (29)$$

Similar inequalities hold by cyclically permuting $A \mapsto B \mapsto C$. This is one of the most important and powerful results of quantum information theory. (It was proved in [23], for more recent proofs see [26, 29].)

3. *Subadditivity:*

$$S(\rho_{AB}) \leq S(\rho_A) + S(\rho_B), \quad (30)$$

where the equality holds if and only if the subsystems A and B are uncorrelated, i.e., $\rho_{AB} = \rho_A \otimes \rho_B$. Note the analogy between (30) and the property $h(X, Y) \leq h(X) + h(Y)$ of the Shannon entropy.

4. *Triangle inequality (or Araki–Lieb inequality):*   For a bipartite system AB in a state ρ_{AB}, one has

$$S(\rho_{AB}) \geq \left| S(\rho_A) - S(\rho_B) \right|. \quad (31)$$

The proof of concavity (28) follows from joint convexity (26): let $\rho^i_1 = \rho^i_{AB}$ and $\rho^i_2 = \rho^i_A \otimes \mathbb{1}/d_B$, where d_B is the dimension of $\mathcal{H}_B$, $\mathbb{1}/d_B$ is the totally mixed state of the system B and $\rho^i_A = \mathrm{Tr}_B \rho^i_{AB}$. Then, with $\rho_{AB} = \sum_i p_i \rho^i_{AB}$ and $\rho_A = \sum_i p_i \rho^i_A$,

$$S\left(\rho_{AB} \| \rho_A \otimes \mathbb{1}/d_B\right) = -S(\rho_{AB}) - \mathrm{Tr}\left(\rho_{AB} \log \rho_A\right) + \log d_B$$

$$= -S(\rho_{AB}) + S(\rho_A) + \log d_B \quad (32)$$

$$\leq \sum_i p_i S\left(\rho^i_{AB} \| \rho^i_A \otimes \mathbb{1}/d_B\right)$$

$$= \sum_i p_i \left(-S(\rho^i_{AB}) - \mathrm{Tr}(\rho^i_{AB} \log \rho^i_A) + \log d_B\right)$$

$$= -\sum_i p_i \left(S(\rho^i_{AB}) - S(\rho^i_A)\right) + \log d_B. \quad (33)$$

Thus, the result follows from (32) and (33).

The proof of inequalities (30) and (31) uses strong subadditivity and the purification technique. As follows: given ρ_{AB} acting in $\mathcal{H}_{AB}$, by adding a suitable ancilla, one purifies it into a pure state projection P_{ABC} on a Hilbert space $\mathcal{H}_{ABC}$. Then, while $S(\rho_{ABC}) = S(P_{ABC}) = 0$, the von Neumann entropies of any two subsystems equal that of the remaining one. Indeed, from the Schmidt decomposition in Lemma 1 one knows that the non-zero eigenvalues of, say, ρ_{AC} and ρ_B are the same and the entropy is determined completely by these eigenvalues. Then, (29) and its cyclic permutations yield

$$
\begin{aligned}
S(\rho_{AB}) &= S(\rho_C) \leq S(\rho_{AC}) + S(\rho_{BC}) = S(\rho_B) + S(\rho_A) \,, \\
S(\rho_{AC}) &= S(\rho_B) \leq S(\rho_{AB}) + S(\rho_{BC}) = S(\rho_C) + S(\rho_A) \,, \qquad (34) \\
S(\rho_{BC}) &= S(\rho_A) \leq S(\rho_{AC}) + S(\rho_{AB}) = S(\rho_B) + S(\rho_C) \,.
\end{aligned}
$$

The first inequality is (30), while the remaining two obtain (31).

Corollary 3 *All separable bipartite states ρ_{AB} have larger entropy than their constituent parties: $S(\rho_{AB}) \geq \max\{S(\rho_A), S(\rho_B)\}$.*

Proof Consider a separable state $\rho_{AB} = \sum_{ij} \lambda_{ij} \rho_A^i \otimes \rho_B^j$; its marginal density matrices are $\rho_A = \sum_i \lambda_A^i \rho_A^i$ and $\rho_B = \sum_j \lambda_B^j \rho_B^j$ where $\lambda_A^i := \sum_j \lambda_{ij}$ and $\lambda_B^j := \sum_i \lambda_{ij}$. Then, by applying the equality in (30), the positivity of the von Neumann entropy yields

$$
S(\rho_{AB}) - S(\rho_A) \geq \sum_{ij} \lambda_{ij} \left(S\left(\rho_A^i \otimes \rho_B^j \right) - S(\rho_A^i) \right) = \sum_{ij} \lambda_{ij} S(\rho_B^j) \geq 0 \,, \quad (35)
$$

and similarly by exchanging A with B.   □

Exercise 3

1. Show that $S(\rho) \geq 0$ and that the equality holds if and only if ρ is a pure state density matrix.
2. Show that, if $U \in M_d(\mathbf{C})$ is a unitary operator, then $S(U^\dagger \rho U) = S(\rho)$.
   Hint: $S(\rho)$ depends only on the eigenvalues of ρ.
3. Let $\rho_1 = \rho$ and $\rho_2 = \mathbb{1}/d$, where $\mathbb{1}$ is the identity operator acting on the Hilbert space $\mathcal{H}$, and show that

$$
S(\rho_1 \| \rho_2) \geq 0 \Rightarrow S(\rho) \leq \log d \,, \qquad (36)
$$

   with the equality holding if and only if $\rho = \mathbb{1}/d$, the completely mixed state.
4. Using the concavity of the function $s(x) := -x \log x$, the spectral decompositions of ρ_i and $\rho := \sum_{i=1}^r p_i \rho_i$ and the inequality

$$
x(\log x - \log y) \geq x - y \qquad \forall 0 \leq x, y \leq 1 \,, \qquad (37)
$$

prove that the quantum relative entropy $S(\rho_1||\rho_2)$ of two density matrices ρ_1 and ρ_2 satisfies $S(\rho_1||\rho_2) \geq 0$ with equality if and only if $\rho_1 = \rho_2$.

5. *By means of the concavity of the function $\eta(x) = -x \log x$ for $0 < x \leq 1$, $\eta(0) = 0$, give another proof that the von Neumann entropy is a concave function of its inputs, i.e., given probabilities $p_i \geq 0$, $\sum_{i=1}^{r} p_i = 1$, and corresponding density operators ρ_i (see Theorem 9 in chapter "Hilbert Space Methods for Quantum Mechanics")*

$$S\left(\sum_{i=1}^{r} p_i \rho_i\right) \geq \sum_{i=1}^{r} p_i S(\rho_i). \tag{38}$$

6. *Let $\rho = \sum_i p_i \rho_i$, where ρ_i are density matrices which have support on orthogonal subspaces. Then prove that*

$$S\left(\sum_i p_i \rho_i\right) = H(\{p_i\}) + \sum_i p_i S(\rho_i), \tag{39}$$

where $H(\{p_i\})$ is the Shannon entropy of the probability distribution $\{p_i\}$.

7. *Consider an ensemble $\{p_j, \rho_j\}$ of density matrices and let $\rho := \sum_j p_j \rho_j$. Prove that for any density matrix ω, the following identity holds:*

$$\sum_j p_j S(\rho_j||\omega) = \sum_j p_j S(\rho_j||\rho) + S(\rho||\omega). \tag{40}$$

This is known as Donald's identity.

3 Entanglement Detection

According to Corollary 1, entangled bipartite pure states are detected by looking at whether their marginal states are also one-dimensional projections. The task becomes a lot more difficult in the case of bipartite mixed states; only in low dimension the question has completely been solved, the main tools being provided by the transposition operation, a linear map which is positive, but not completely positive. We recall some definitions (see the Introduction to this chapter, chapter "Hilbert Space Methods for Quantum Mechanics", Sect. 2.4 and chapter "Quantum Probability and Quantum Information Theory", Sect. 5 and [1] for further details).

Definition 2 A linear map $\Lambda : M_{d_1}(\mathbf{C}) \mapsto M_{d_2}(\mathbf{C})$ is positive if it maps positive matrices into positive matrices:

$$M_{d_1}(\mathbf{C}) \ni X \geq 0 \longmapsto M_{d_2}(\mathbf{C}) \ni \Lambda[X] \geq 0. \tag{41}$$

A linear map $\Lambda : M_{d_1}(\mathbf{C}) \mapsto M_{d_2}(\mathbf{C})$ is completely positive if

$$\Lambda \otimes \mathrm{id}_n : M_{d_1}(\mathbf{C}) \otimes M_n(\mathbf{C}) \mapsto M_{d_1}(\mathbf{C}) \otimes M_n(\mathbf{C}) \tag{42}$$

is positive for all n.

Remark 3 The matrix algebra $M_{d_1}(\mathbf{C}) \otimes M_n(\mathbf{C})$ describes a composite system $S + A$, where S is typically a d_1-level system of interest and A an n-level ancilla statistically coupled to it via the states of $\mathcal{S}(S + A)$. Linear maps of the form $\Lambda \otimes \mathrm{id}_n$, where $\Lambda :$ $M_{d_1}(\mathbf{C}) \mapsto M_{d_2}(\mathbf{C})$, are called *local* for they affect only the system of interest while leaving the ancilla unaffected. The maps Λ are usually defined on the observables of a system; then, because of linearity, their action is transferred onto the space of states $\mathcal{S}(d_1)$ by *duality*; in the case of a positive map $\Lambda : M_{d_1}(\mathbf{C}) \mapsto M_{d_2}(\mathbf{C})$,

$$\mathrm{Tr}\left(\rho \, \Lambda[X]\right) = \mathrm{Tr}\left(\Lambda^*[\rho] \, X\right), \tag{43}$$

where $\rho \in \mathcal{S}(d_2)$ is a density matrix for a d_2-level system, $X \in M_{d_1}(\mathbf{C})$, and $\Lambda^*[\rho] \in \mathcal{S}(d_1)$ is a density matrix of the d_1-level system: $\Lambda^* : \mathcal{S}(S_{d_2}) \mapsto \mathcal{S}(S_{d_1})$. As seen in chapter "Hilbert Space Methods for Quantum Mechanics", Sect. 2.4 and chapter "Quantum Probability and Quantum Information Theory", Sect. 5 quantum operations and quantum channels are usually assumed to preserve the trace of the density matrices they act upon; by duality, one sees that trace-preserving maps, $\mathrm{Tr}\left(\Lambda^*[\rho]\right) = 1$, on the space of states correspond to identity-preserving (or unital) maps, $\Lambda[\mathbb{1}_{d_1}] = \mathbb{1}_{d_2}$, on the algebra of operators.

By the theorems of Stinespring [31] and Kraus [22] (see chapters "Hilbert Space Methods for Quantum Mechanics" and "Quantum Probability and Quantum Information Theory"), completely positive maps $\Lambda : M_{d_1}(\mathbf{C}) \mapsto M_{d_2}(\mathbf{C})$ are completely characterized (see chapter "Quantum Probability and Quantum Information Theory", Sect. 5.4); they are all and only those linear maps that can be cast in the form

$$\Lambda[X] = \sum_i L_i \, X \, L_i^\dagger, \tag{44}$$

where $X \in M_{d_1}(\mathbf{C})$ and $L_i : \mathbf{C}^{d_1} \mapsto \mathbf{C}^{d_2}$ are $d_1 \times d_2$ matrices such that $\sum_i L_i^\dagger L_i$ converges in the norm-topology of $M_{d_1}(\mathbf{C})$ ($\sum_i L_i^\dagger L_i = \mathbb{1}$ if Λ is unital).

By means of (44), one checks whether a linear map $\Lambda : M_{d_1}(\mathbf{C}) \mapsto M_{d_2}(\mathbf{C})$ is completely positive by looking at the so-called *Choi matrix* constructed with the symmetric projection (15) [1, 8]:

$$M_{d_2}(\mathbf{C}) \otimes M_{d_1}(\mathbf{C}) \ni C_\Lambda := \Lambda \otimes \mathrm{id}_{d_1}[P_+^{(d_1)}] = \frac{1}{d_1} \sum_{i,j=1}^{d_1} \Lambda[|i\rangle\langle j|] \otimes |i\rangle\langle j|. \tag{45}$$

The map Λ is completely determined by its actions on the d_1^2 operators $|i\rangle\langle j| \in M_{d_1}(\mathbf{C})$, the so-called *matrix units*; namely, one knows Λ if one knows the $d_1^2 \times d_2^2$ complex values

$$\langle \alpha | \, \Lambda[|i\rangle\langle j|] \, |\beta\rangle \, , \tag{46}$$

where $\{|\alpha\rangle\}_{\alpha=1}^{d_2}$ is an orthonormal basis in $\mathbf{C}^{d_2}$.

Remark 4 This construction associates with a linear map $\Lambda : M_{d_1}(\mathbf{C}) \mapsto M_{d_2}(\mathbf{C})$ a matrix $C_\Lambda \in M_{d_2}(\mathbf{C}) \otimes M_{d_1}(\mathbf{C})$ whose entries are

$$\langle \alpha, i | \, C_\Lambda \, |\beta, j\rangle = \langle \alpha | \, \Lambda[|i\rangle\langle j|] \, |\beta\rangle \, . \tag{47}$$

Vice versa, given a matrix $C \in M_{d_2}(\mathbf{C}) \otimes M_{d_1}(\mathbf{C})$ with entries given by the left-hand side of the previous relation, then its right-hand side associates with it a linear map $\Lambda_C : M_{d_1}(\mathbf{C}) \mapsto M_{d_2}(\mathbf{C})$.

Lemma 2 $\Lambda : M_{d_1}(\mathbf{C}) \mapsto M_{d_2}(\mathbf{C})$ *is completely positive if and only if its Choi matrix C_Λ is positive.*

Proof If $\Lambda : M_{d_1}(\mathbf{C}) \mapsto M_{d_2}(\mathbf{C})$ is completely positive, then $\Lambda \otimes \mathrm{id}_{d_1}$ is positive on $M_{d_1}(\mathbf{C}) \otimes M_{d_1}(\mathbf{C})$, whence $C_\Lambda \geq 0$. If $C_\Lambda \geq 0$, consider its spectral decomposition

$$C_\Lambda = \sum_{k=1}^{d_1 d_2} \ell_k \, |\Psi_k\rangle\langle\Psi_k| \, , \tag{48}$$

where $\ell_k \geq 0$ and $|\Psi_k\rangle \in \mathbf{C}^{d_2} \otimes \mathbf{C}^{d_1}$. Using the completely symmetric state $|\Psi_+^{(d_1)}\rangle$ in (14) and by expanding $|\Psi_k\rangle = \sum_{\alpha=1}^{d_2} \sum_{i=1}^{d_1} \Psi_{\alpha i} |\alpha\rangle \otimes |i\rangle$ with respect to the orthonormal bases $\{|i\rangle\}_{i=1}^{d_1}$ in $\mathbf{C}^{d_1}$ and $\{|\alpha\rangle\}_{\alpha=1}^{d_2}$ in $\mathbf{C}^{d_2}$, one rewrites

$$|\Psi_k\rangle = V_k \otimes \mathbb{1}_{d_1} \, |\Psi_+^{(d_1)}\rangle \, , \tag{49}$$

by means of the $d_2 \times d_1$ matrices $V_k : \mathbf{C}^{d_1} \mapsto \mathbf{C}^{d_2}$ such that

$$V_k|i\rangle = d_1 \sum_{\alpha=1}^{d_2} \Psi_{\alpha i} |\alpha\rangle \, . \tag{50}$$

Setting $L_k = \sqrt{\ell_k} \, V_k$, the Choi matrix can be recast as

$$C_\Lambda = \sum_{k=1}^{d_1 d_2} L_k \otimes \mathbb{1}_{d_1} \, P_+^{(d_1)} \, L_k^\dagger \otimes \mathbb{1}_{d_1} \, , \tag{51}$$

where $P_+^{(d_1)}$ projects onto $|\Psi_+^{(d_1)}\rangle$.

Then, Λ and $\mathbb{L} : M_{d_1}(\mathbf{C}) \mapsto M_{d_2}(\mathbf{C})$ defined by $\mathbb{L}[X] = \sum_{k=1}^{d_1 d_2} L_k \, X \, L_k^\dagger$ have the same elements (46) and thus coincide, whence Λ can be written in the form (44) and is thus completely positive. $\qquad \square$

Though positive maps still lack a full characterization like that of completely positive maps, the Choi matrix allows one to sort them out as follows.

Lemma 3 *A linear map* $\Lambda : M_{d_1}(\mathbf{C}) \mapsto M_{d_2}(\mathbf{C})$ *is positive if and only if the Choi matrix* C_Λ *is block-positive, that is, if and only if*

$$\langle \phi \otimes \psi | \, \Lambda \otimes \mathrm{id}_{d_1}[P_+^{(d_1)}] \, | \phi \otimes \psi \rangle \geq 0 \tag{52}$$

for all $\phi \in \mathbf{C}^{d_2}$ *and* $\psi \in \mathbf{C}^{d_1}$.

Proof Using the explicit form of the symmetric projection $P_+^{(d_1)}$ the inequality in the statement for the Lemma reads

$$\langle \phi | \, \Lambda[|\psi^*\rangle\langle\psi^*|] \, | \phi \rangle \geq 0 \,, \tag{53}$$

where $|\psi^*\rangle$ is the conjugate of $|\psi\rangle$ with respect to the chosen orthonormal basis in $\mathbf{C}^{d_1}$. Then, the map $\Lambda : M_{d_1}(\mathbf{C}) \mapsto M_{d_2}(\mathbf{C})$ is positive as it preserves the positivity of all projections $|\psi\rangle\langle\psi| \in M_{d_1}(\mathbf{C})$. Vice versa, if the Choi matrix C_Λ is block-positive, by means of (21) one rewrite

$$0 \leq \langle \phi \otimes \psi^* | \, C_\Lambda \, | \phi \otimes \psi^* \rangle = \frac{1}{d_1} \sum_{i,j=1}^{d_1} \langle \phi | \, \Lambda[|i\rangle\langle j|] \, |\phi\rangle \, \langle \psi^* i | \rangle \, \langle j | \psi^* \rangle \tag{54}$$

$$= \langle \phi | \, \Lambda[|\psi\rangle\langle\psi|] \, |\phi\rangle \,.$$

$\square$

The Choi matrix C_Λ results from the action of $\Lambda \otimes \mathrm{id}_{d_1}$ on the pure state $P_+^{(d_1)}$ of the composite system consisting of two d_1-level systems S; thence, the physical meaning of the previous two lemmas is as follows. A completely positive map $\Lambda : M_{d_1}(\mathbf{C}) \mapsto M_{d_2}(\mathbf{C})$ is identified by the fact that it maps the state $P_+^{(d_1)}$ into another state (after normalization) over the algebra $M_{d_2}(\mathbf{C}) \otimes M_{d_1}(\mathbf{C})$, namely, the Choi matrix $C_\Lambda \geq 0$. On the contrary, if Λ is only positive, but not completely positive, $\Lambda \otimes \mathrm{id}_{d_1}$ maps $P^{(d_1)}$ outside the space of states. Such a "physical interpretation" can be generalized to all bipartite states.

Suppose $\Lambda : M_{d_1}(\mathbf{C}) \mapsto M_{d_2}(\mathbf{C})$ is a positive map, then $\Lambda \otimes \mathrm{id}_n$ preserves the positivity of all separable states of the composite system $S+A$, A an n-level ancilla; that is of all states of the form (1) where $\rho_1^i \in \mathcal{S}(S)$ and $\rho_2^j \in \mathcal{S}(A)$. Indeed,

$$\Lambda \otimes \mathrm{id}_n[\rho] = \sum_{ij} \lambda_{ij} \, \Lambda[\rho_1^i] \otimes \rho_2^j \geq 0 \,, \tag{55}$$

for $\rho_{1,2}^{i,j} \geq 0$ as well as $\Lambda[\rho_1^i] \geq 0$.

Therefore, if $\Lambda : M_{d_1}(\mathbf{C}) \mapsto M_{d_1}(\mathbf{C})$ is a positive map and ρ is a state of a bipartite system $S_1 + S_2$, S_i, $i = 1, 2$, a d_i-level system, and $\Lambda \otimes \mathrm{id}_{d_2}[\rho]$ is not positive semi-definite, then ρ must be entangled. The positive map Λ which exposes the entanglement of ρ is called an *entanglement witness* [20, 13, 8]. Notice that, in order to work as an entanglement witness, Λ must be positive, but not completely positive.

When Λ is taken as the transposition, $\Lambda = T : M_{d_1}(\mathbf{C}) \mapsto M_{d_1}(\mathbf{C})$, the action of $T \otimes \mathrm{id}_{d_2}$ is called *partial transposition* and the previous observation is the content of

Lemma 4 (Peres criterion) [28] *A state $\rho \in \mathcal{S}(d_1 d_2)$ of a bipartite system $S_1 + S_2$ consisting of a d_1- and a d_2-level system S_1 and S_2 is entangled if it does not remain positive under partial transposition:*

$$T \otimes \mathrm{id}_{d_2}[\rho] \ngeq 0 \implies \rho \text{ entangled} . \tag{56}$$

Definition 3 The bipartite states of $S_1 + S_2$ which remain positive under partial transposition $T \otimes \mathrm{id}_{d_2}$ are called PPT states and those which do not are known as NPT states.

Notice that, while transposition depends on the chosen representation, that is it needs an orthonormal basis to be performed, the spectrum of a matrix does not. Therefore, it is enough to check whether a bipartite state is PPT or NPT with respect to the most convenient orthonormal basis.

Example 1

1. The action of $T \otimes \mathrm{id} : M_d(\mathbf{C}) \otimes M_d(\mathbf{C}) \mapsto M_d(\mathbf{C}) \otimes M_d(\mathbf{C})$ turns the projection (15) into

$$T \otimes \mathrm{id}[P_+^{(d)}] = \frac{1}{\sqrt{d}} \sum_{i,j=1}^{d} |j\rangle\langle i| \otimes |i\rangle\langle j| =: \frac{1}{d} V , \tag{57}$$

where $V : \mathbf{C}^d \otimes \mathbf{C}^d \mapsto \mathbf{C}^d \otimes \mathbf{C}^d$ is the so-called flip operator. Indeed,

$$V|\psi \otimes \varphi\rangle = |\varphi \otimes \psi\rangle \qquad \forall \psi, \varphi \in \mathbf{C}^d . \tag{58}$$

Since $V = V^\dagger$ and $V^2 = 1$, V has eigenvalues 1 and -1. Therefore, $T \otimes \mathrm{id}[P_+^{(d)}]$ is NPT and $P_+^{(d)}$ entangled. Since $P_+^{(d)}$ is a pure state, this is not a surprise since it is maximally entangled ($\mathrm{Tr}_2(P_+^{(d)}) = 1/d$). More interesting is the following example.

2. Consider the family of states, called isotropic, of the form [19]

$$\rho_F = \alpha \mathbb{1}_{d^2} + \beta P_+^{(d)} . \tag{59}$$

Positivity, $\rho_F \geq 0$, and normalization, $\mathrm{Tr}(\rho_F) = 1$, yield

$$\alpha \geq 0 , \ \alpha d^2 + \beta = 1 , \ 0 \leq F := \mathrm{Tr}(\rho_F P_+^{(d)}) = \alpha + \beta \leq 1 . \tag{60}$$

Isotropic states are thus mixtures of the totally mixed state $\mathbb{1}/d^2$ and the totally symmetric state (14),

$$\rho_F = \frac{d^2(1-F)}{d^2-1}\frac{\mathbb{1}_{d^2}}{d^2} + \frac{d^2F-1}{d^2-1}P_+^{(d)} . \tag{61}$$

Since $\langle \psi \otimes \phi | P_+^{(d)} | \psi \otimes \phi \rangle = |\langle \psi | \phi^* \rangle|^2$, where ψ^* is the vector in $\mathbf{C}^d$ with complex conjugate components with respect to ψ, if ρ_F is separable, then

$$F = \mathrm{Tr}\,(\rho_F\, P_+^{(d)}) = \frac{1}{d}\sum_{ij}\mu_{ij}|\langle \psi_i^1|(\phi_j^2)^*\rangle|^2 \le \frac{1}{d} . \tag{62}$$

The condition $0 \le F \le 1/d$ is necessary and also sufficient for separability. The reason is that isotropic states are all and only those $d^2 \times d^2$ density matrices which commute with all local unitaries of the form $U \otimes U^*$, where U is any unitary matrix in $M_d(\mathbf{C})$ and U^* denotes its complex conjugate (not its adjoint). Moreover, since $(U \otimes U^*)\,P_+^{(d)}\,(U^\dagger \otimes U^{\mathrm{T}}) = P_+^{(d)}$, any isotropic ρ_F arises from a twirling of the form

$$\rho_F = \int_{\mathcal{U}} dU\,(U \otimes U^*)\,\rho\,(U^\dagger \otimes U^{\mathrm{T}}), \tag{63}$$

where U^{T} denotes the transposition of U and ρ is such that $\mathrm{Tr}\,(\rho\,P_+^{(d)}) = F$. If $Fd \le 1$ and $\rho = |\psi\rangle\langle\psi| \otimes |\phi\rangle\langle\phi|$, with $|\phi\rangle = \sqrt{dF}\,|\psi\rangle + \sqrt{1-dF}\,|\psi^\perp\rangle$, then one can show that ρ_F can be obtained by twirling a separable state and is thus itself separable. The above necessary and sufficient conditions for separability coincides with the isotropic states being positive under partial transposition. Indeed,

$$\mathrm{T} \otimes \mathrm{id}_d[\rho_F] = \frac{1-F}{d^2-1}\mathbb{1}_{d^2} + \frac{d^2F-1}{d(d^2-1)}V \tag{64}$$

has positive eigenvalues $(dF+1)/(d^2+d) \ge 0$ and $(1-dF)/(d^2-d)$ if and only if $0 \le F \le 1/d$.

3. With reference to Corollary 3, consider the case of two qubits and set $d = 2$ in the previous example. Then, if $F > 1/2$ the isotropic state

$$\rho_F = \frac{(1-F)}{3}\mathbb{1}_4 + \frac{4F-1}{3}P_+^{(2)} \tag{65}$$

is entangled. However, Fig. 1 shows that some of them have von Neumann entropy,

$$S(\rho_F) = -(1-F)\log\frac{1-F}{3} - F\log F , \tag{66}$$

larger than that of their reduced density matrices $\mathrm{Tr}_{A,B}\rho_F = \mathbb{1}/2$.

Fig. 1 $S(\rho_F) - S(\rho_{A,B}) = S(\rho_F) - \log 2$

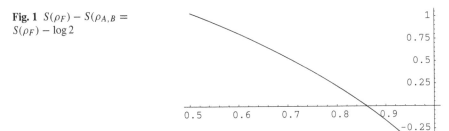

4. This chapter is entirely devoted to bipartite-entangled states; a flavor of the intricacies of multipartite entanglement [20] can be conveyed by means of the GHZ states of three qubits that, with respect to the standard basis, read

$$|GHZ\rangle = \frac{|000\rangle + |111\rangle}{\sqrt{2}} \, . \tag{67}$$

Because of the symmetry there is no restriction in distinguishing between one of the three qubits and the other two. Partial transposition with respect to the third one reveals that there is entanglement between the latter and the first two qubits; indeed,

$$Z := \mathrm{id}_{12} \otimes T_3[|GHZ\rangle\langle GHZ|] = \frac{1}{2}\Big(|000\rangle\langle 000| + |001\rangle\langle 110| \\ + |110\rangle\langle 001| + |111\rangle\langle 111|\Big) \tag{68}$$

is such that

$$Z = \frac{1}{\sqrt{2}}\Big(|001\rangle - |110\rangle\Big) = -\frac{1}{\sqrt{2}}\Big(|001\rangle - |110\rangle\Big) \, . \tag{69}$$

Since Z has an eigenvalue -1, partial transposition shows there is entanglement between the first two qubits and third one; however, the partial trace with respect to the latter gives the separable totally mixed state:

$$\mathrm{Tr}_3\Big(|GHZ\rangle\langle GHZ|\Big) = \frac{\mathbb{1}_{12}}{4} \, . \tag{70}$$

Indeed, the trace operation is a state transformation with extreme decohering effects. However, can localize the entanglement on the pair consisting of qubits 1 and 2 by projecting the third qubit onto a suitable state $|\psi\rangle = \alpha|0\rangle + \beta|1\rangle$ (instead of tracing it out):

$$\langle \psi | GHZ \rangle \langle GHZ | \psi \rangle = \frac{1}{2} \Big(|\alpha|^2 \, |00\rangle\langle 00| + \alpha\beta^* \, |00\rangle\langle 11|$$
$$+ \alpha^*\beta \, |11\rangle\langle 00| + |\beta|^2 \, |11\rangle\langle 11| \Big) . \tag{71}$$

After normalization, the resulting state of qubits 1 and 2 may be entangled, for instance, when $\alpha = \beta = 1/\sqrt{2}$.

Lemma 4 provides a sufficient condition for a bipartite mixed state to be entangled; using Lemmas 2 and 3, one derives a necessary, albeit non-constructive condition[19].

Proposition 1 (Peres–Horodecki criterion) *A bipartite state ρ of a composite system $S_1 + S_2$ described by the algebra $M_{d_1}(\mathbf{C}) \otimes M_{d_2}(\mathbf{C})$ is entangled if and only there exists a positive map $\Lambda : M_{d_2}(\mathbf{C}) \mapsto M_{d_1}(\mathbf{C})$ such that, under the action of the dual map, $\Lambda^* \otimes \mathrm{id}_{d_2}[\rho] \ngeq 0$.*

Proof As the space of states itself, also the set of separable states over $M_{d_1}(\mathbf{C}) \otimes M_{d_2}(\mathbf{C})$, $\mathcal{S}_{\mathrm{sep}}(S_1 + S_2)$, is convex and closed (with respect to the trace norm, see Exercises 1.1 and 1.2). By the Hahn–Banach theorem, $\mathcal{S}(S_1+S_2)$ can thus be strictly separated from any entangled state ρ_{ent} by a hyperplane, that is, by a continuous linear functional $\mathcal{R} : \mathcal{S}(S_1 + S_2) \mapsto \mathbb{R}$ and a real constant a such that $\mathcal{R}(\rho_{\mathrm{ent}}) < a \leq \mathcal{R}(\rho_{\mathrm{sep}})$. As the trace norm, $\|X\|_{\mathrm{Tr}} = \mathrm{Tr}\sqrt{X^\dagger X}$, and the Hilbert–Schmidt norm, $\|X\|_{\mathrm{HS}} = \sqrt{\mathrm{Tr}(X^\dagger X)}$, define equivalent topologies in finite dimension, the action of $\mathcal{R}$ can be represented by means of $R = R^\dagger \in M_{d_1}(\mathbf{C}) \otimes M_{d_2}(\mathbf{C})$ such that $\mathcal{R}(\rho) = \mathrm{Tr}(R\,\rho)$. Setting $S := R' - a\mathbb{1}$, it thus follows that $\rho \in M_{d_1}(\mathbf{C}) \otimes M_{d_2}(\mathbf{C})$ is entangled if and only if there exists $S \in M_{d_1}(\mathbf{C}) \otimes M_{d_2}(\mathbf{C})$ such that $\mathrm{Tr}(S\,\rho) < 0$ while $\mathrm{Tr}(S\,\rho_{\mathrm{sep}}) \geq 0$ for all $\rho_{\mathrm{sep}} \in \mathcal{S}_{\mathrm{sep}}(S_1 + S_2)$.

Furthermore, according to Remark 4, any such matrix is the Choi matrix of a map $\Lambda_S : M_{d_2}(\mathbf{C}) \mapsto M_{d_1}(\mathbf{C})$: $S = \Lambda_S \otimes \mathrm{id}_{d_2}[P_+^{(d_2)}]$. Since S is block-positive (see Lemma 3), Λ is a positive map. Let Λ_S^* be its dual such that

$$\mathrm{Tr}(S\,\rho) = \mathrm{Tr}\left(P_+^{(d_2)} \Lambda_S^* \otimes \mathrm{id}_{d_2}[\rho] \right) \tag{72}$$

for all $\rho \in \mathcal{S}(S_1+S_2)$. If ρ is an entangled state, then $\mathrm{Tr}(S\,\rho) < 0$ and $\Lambda_S^* \otimes \mathrm{id}_{d_2}[\rho]$ cannot be positive definite. Vice versa, if $\Lambda^* \otimes \mathrm{id}_{d_2}[\rho] \geq 0$ for all positive $\Lambda : M_{d_2}(\mathbf{C}) \mapsto M_{d_1}(\mathbf{C})$, then ρ must belong to $\mathcal{S}_{\mathrm{sep}}(S_1 + S_2)$. □

The previous argument states that a positive map $\Lambda : M_{d_2}(\mathbf{C}) \mapsto M_{d_1}(\mathbf{C})$ may witness the entanglement of a state $\rho \in \mathcal{S}(S_1 + S_2)$ if $\Lambda^* \otimes \mathrm{id}_{d_1}$ turns ρ into a non-positive matrix. Therefore, Λ cannot be a CP map, otherwise $\Lambda^* \otimes \mathrm{id}_{d_1}$ would automatically be a positive map. Unlike for CP maps that are completely determined by their form (44), positive, but not completely positive, maps still lack a complete characterization. Consequently, given an entangled state $\rho \in \mathcal{S}(S_1 + S_2)$, the problem of seeking its entanglement witness is in general an extremely hard problem.

A particular class of positive maps is the following one.

Definition 4 A map $\Lambda : M_{d_1}(\mathbf{C}) \mapsto M_{d_2}(\mathbf{C})$ is decomposable if it is positive and $\Lambda = \Lambda_1 + \Lambda_2 \circ T$, where $\Lambda_{1,2} : M_{d_1}(\mathbf{C}) \mapsto M_{d_2}(\mathbf{C})$ are CP maps and T is the transposition on $M_{d_1}(\mathbf{C})$ with respect to a fixed orthonormal basis in $\mathbf{C}^{d_1}$.

Example 2 The so-called *reduction map* [18] $\Lambda : M_d(\mathbf{C}) \mapsto M_d(\mathbf{C})$,

$$\Lambda[X] = \mathrm{Tr}\,(X)\,\mathbb{1}_d - X\,, \qquad X \in M_d(\mathbf{C}) \tag{73}$$

is decomposable. First, it is positive: in fact, if $X \geq 0$, the spectrum of $\Lambda[X]$ consists of eigenvalues $\sum_{i=1,i\neq j}^d x_i \geq 0$, $1 \leq j \leq d$, where $x_i \geq 0$ are the eigenvalues of X. Second, it is not completely positive: indeed, its Choi matrix

$$\Lambda \otimes \mathrm{id}_d[P_+^{(d)}] = \frac{\mathbb{1}_{d^2}}{d} - P_+^{(d)} \tag{74}$$

has one negative eigenvalue $1/d - 1$ and $1/d$ as a $(d^2 - 1)$-degenerate positive eigenvalue. However, $\widetilde{\Lambda} = \Lambda \circ T$ is completely positive, since

$$\widetilde{\Lambda} \otimes \mathrm{id}_d[P_+^{(d)}] = \frac{1}{d}\left(\mathbb{1}_{d^2} - V\right), \tag{75}$$

where V is the flip operator of Example 1.1, has eigenvalues $1/d$ ($d(d-1)/2$-degenerate) and 0 ($d(d+1)/2$-degenerate). Since, as $T \circ T = \mathrm{id}$, it follows that Λ can be decomposed as $\Lambda = \widetilde{\Lambda} \circ T$.

Let $(d_1, d_2) = (2, 2), (2, 3), (3, 2)$, then a theorem of Woronowicz [32, 36] asserts that all positive maps $\Lambda : M_{d_1}(\mathbf{C}) \mapsto M_{d_2}(\mathbf{C})$ are decomposable. This fact makes transposition an exhaustive entanglement witness in low dimension; in other words, those states that remain *positive under partial transposition* are separable and vice versa.

Corollary 4 *If* $(d_1, d_2) = (2, 2), (2, 3), (3, 2)$, *then,* $\rho \in \mathcal{S}(S_1 + S_2)$ *is entangled if and only if* $T \otimes \mathrm{id}_{d_2}[\rho] \ngeq 0$.

Proof If $\rho \in \mathcal{S}(S_1 + S_2)$ is separable then $T \otimes \mathrm{id}_{d_2}[\rho] \geq 0$ from Lemma 4. Vice versa, because of the assumed dimensions, Woronowicz theorem ensures that any positive map is decomposable. Therefore, if $T \otimes \mathrm{id}_{d_2}[\rho] \geq 0$, it turns out that, for all positive $\Lambda : M_{d_1}(\mathbf{C}) \mapsto M_{d_1}(\mathbf{C})$,

$$\Lambda \otimes \mathrm{id}_{d_2}[\rho] = \Lambda_1 \otimes \mathrm{id}_{d_1}[\rho] + \Lambda_2 \otimes \mathrm{id}_{d_2}\left[T \otimes \mathrm{id}_{d_2}[\rho]\right] \geq 0\,, \tag{76}$$

as $\Lambda_{1,2}$ are completely positive maps.   □

Woronowicz theorem does not extend to higher dimension; there are instances of non-decomposable positive maps already for $d_1 = d_2 = 3$ [17]; as a consequence partial transposition is not an exhaustive entanglement witness in higher dimension. In other words, all NPT states are entangled, but there can exist PPT-entangled states. The entanglement witnesses of PPT-entangled states $\rho \in \mathcal{S}(S_1 + S_2)$ must be non-decomposable positive maps, otherwise as in the proof of the previous lemma,

$$\Lambda \otimes \mathrm{id}_{d_2}[\rho] = \Lambda_1 \otimes \mathrm{id}_{d_2}[\rho] + \Lambda_2 \otimes \mathrm{id}_{d_2}\left[\mathrm{T} \otimes \mathrm{id}_{d_2}[\rho]\right] \geq 0 . \tag{77}$$

We shall consider concrete examples of these states and their detection via non-decomposable positive maps in Sect. 4.2; here we will just make the following observation.

Lemma 5 *Suppose* $\rho \in M_{d_1}(\mathbf{C}) \otimes M_{d_2}(\mathbf{C})$ *is a PPT state. If there exists a positive map* $\Lambda : M_{d_2}(\mathbf{C}) \mapsto M_{d_1}(\mathbf{C})$ *such that*

$$F_\rho(\Lambda) = \mathrm{Tr}\left(\rho \, \Lambda \otimes \mathrm{id}_{d_2}[P_+^{(d_2)}]\right) < 0 , \tag{78}$$

then, ρ *is PPT entangled and* Λ *non-decomposable.*

Proof If ρ were separable, then $F_\rho(\Lambda) \geq 0$ as Λ is a positive map (see the discussion before Lemma 4). The same would be true if Λ were decomposable, since ρ is assumed to be PPT. □

Example 3 No state vector $|\Psi\rangle \in \mathbf{C}^{d_1} \otimes \mathbf{C}^{d_2}$ can be PPT entangled. Indeed, consider the Schmidt decomposition of an entangled vector state

$$|\Psi\rangle = \sum_{j=1}^{d} \sqrt{\lambda_j} |\psi_j^{(1)}\rangle \otimes |\psi_j^{(2)}\rangle , \tag{79}$$

where $d := \min\{d_1, d_2\}$, $\{|\psi_j^{(1,2)}\rangle\}_{j=1}^{d}$ are orthonormal sets in the Hilbert spaces $\mathbf{C}^{d_{1,2}}$ and the Schmidt coefficients $\lambda_j > 0$ for at least two indices. Then, the partial transposition with respect to the orthonormal basis having $\{|\psi_j^{(1)}\rangle\}_{j=1}^{d}$ among its elements yields

$$R := \mathrm{T} \otimes \mathrm{id}_{d_2}[|\Psi\rangle\langle\Psi|] = \sum_{i,j=1}^{d} \sqrt{\lambda_i \lambda_j} |\psi_j^{(1)}\rangle\langle\psi_i^{(1)}| \otimes |\psi_i^{(2)}\rangle\langle\psi_j^{(2)}| . \tag{80}$$

If $\lambda_{1,2} > 0$, let $|\Phi\rangle = |\psi_1^{(1)}\psi_2^{(2)}\rangle - |\psi_2^{(1)}\psi_1^{(2)}\rangle/\sqrt{2}$, then $R|\Phi\rangle = -\sqrt{\lambda_1\lambda_2}\,|\Phi\rangle$, whence $\mathrm{T} \otimes \mathrm{id}_{d_2}[|\Psi\rangle\langle\Psi|]$ is not positive-definite.

Exercise 4

1. *Using the Pauli matrices* $\sigma_{1,2,3}$ *plus* $\sigma_0 = \mathbb{1}_2$, *show that one can express the trace operation* $X \mapsto \widehat{\mathrm{Tr}}[X] = \mathrm{Tr}(X)\mathbb{1}$ *in the following way:*

$$\widehat{\mathrm{Tr}}(X) = \frac{1}{2} \sum_{\mu=0} \sigma_\mu X \sigma_\mu \qquad \forall X \in M_2(\mathbf{C}) . \tag{81}$$

Hint: Consider $X = \sigma_\mu$, $\mu = 0, 1, 2, 3$.

2. *Using the Pauli matrices $\sigma_{1,2,3}$ plus $\sigma_0 = \mathbb{1}_2$, show that one can express the transposition in the following way:*

$$\mathrm{T}[X] = \frac{1}{2} \sum_{\mu=0} \varepsilon_\mu \sigma_\mu \, X \, \sigma_\mu \qquad \forall X \in M_2(\mathbf{C}) , \tag{82}$$

   where $\varepsilon_0 = \varepsilon_1 = \varepsilon_3 = -\varepsilon_2 = 1$.
   Hint: Consider $X = \sigma_\mu$, $\mu = 0, 1, 2, 3$.

3. *Find the subspaces of $\mathbf{C}^d \otimes \mathbf{C}^d$ corresponding to the eigenvalues ± 1 of the flip operator V of Example 1.1 and their degeneracy.*
   Hint: Consider the symmetric and anti-symmetric vectors in $\mathbf{C}^d \otimes \mathbf{C}^d$.

4. *Let $\Lambda_1 : M_{d_1}(\mathbf{C}) \mapsto M_d(\mathbf{C})$ and $\Lambda_2 : M_d(\mathbf{C}) \mapsto M_{d_2}(\mathbf{C})$ be two CP maps; show that their composition $\Lambda_2 \circ \Lambda_1 : M_{d_1}(\mathbf{C}) \mapsto M_{d_2}(\mathbf{C})$ is also CP.*
   Hint: Use the Stinespring representation *(44)*.

5. *With reference to Example 1.2, consider the family of states, called Werner states,*

$$\rho_W = \alpha \mathbb{1}_{d^2} + \beta \, V , \tag{83}$$

   where V is the flip operator of Example 1.1 and $W := \mathrm{Tr}\,(\rho_W \, V)$. Use the spectral properties of V to write

$$\rho_W = \frac{d(d-W)}{d^2-1} \frac{\mathbb{1}_{d^2}}{d^2} + \frac{dW-1}{d(d^2-1)} V , \quad -1 \leq W \leq 1 . \tag{84}$$

   Then, use the relation between the totally symmetric state $P_+^{(d)}$ and V to show that the Werner states are separable if and only if $-1 \leq W < 0$.
   Hint: Relate the isotropy parameter F to the Werner parameter W.

6. *Let $\rho \in M_n(\mathbf{C})$ be a density matrix and show that any choice of $M_n(\mathbf{C}) \ni X_i$, $i \in I$ such that $\sum_{i \in I} X_i^\dagger X_i = \mathbb{1}_n$ can be used to decompose $\rho = \sum_{i \in I} \lambda_i \, \rho_i$.*
   Hint: Write $\rho = \sqrt{\rho} \, \mathbb{1}_d \, \sqrt{\rho}$ with the identity written in terms of the X_is.

7. *Show that the flip operator introduced in Example 1.1 satisfies*

$$V A \otimes B V = B \otimes A \qquad \forall A, B \in M_d(\mathbf{C}) . \tag{85}$$

3.1 Entanglement Manipulation

Entangled states are a resource that can be used to perform classically forbidden tasks, appropriately termed *quantum novelties* in chapter "Quantum Probability and Quantum Information Theory", Sect. 7, like *teleportation* and *super-dense coding*, like *quantum information transmission* discussed in detail in chapter "Physical Realization of Quantum Information", *quantum cryptography* in chapter "Quantum Cryptography", and *quantum computation* in chapter "Quantum Algorithms".

Maximally entangled two qubit states like the singlet state $|\Psi_{11}\rangle$ in (9) provide optimal resources for these tasks. However, one not always has maximally entangled states at disposal, rather of the form

$$|\psi\rangle = \alpha\,|00\rangle + \beta\,|01\rangle + \gamma\,|10\rangle + \delta\,|11\rangle \qquad (86)$$

with $|\alpha|^2 + |\beta|^2 + |\gamma|^2 + |\delta|^2 = 1$, or mixed states. One would then like to quantify the entanglement content of a bipartite state and find methods to manipulate it. Of course, taking into account the non-local character of entanglement, the class of allowed operations must be restricted in order to avoid that entanglement be augmented by non-local operations involving both parties at the same time.

Since the beginning of quantum information theory, an important issue has been the so-called *distillation* [25, 12, 13, 20] of maximally entangled states like $|\Psi_{11}\rangle$ out of non-maximally entangled states, by means of *local quantum operations*, that is, by completely positive maps acting locally on the parties and *classical communication* among the parties: these are known as *LOCC operations*.

Consider two parties A and B sharing N copies of a bipartite system each one of them described by a mixed state ρ so that they act upon the tensor product state

$$\rho^{\otimes N} = \underbrace{\rho \otimes \rho \otimes \cdots \otimes \rho}_{N\ \text{times}}\,. \qquad (87)$$

If $\mathcal{D} = \{\mathcal{D}_N\}$ denotes a particular sequence of LOCC protocols that A and B apply to their local parties in the states $\rho^{\otimes N}$ outputting M_D singlets in the state vector $|\Psi_{11}\rangle\langle\Psi_{11}|^{\otimes M_D}$, then the optimal asymptotic yield

$$E_D[\rho] = \sup_{\mathcal{D}} \lim_{N\to+\infty} \frac{M_D}{N}\,, \qquad (88)$$

called *entanglement of distillation*, represents the optimal fraction of singlet qubit states that one can distill by means of LOCC out of the mixed state ρ.

Remark 5 Since successful local manipulations at some point must yield a singlet as output and the latter is not positive under partial transposition, it turns out that distillability of a bipartite state ρ is only possible if the initial state is entangled but also only if it is not PPT [13, 20]. Indeed, all pure bipartite states are distillable [13, 20] (see Example 3.2). Instead, while PPT states are not distillable, it is still unknown whether all NPT states are distillable (see Definition 3); the entanglement which cannot be distilled is called bound entanglement.

The distillation of maximally entangled states out of a mixed state has a dual operation, called *entanglement dilution*, whereby one uses a certain number M_F of singlet states to *form*, by means of a sequence of suitable LOCC protocols $\mathcal{F} = \{\mathcal{F}_N\}_N$, N copies of a bipartite mixed state ρ. The asymptotic optimal yield

$$E_C[\rho] = \inf_{\mathcal{F}} \lim_{N \to +\infty} \frac{M_F}{N} , \qquad (89)$$

called *entanglement cost*, represents the minimal fraction of singlet qubit states that are needed to form a mixed bipartite state ρ by means of *LOCC*.

Consider (86); as seen in Sect. 2, if it is separable, the reduced density matrix

$$\rho_1 = \mathrm{Tr}_2(|\psi\rangle\langle\psi|) = \begin{pmatrix} |\alpha|^2 + |\beta|^2 & \alpha\gamma^* + \beta\delta^* \\ \alpha^*\gamma + \beta^*\delta & |\gamma|^2 + |\delta|^2 \end{pmatrix} \qquad (90)$$

is a projection and vice versa. Therefore, the closer ρ_1 is to a projection, the less entangled is $|\psi\rangle$. A natural *entanglement measure* [13] is thus the von Neumann entropy of ρ_1 (see Exercise 1.2):

$$S(\rho_1) = -\frac{1-p}{2} \log \frac{1-p}{2} - \frac{1+p}{2} \log \frac{1+p}{2} , \qquad (91)$$

$$p = \sqrt{1 - C^2} , \qquad 0 \le C = 2\,|\alpha\delta - \beta\gamma| \le 1 . \qquad (92)$$

Indeed, it turns out that for pure bipartite states $|\Psi\rangle\langle\Psi|$

$$E_D[|\Psi\rangle\langle\Psi|] = E_C[|\Psi\rangle\langle\Psi|] = S(\rho_1) , \qquad (93)$$

namely, the von Neumann entropy of the marginal states of $|\Psi\rangle$ is a useful operational measure of the entanglement content of $|\Psi\rangle$.

Furthermore, for pure bipartite states, the entanglement distillation and dilution processes are the inverse of each other. In general, one can only say that $E_C[\rho] \ge E_D[\rho]$; otherwise, starting from $E_C[\rho]$ singlets, one could form one copy of ρ and then distill $E_D[\rho] > E_C[\rho]$ singlets thus increasing the initial amount of entanglement by means of local operations which is impossible. The general irreversibility of entanglement manipulation comes about because of the existence of bound entanglement, that is of entanglement, like that of PPT entangled states, which cannot be distilled: The formation of any such state ρ requires an entanglement cost that cannot be retrieved, $E_C[\rho] > E_D[\rho] = 0$.

In the case of pure bipartite states, reversibility is guaranteed by the fact that they can always be distilled (see Remark 5), whence $E_C[|\Psi\rangle\langle\Psi|] \le E_D[|\Psi\rangle\langle\Psi|]$. In fact, if $E_D[|\Psi\rangle\langle\Psi|] > E_C[|\Psi\rangle\langle\Psi|]$, then one could use one singlet to form $1/E_C[|\Psi\rangle\langle\Psi|]$ states $|\Psi\rangle$ and then distill $E_D[|\Psi\rangle\langle\Psi|]/E_C[|\Psi\rangle\langle\Psi|] > 1$ singlets, again increasing the given amount of entanglement by local operations and classical communication.

3.2 Concurrence

Bipartite density matrices ρ can always be decomposed into infinitely many convex combinations of pure bipartite states; (see Exercise 4.6); thus, a natural replacement of (89) is [9]

$$E_F[\rho] = \inf \left\{ \sum_i \lambda_i S(\rho_1^i) \; : \; \rho = \sum_i \lambda_i |\psi^i\rangle\langle\psi^i| \right\} , \tag{94}$$

that is the smallest convex combination of the von Neumann entropies $S(\rho_1^i)$ of the reduced density matrices $\text{Tr}_2(|\psi^i\rangle\langle\psi^i|)$ of the pure states $|\psi^i\rangle\langle\psi^i|$ in terms of which ρ can be convexly expanded as $\sum_i \lambda_i |\psi^i\rangle\langle\psi^i|$, $\lambda_i \geq 0$, $\sum_i \lambda_i = 1$. This quantity is known as *entanglement of formation* [13].

Remark 6 The entanglement cost and the entanglement of formation are related by [13]

$$E_C[\rho] = \lim_{N\to\infty} \frac{E_F[\rho^{\otimes N}]}{N} . \tag{95}$$

Consider $N = 2$; since a protocol that yields ρ can be used to form $\rho \otimes \rho$, one in general has $E_F[\rho \otimes \rho] \leq 2 E_F[\rho]$ and thus $E_C[\rho] < E_F[\rho]$. A recent result [16] (see chapter "Quantum Entropy and Information", Sect. 7.5) has proved that the entanglement of formation is not additive and thus the entanglement cost is not in general equal to the entanglement of formation.

Surprisingly [35], in the case of two qubits, the variational quantity (94) can be expressed as in (91) with C in (92) substituted by the so-called *concurrence*

$$C = \max \{0, \lambda_1 - \lambda_2 - \lambda_3 - \lambda_4\} , \tag{96}$$

where $\lambda_1 \geq \lambda_2 \geq \lambda_3 \geq \lambda_4$ are the square roots of the (positive) eigenvalues of the 4×4 matrix $R = \rho \sigma_2^{(1)}\sigma_2^{(2)} \rho^* \sigma_2^{(1)}\sigma_2^{(2)}$, where ρ^* is the density matrix with complex conjugate entries with respect to those of ρ (in a fixed representation).

Example 4

1. Given the two-qubit state vector

$$|\tilde{\psi}\rangle = \sigma_2 \otimes \sigma_2 |\psi^*\rangle = -\alpha^* |11\rangle + \beta^* |10\rangle + \gamma^* |10\rangle - \delta^* |00\rangle , \tag{97}$$

where $|\psi^*\rangle$ is the conjugate vector of $|\psi\rangle$ with respect to the basis $|0\rangle, |1\rangle$, it is immediate to check that the 4×4 matrix

$$R = |\psi\rangle\langle\psi|\tilde{\psi}\rangle\langle\tilde{\psi}| \tag{98}$$

has $C^2 = 4 |\alpha\delta - \beta\gamma|^2$ as positive eigenvalue. The concurrence of the pure state $|\psi\rangle$ is thus given by the square root of the latter: when it is maximal, $C = 1$, then $p = 1$ and $S(\rho_1) = \log 2$ is maximal, in which case $|\psi\rangle$ is maximally entangled; otherwise, when $C = 0$ is minimal, $p = 1$, $S(\rho_1) = 0$, and $|\psi\rangle$ is separable.

2. Consider the two-qubit density matrix

$$\rho = \begin{pmatrix} a & 0 & 0 & 0 \\ 0 & b & c & 0 \\ 0 & c & d & 0 \\ 0 & 0 & 0 & e \end{pmatrix}, \quad a, b, d, e \geq 0, \ a+b+d+e = 1, \ bd \geq c^2, \tag{99}$$

written with respect to the standard basis $|ij\rangle$, $i, j = 0, 1$. One readily computes

$$\tilde{\rho} = \begin{pmatrix} e & 0 & 0 & 0 \\ 0 & d & c & 0 \\ 0 & c & b & 0 \\ 0 & 0 & 0 & a \end{pmatrix}, \quad R = \rho\tilde{\rho} = \begin{pmatrix} ae & 0 & 0 & 0 \\ 0 & c^2+bd & 2bc & 0 \\ 0 & 2bc & c^2+bd & 0 \\ 0 & 0 & 0 & ae \end{pmatrix}. \tag{100}$$

The eigenvalues of R are $ae \geq 0$ doubly degenerate and $(\sqrt{bd} - c)^2$, whence the concurrence is

$$C(\rho) = \max\left\{0, 2(|c| - \sqrt{ae})\right\} \tag{101}$$

and is thus proportional to the negative eigenvalue of the partially transposed ρ.

4 Complete Positivity, Open Quantum Systems, and Entanglement

The standard time evolution of the observables $X = X^\dagger \in M_d(\mathbf{C})$ of a finite-level system S, $X \mapsto X_t$, forms a group of linear maps, $\mathbb{U}_t : M_d(\mathbf{C}) \mapsto M_d(\mathbf{C})$

$$X_t := \mathbb{U}_t[X] = U_t^\dagger \, X \, U_t, \quad U_t = e^{-iHt} \quad (\hbar = 1), \tag{102}$$

where $H = H^\dagger \in M_d(\mathbf{C})$ is the Hamiltonian of the system, while the dual time evolution (see (43)) of its states,

$$\mathbb{U}_t^*[\rho] = U_t \, \rho \, U_t^\dagger, \tag{103}$$

is a solution to the *Liouville–von Neumann* equation

$$\partial_t \rho_t = -i\left[H, \rho_t\right]. \tag{104}$$

The time evolution generated by this equation is reversible and is typical of *closed quantum systems*, namely, of quantum systems which can be considered isolated from the environment in which they are immersed.

When the interactions between system S and environment E, typically a heat bath in equilibrium at a certain temperature, cannot be neglected, one treats the subsystem S of interest as an *open quantum system* and tries to recover for it a so-called *reduced dynamics* by tracing away the (usually infinitely many) degrees of freedom of the environment. Practically speaking, one starts from a total Hamiltonian $H_T = H_S + H_E + \lambda H_I$, where $H_{S,E}$ are the Hamiltonian of the system and environment, respectively, while H_I is an interaction term with λ an a-dimensional coupling constant. The formal time evolution of any system S observable of the form $X_S \otimes \mathbb{1}_E$ is given by

$$X_S \otimes \mathbb{1}_E \mapsto e^{itH_T} X_S \otimes \mathbb{1}_E e^{-itH_T}, \tag{105}$$

and the reduced dynamics in the Heisenberg picture is obtained by means of the expectation values

$$X_S \mapsto X_S(t) = \omega_E \left(e^{it H_T} X_S \otimes \mathbb{1}_E e^{-it H_T} \right) \tag{106}$$

with respect to an environment state ω_E which is left invariant by the environment time evolution. The resulting reduced dynamics is highly complicated; however, if the coupling is sufficiently weak ($\lambda \ll 1$), a number of techniques allow one to disentangle the time evolution of S from that of $S + E$ and to approximate it by a semigroup of completely positive maps γ_t, $t \geq 0$, on the system density matrices satisfying the forward in time composition law

$$\gamma_{t+s} = \gamma_t \circ \gamma_s = \gamma_s \circ \gamma_t, \quad s, t \geq 0. \tag{107}$$

The presence of the environment E manifests itself as a source of dissipation and noise [2, 11], described by the generator of the semigroup, namely, by the left-hand side of the *master equation*

$$\partial_t \rho_t = -i\left[H, \rho_t\right] + \mathbb{D}[\rho_t] =: \mathbb{L}[\rho], \tag{108}$$

where $\rho \mapsto \mathbb{D}[\rho]$ is a linear map of the form [15]

$$\mathbb{D}[\rho] = \sum_{i,j=1}^{d^2-1} K_{ij} \left(F_i \rho F_j^\dagger - \frac{1}{2}\left\{ F_j^\dagger F_i, \rho \right\} \right), \tag{109}$$

with $K = [K_{ij}]$ a positive $(d^2 - 1) \times (d^2 - 1)$ matrix, known as *Kossakowski matrix*, and the F_i $d \times d$ matrices that together with $F_{d^2} = \mathbb{1}/d^2$ form an orthonormal basis in $M_d(\mathbf{C})$ with respect to the Hilbert–Schmidt scalar product $\mathrm{Tr}\,(F_j^\dagger F_i) = \delta_{ij}$ (see Exercise 4.6).

Remark 7 Semigroups generated by master equations of the form (108) are the standard way to describe decoherence effects suffered by few degrees of freedom as a result of the action of many degrees of freedoms weakly coupled to them. Usually, the few degrees of freedom are those of a finite-level open quantum system immersed in a large thermal reservoir with which it weakly interacts and which acts as a source of dissipation and noise. The semigroup property corresponds to the irreversibility of the dynamics due to the non-standard piece $\mathbb{D}$ which describes the noisy and dissipative effects caused by the presence of the environment and are such that $\gamma_{-t}[\rho]$ for $t > 0$ cannot be a solution to (109). Since $K \geq 0$, by diagonalizing it, the term $\sum_{i,j=1}^{d^2-1} K_{ij} F_i \rho F_j^\dagger$ can be cast in Kraus–Stinespring form (44), $\sum_{\ell=1}^{d^2-1} D_\ell \rho D_\ell^\dagger$; this identifies it as a completely positive map that generically transforms pure states into mixtures and may thus be interpreted as a noise term. Further, the anti-commutator corresponds to a damping term

$$\partial_t \rho_t = -i \left(-\frac{i}{2}\Gamma\right) \rho_t + i \rho_t \left(\frac{i}{2}\Gamma\right) = -\frac{1}{2}\{\Gamma, \rho_t\}, \tag{110}$$

where $\Gamma = \sum_{i,j=1}^{d^2-1} K_{ij} F_j^\dagger F_i \geq 0$. Altogether, these two contributions to the dissipative part of the generator contribute to preserve the overall probability: $\partial_t \mathrm{Tr}(\rho_t) = 0$.

In other words, the master equation (108) can be taken as quantum version of a classical Fokker–Planck equation with the completely positive map in (109) corresponding to the classical white-noise diffusive term and the anti-commutator to the friction term.

The reversible time evolution generated by (103) is automatically completely positive for it consists of linear maps of the form (44) with only one (unitary) Kraus operator; instead, because of the presence of the dissipative term $\mathbb{D}[\rho]$ this is not in general true for the irreversible maps γ_t.

Theorem 1 *The semigroup $\{\gamma_t\}_{t\geq 0}$ generated by (108) consists of completely positive maps if and only if the Kossakowski matrix $K = [K_{ij}] \geq 0$* [15, 26].

Proof We give a sketch of the proof following [15]. If γ_t is completely positive, then, from Lemma 2, $\gamma_t \otimes \mathrm{id}_d[P_+^{(d)}] \geq 0$. Let $|\psi\rangle \in \mathbf{C}^{d^2}$ be such that $\langle\psi|\Psi_+^{(d)}\rangle = 0$; by expanding in t, one finds

$$0 \leq \langle\psi| \gamma_t \otimes \mathrm{id}_d[P_+^{(d)}] |\psi\rangle = t \langle\psi| \mathbb{L} \otimes \mathrm{id}_d[P_+^{(d)}] |\psi\rangle + o(t)$$

$$= t \sum_{i,j=1}^{d^2-1} K_{ij} \langle\psi| F_i \otimes \mathbb{1}_d P_+^{(d)} F_j^\dagger \otimes \mathbb{1}_d |\psi\rangle + o(t), \tag{111}$$

whence, for all $|\psi\rangle \in \mathbf{C}^{d^2}$, $\sum_{i,j=1}^{d^2-1} K_{ij} \langle\psi| F_i \otimes \mathbb{1}_d P_+^{(d)} F_j^\dagger \otimes \mathbb{1}_d |\psi\rangle \geq 0$. Let $\Psi = [\Psi_{ij}]$ be the $d \times d$ matrix with entries given by the components of $|\psi\rangle$ with respect to a chosen orthonormal basis; the orthogonality condition $\langle\psi|\Psi_+^{(d)}\rangle = \sum_{i=1}^d \psi_{ii} = 0$ makes Ψ a traceless matrix that can be expanded as $\Psi = \sum_{i=1}^{d^2-1} x_i F_i$ using the Hilbert–Schmidt basis appearing in generator (109). By explicit computation, $\sqrt{d}\langle\Psi_+^{(d)}| F_j^\dagger \otimes \mathbb{1}_d |\psi\rangle = \mathrm{Tr}(F_j^\dagger \Psi) = x_j$ (see Exercise 4.6), thus

$$\sum_{i,j=1}^{d^2-1} K_{ij} \langle \psi | F_i \otimes \mathbb{1}_d \, P_+^{(d)} \, F_j^\dagger \otimes \mathbb{1}_d \, | \psi \rangle = \sum_{i,j=1}^{d^2-1} x_i^* \, K_{ij} \, x_j = \langle x | \, K \, | x \rangle \geq 0 \,. \quad (112)$$

Since vectors $|x\rangle \in \mathbf{C}^{d^2-1}$, traceless $d^2 \times d^2$ matrices, and $|\psi\rangle \perp |\Psi_+^{(d)}\rangle$ are in one-to-one correspondence, by varying the latter one gets that $\langle x | \, K \, | x \rangle \geq 0$ for all $|x\rangle \in \mathbf{C}^{d^2-1}$, whence $K \geq 0$.

In order to prove the converse, consider two orthogonal vectors $|\psi\rangle, |\phi\rangle \in \mathbf{C}^{d^2}$ and, as in the argument of above, the corresponding $d^2 \times d^2$ matrices $\Psi = [\psi_{ij}]$, $\Phi = [\phi_{ij}]$. Then, $\langle \psi | \phi \rangle = 0$ results in $\mathrm{Tr}\,(\Psi^\dagger \Phi) = \mathrm{Tr}\,(\Psi \Phi^\dagger) = 0$; as previously, one expands $\Psi \Phi^\dagger = \sum_{i=1}^{d^2-1} x_i \, F_i$ and computes, up to first order in t,

$$\langle \psi | \, \gamma_t \otimes \mathrm{id}_d [|\psi\rangle\langle\psi|] \, |\phi\rangle \simeq t \, \langle \psi | \, \gamma_t \otimes \mathrm{id}_d [|\psi\rangle\langle\psi|] \, |\phi\rangle$$

$$\simeq t \sum_{i,j=1}^{d-1} K_{ij} \, (\mathrm{Tr}\,(F_i^\dagger \Psi \Phi^\dagger))^* \, \mathrm{Tr}\,(F_j^\dagger \Psi \Phi^\dagger) \quad (113)$$

$$\simeq t \, \langle x | \, K \, | x \rangle \,.$$

The positivity of the Kossakowski matrix implies $\langle \phi | \, \mathbb{L} \otimes \mathrm{id}_d [|\psi\rangle\langle\psi|] \, |\phi\rangle \geq 0$ for all $\langle \psi | \phi \rangle = 0$ so that the leftmost side of the previous expression cannot become negative. Suppose there exists an initial state ρ such that $\gamma_t \otimes \mathrm{id}_d[\rho]$ assumes a zero eigenvalue at a certain time t^* and a negative eigenvalue at $t = t^* + \varepsilon$ as soon as $\varepsilon > 0$. By expanding in ε, one finds

$$0 > \langle \phi | \, \gamma_{t^*+\varepsilon} \otimes \mathrm{id}_d[\rho] \, |\phi\rangle \simeq \langle \phi | \, \gamma_{t^*}[\rho] \, |\phi\rangle + \varepsilon \, \langle \phi | \, \mathbb{L} \otimes \mathrm{id}_d[\gamma_{t^*} \otimes \mathrm{id}_d[\rho]] \, |\phi\rangle \,. \quad (114)$$

Since ε can be taken arbitrarily small, this cannot be true if $|\phi\rangle$ is orthogonal to the eigenvectors $|\psi\rangle$ of $\gamma_{t^*} \otimes \mathrm{id}_d[\rho]$ and the contribution of order 1 in $\varepsilon > 0$ is strictly positive, $\langle \phi | \, \mathbb{L} \otimes \mathrm{id}_d [|\psi\rangle\langle\psi|] \, |\phi\rangle > 0$ for all $|\psi\rangle$. A more refined analysis extends this argument to the case when $\langle \phi | \, \mathbb{L} \otimes \mathrm{id}_d [|\psi\rangle\langle\psi|] \, |\phi\rangle \geq 0$ [21]. Thus if the Kossakowski matrix is positive, $\gamma_t \otimes \mathrm{id}_d$ preserves positivity whence γ_t is completely positive. □

If K is not positive definite, then the physical consistency of γ_t is doubtful as negative probabilities may arise by the action of $\gamma_t \otimes \mathrm{id}_d$ on the entangled states of the bipartite system $S + S$, surely on the symmetric state $P_+^{(2)}$ because of Lemma 2. On the other hand, if there were in nature no entangled states of $S + S$, then the positivity of γ_t would be enough to make it a possible mathematical description of a physical process.

Example 5

1. Consider the one-qubit purely dissipative master equation

$$\partial_t \rho_t = \mathbb{L}_1[\rho_t] = \frac{1}{2} \sum_{i=1}^{3} (\sigma_i \, \rho_t \, \sigma_i - \rho_t) \,. \quad (115)$$

Comparing with (108), $F_i = \sigma_i/\sqrt{2}$, $\sigma_{1,2,3}$ the Pauli matrices, and the Kossakowski matrix $K = \mathrm{diag}(1, 1, 1)$ is positive. Using Exercises 5.2 and 4.1 and the Bloch representation of the initial state, $\rho = \frac{1}{2}(\mathbb{1}_2 + \boldsymbol{r} \cdot \boldsymbol{\sigma})$, one computes

$$
\begin{aligned}
\gamma_t^{(1)}[\rho] &= \frac{1}{2}(\mathbb{1}_2 + \lambda_t \boldsymbol{r} \cdot \boldsymbol{\sigma}) = \frac{1 - \lambda_t}{2}\mathbb{1}_2 + \lambda_t \rho \\
&= \left(\lambda_t \mathrm{id}_2 + \frac{1 - \lambda_t}{2}\widehat{\mathrm{Tr}}\right)[\rho] = \frac{1 + 3\lambda_t}{4}\rho + \frac{1 - \lambda_t}{4}\sum_{i=1}^{3}\sigma_i \rho \sigma_i ,
\end{aligned}
\tag{116}
$$

where $\lambda_t = \exp(-2t)$ and we have used that $\mathrm{Tr}(\rho) = 1$. Therefore, the dissipative time evolution $\gamma_t^{(1)}$ is written in Kraus–Stinespring form with Kraus operators given by the Pauli matrices; it is thus completely positive in accordance with Theorem 1.

2. Consider the one-qubit purely dissipative master equation

$$
\partial_t \rho_t = \mathbb{L}_2[\rho_t] = \frac{1}{4}\sum_{i=1}^{3}\varepsilon_i(\sigma_i \rho_t \sigma_i - \rho_t) ,
\tag{117}
$$

where, as in Exercise 4.2, $\varepsilon_1 = 1 = \varepsilon_3 = -\varepsilon_2$. Comparing with (108), the Kossakowski matrix $K = \mathrm{diag}(1, 1, 1)$ is not positive. Integration yields

$$
\begin{aligned}
\gamma_t^{(2)}[\rho] &= \frac{1}{2}(\mathbb{1}_2 + r_1\sigma_1 + \lambda_t r_2\sigma_2 + r_3\sigma_3) = \rho + \frac{\lambda_t - 1}{2}r_2\sigma_2 = \\
&= \left(\frac{1 + \lambda_t}{2}\mathrm{id}_2 + \frac{1 - \lambda_t}{2}\mathrm{T}\right)[\rho] = \frac{3 + \lambda_t}{4}\rho + \frac{1 - \lambda_t}{4}\sum_{i=1}^{3}\varepsilon_i\sigma_i \rho \sigma_i ,
\end{aligned}
\tag{118}
$$

where it has been used that $r_2\sigma_2 = \rho - \mathrm{T}[\rho]$ and it has been set $\lambda_t = \exp(-2t)$. As in the previous example, the Kraus operators are the Pauli matrices; however the time evolution is not in Kraus–Stinespring form as one of the terms is multiplied by $\varepsilon_2 = -1$. Because of Theorem 1, it cannot be completely positive; indeed, the corresponding Choi matrix is easily computed by using Exercises 5.2 and 5.3

$$
\begin{aligned}
\gamma_t^{(2)} \otimes \mathrm{id}_2[P_+^{(2)}] &= \frac{1}{4}\left(\mathbb{1}_4 + \sigma_1 \otimes \sigma_1 - \lambda_t \sigma_2 \otimes \sigma_2 + \sigma_3 \otimes \sigma_3\right) \\
&= \frac{1}{2}\begin{pmatrix} 1 & 0 & 0 & (1 + \lambda_t)/2 \\ 0 & 0 & (1 - \lambda_t)/2 & 0 \\ 0 & (1 - \lambda_t)/2 & 0 & 0 \\ (1 + \lambda_t)/2 & 0 & 0 & 1 \end{pmatrix}
\end{aligned}
\tag{119}
$$

and has a negative eigenvalue $-(1 - \lambda_t)/2$ for all $t > 0$.

3. Complete positivity is stronger than simple positivity [4]; in order to compare the physical constraints imposed on the dissipative dynamics by these two requests, consider the following one-parameter family $\{\rho_t\}_{t\geq 0}$ of density matrices of a single two-level system

$$\rho_t = \frac{1}{2}\begin{pmatrix} 1 + e^{-at}\, r_3 & e^{-bt}(r_1 - i\, r_2) \\ e^{-bt}(r_1 + i\, r_2) & 1 - e^{-at}\, r_3 \end{pmatrix}, \tag{120}$$

with $a, b \geq 0$ and $r_i \in \mathbb{R}$, $i = 1, 2, 3$, such that $\sum_{i=1}^{3} r_i^2 \leq 1$. These conditions guarantee that the matrices ρ_t are positive semi-definite; indeed, for 2×2 density matrices, positivity semi-definiteness is equivalent to the non-negativity of both trace and determinant. Now, $\mathrm{Tr}\,(\rho_t) = 1$ while

$$\mathrm{Det}(\rho_t) = \frac{1 - \left(e^{-2at} r_3^2 + e^{-2bt}(r_1^2 + r_2^2)\right)}{4} \geq \frac{1 - \left(r_1^2 + r_2^2 + r_3^2\right)}{4} \geq 0\,.$$

By using the Bloch vector representation (see Sect. 2.1 in chapter "Hilbert Space Methods for Quantum Mechanics"), one writes the time evolution (120) as

$$\rho = \frac{\mathbb{1}_2 + \boldsymbol{r}\cdot\boldsymbol{\sigma}}{2}\,, \quad \boldsymbol{r} = \begin{pmatrix} r_1 \\ r_2 \\ r_3 \end{pmatrix} \longmapsto \rho_t = \frac{\mathbb{1}_2 + \boldsymbol{r_t}\cdot\boldsymbol{\sigma}}{2}\,, \quad \boldsymbol{r_t} = \begin{pmatrix} e^{-bt} r_1 \\ e^{-bt} r_2 \\ e^{-at} r_3 \end{pmatrix}.$$
$$\tag{121}$$

In fact, ρ_t is obtained as the action of a linear map γ_t on ρ,

$$\rho_t = \frac{\mathbb{1}_2 + \sum_{i=1}^{3} r_i \gamma_t[\sigma_i]}{2}\,, \quad \gamma_t[\sigma_{1,2}] = e^{-bt}\,\sigma_{1,2}\,, \quad \gamma_t[\sigma_3] = e^{-at}\,\sigma_3\,. \tag{122}$$

By taking the time derivative and using Exercise 5.2, one finds

$$\dot{\sigma}_{1,2} = -b\,\sigma_{1,2} = \mathbb{D}[\sigma_{1,2}]\,, \quad \dot{\sigma}_3 = -a\,\sigma_3 = \mathbb{D}[\sigma_3]\,, \tag{123}$$

where

$$\mathbb{D}[\rho] = \frac{a}{4}\left(\sigma_1\,\rho\,\sigma_1 + \sigma_2\,\rho\,\sigma_2 - 2\rho\right) + \left(\frac{b}{2} - \frac{a}{4}\right)\left(\sigma_3\,\rho\,\sigma_3 - \sigma_3\right). \tag{124}$$

Therefore, the positive, trace-preserving maps $\gamma_t = e^{t\mathbb{D}}$ form a semigroup that describe an irreversible process where the diagonal and off-diagonal elements of any initial density matrix decay with different rates a, respectively, b, to those of the totally mixed state $\rho_\infty = \frac{1}{2}$. Comparing with (109) one sees that the matrices $F_i = \sigma_i/\sqrt{2}$, while the Kossakowski matrix reads

$$K = \frac{1}{2} \begin{pmatrix} a & 0 & 0 \\ 0 & a & 0 \\ 0 & 0 & b - \frac{a}{2} \end{pmatrix}. \tag{125}$$

Such a matrix is positive, and thus the map γ_t is completely positive, if and only if $a \geq 2b$. Indeed, by using Exercise 5.3, the Choi matrix (45) of γ_t results:

$$\gamma_t \otimes \mathrm{id}_2[P_+^{(2)}] = \frac{1}{4}\Big(\mathbb{1} \otimes \mathbb{1} + e^{-bt}\left(\sigma_1 \otimes \sigma_1 - \sigma_2 \otimes \sigma_2\right) + e^{-at}\,\sigma_3 \otimes \sigma_3\Big)$$

$$= \frac{1}{4} \begin{pmatrix} 1 + e^{-at} & 0 & 0 & 2e^{-bt} \\ 0 & 1 - e^{-at} & 0 & 0 \\ 0 & 0 & 1 - e^{-at} & 0 \\ 2e^{-bt} & 0 & 0 & 1 + e^{-at} \end{pmatrix}. \tag{126}$$

This matrix is positive if and only if $1 + e^{-at} \geq 2e^{-bt}$: when $b \geq a/2$ this is always true, whereas if $b < a/2$, an expansion at small times yields

$$1 + e^{-at} - 2e^{-bt} \simeq (2b - a)\,t < 0. \tag{127}$$

4.1 Positive and Completely Positive Semigroups

The last example shows that complete positivity implies a hierarchy between the decay rates of the matrix elements of the system S that simple positivity would not ask for. This is due to the fact that one assumes that there might exist entangled states of bipartite systems evolving according to a dissipative time evolution of the form $\gamma_t \otimes \mathrm{id}$, where only one party is affected by the dynamics. While the existence of entangled states is hardly questionable, the form of the time evolution might look too abstract, physically speaking. For instance, instead of thinking of the inert party as a distant quantum system that happened to be entangled with the other one before the dynamics γ_t started, one could think of two systems of the same kind, not interacting, but immersed in a same environment or being affected by two independent dissipative dynamics: In the first case, the mathematical description of the actual time evolution would use a semigroup consisting of maps $\Gamma_t = \gamma_t \otimes \gamma_t$, in the second one of maps of the form $\Gamma_t = \gamma_t^{(1)} \otimes \gamma_t^{(2)}$.

A minimal request of physical consistency is that the maps Γ_t preserve the positivity of all states of $S + S$ on which they act. The following two results show that positivity of $\gamma_t \otimes \gamma_t$ is only guaranteed if γ_t is completely positive [6], while that of $\gamma_t^{(1)} \otimes \gamma_t^{(2)}$, $\gamma_t^{(1)} \neq \gamma_t^{(2)}$, is less constraining [5, 6].

Proposition 2 *Given a semigroup of maps γ_t generated by* (108), *the map $\Gamma_t = \gamma_t \otimes \gamma_t$, $t \geq 0$, is positive if and only it γ_t is completely positive [7].*

Proof If γ_t is completely positive, both $\gamma_t \otimes \mathrm{id}_d$ and $\mathrm{id}_d \otimes \gamma_t$ are positive and such is their composition Γ_t.

The maps Γ_t form a semigroup with generator $\mathbb{L} \otimes \mathrm{id}_d + \mathrm{id}_d \otimes \mathbb{L}$; if they are positive, then, as in the proof of Theorem 1, for all orthogonal $\psi, \phi \in \mathbf{C}^d \otimes \mathbf{C}^d$, it must hold that

$$
0 \le \langle \phi | \, (\mathbb{L} \otimes \mathrm{id}_d + \mathrm{id}_d \otimes \mathbb{L})[|\psi\rangle\langle\psi|] \, |\phi\rangle
$$

$$
= \sum_{i,j=1}^{d^2-1} K_{ij} \left((\mathrm{Tr}\,(F_i^\dagger \Psi \Phi^\dagger))^* \, \mathrm{Tr}\,(F_j^\dagger \Psi \Phi^\dagger) + (\mathrm{Tr}\,(F_i^\dagger (\Psi^\dagger \Phi)^\mathrm{T}))^* \, \mathrm{Tr}\,(F_j^\dagger (\Psi^\dagger \Phi)^\mathrm{T}) \right),
$$

$$(128)$$

where $\Psi = \left[\psi_{ij} \right]$ and $\Phi = \left[\phi_{ij} \right]$ are $d \times d$ matrices consisting of the coefficients of the expansions of $|\psi\rangle$ and $|\phi\rangle$ with respect to a chosen orthonormal basis $\{|ij\rangle\}_{i,j=1}^d$ in $\mathbf{C}^d \otimes \mathbf{C}^d$ and $(X)^\mathrm{T}$ denotes transposition of a matrix $X \in M_d(\mathbf{C})$ with respect to the chosen basis. Notice that, since ψ and ϕ are assumed orthogonal, it follows that $\mathrm{Tr}\,(\psi^\dagger \Phi) = 0$. Let $M_d(\mathbf{C}) \ni X = \sum_{i=1}^{d^2-1} x_i F_i$, $x_i = \mathrm{Tr}\,(F_i^\dagger X)$, be a generic traceless matrix; because of their same spectral structure, X and its transposed X^T are similar: $X^\mathrm{T} = \Phi^{-1} X \Phi$, $\Phi \in M_d(\mathbf{C})$. Set $\Psi^\dagger = \Phi^{-1} X$; then, $X = \Phi \Psi^\dagger$ and $(\Psi^\dagger \Phi)^\mathrm{T} = (X^\mathrm{T})^\mathrm{T} = X$, whence

$$
\langle \phi | \, (\mathbb{L} \otimes \mathrm{id}_d + \mathrm{id}_d \otimes \mathbb{L})[|\psi\rangle\langle\psi|] \, |\phi\rangle = 2\langle x | \, K \, | x \rangle \ge 0, \tag{129}
$$

for all $|x\rangle \in \mathbf{C}^{d^2-1}$; thus, $K = [K_{ij}] \ge 0$ and γ_t is completely positive by Theorem 1. $\qquad\square$

Proposition 3 *Consider two semigroups $\gamma_t^{(1)}$ and $\gamma_t^{(2)}$ generated by two master equations of the form* (108), *with dissipative terms*

$$
\mathbb{D}_1[\rho] = \sum_{i=1}^{d^2-1} f_i^{(1)} \left(F_i^{(1)} \rho F_i^{(1)} - \frac{1}{2} \left\{ (F_i^{(1)})^2, \rho \right\} \right), \tag{130}
$$

$$
\mathbb{D}_2[\rho] = \sum_{i=1}^{d^2-1} f_i^{(2)} \left(F_i^{(2)} \rho F_i^{(2)} - \frac{1}{2} \left\{ (F_i^{(2)})^2, \rho \right\} \right), \tag{131}
$$

diagonal with respect to a Hilbert–Schmidt basis of traceless, Hermitian matrices $F_i^{1,2} \in M_d(\mathbf{C})$. If all coefficients $f_i^{(1,2)} \ge 0$, but for one fixed $f_k^{(2)}$ and $f_i^{(1)} \ge |f_k^{(2)}|$ for all i, $f_i^{(2)} \ge |f_k^{(2)}|$ for all $i \ne k$, then the maps $\Gamma_t = \gamma_y^{(1)} \otimes \gamma_t^{(2)}$ are positive for all $t \ge 0$ [6].

Proof As in the proof of Theorem 1, we show that for all orthogonal $\psi, \phi \in \mathbf{C}^d \otimes \mathbf{C}^d$,

$$\langle \phi | (\mathbb{L}_1 \otimes \text{id}_d + \text{id}_d \otimes \mathbb{L}_2)[|\psi\rangle\langle\psi|] |\phi\rangle = \sum_{i=1}^{d^2-1} f_i^{(1)} \left| \text{Tr}\,(F_i^{(1)} \Psi \Phi^\dagger) \right|^2$$

$$+ \sum_{1=i\neq k}^{d^2-1} f_i^{(2)} \left| \text{Tr}\,(F_i^{(2)} (\Psi^\dagger \Phi)^\mathrm{T}) \right|^2 - |f_k^{(2)}| \left| \text{Tr}\,(F_k^{(2)} (\Psi^\dagger \Phi)^\mathrm{T}) \right|^2 \geq 0 \,.$$

(132)

Because of the assumptions on the Hilbert–Schmidt bases $\{F_i^{(1,2)}\}_{i=1}^{d^2}$, one has (see Exercise 5.4)

$$\sum_{i=1}^{d^2-1} \left(\text{Tr}\,(F_i^{(1)} \Psi \Phi^\dagger) \right)^2 = \text{Tr}\left(\Psi \Phi^\dagger \Psi \Phi^\dagger \right) = \text{Tr}\left((\Phi^\dagger \Psi)^\mathrm{T} (\Phi^\dagger \Psi)^\mathrm{T} \right)$$

$$= \sum_{i=1}^{d^2-1} \left(\text{Tr}\,(F_i^{(2)} (\Phi^\dagger \Psi)^\mathrm{T}) \right)^2 \,.$$

(133)

Therefore, one can bound $\left| \text{Tr}\,(F_k^{(2)} (\Phi^\dagger \Psi)^\mathrm{T}) \right|^2$ from above by

$$\sum_{1=i}^{d^2-1} f_i^{(2)} \left| \text{Tr}\,(F_i^{(2)} (\Psi^\dagger \Phi)^\mathrm{T}) \right|^2 - \sum_{1=i\neq k}^{d^2-1} f_i^{(2)} \left| \text{Tr}\,(F_i^{(2)} (\Psi^\dagger \Phi)^\mathrm{T}) \right|^2 \,,$$

(134)

whence it follows that

$$\langle \phi | (\mathbb{L}_1 \otimes \text{id}_d + \text{id}_d \otimes \mathbb{L}_2)[|\psi\rangle\langle\psi|] |\phi\rangle \geq \sum_{i=1}^{d^2-1} \left(f_i^{(1)} - |f_k^{(2)}| \right) \left| \text{Tr}\,(F_i^{(1)} \Psi \Phi^\dagger) \right|^2$$

$$+ \sum_{1=i\neq k}^{d^2-1} \left(f_i^{(2)} - |f_k^{(2)}| \right) \left| \text{Tr}\,(F_i^{(2)} (\Psi^\dagger \Phi)^\mathrm{T}) \right|^2 \geq 0 \,.$$

(135)

$\square$

Example 6

1. The semigroups of Examples 5.1 and 5.2 fulfill the hypothesis of Proposition 2; therefore, the maps

$$\Gamma_t = \gamma_t^{(1)} \otimes \gamma_t^{(2)} = \lambda_t \frac{1+\lambda_t}{2} \text{id}_{12} + \frac{1-\lambda_t^2}{4} \widehat{\text{Tr}}_1$$

$$+ \lambda_t \frac{1-\lambda_t}{2} T_2 + \frac{(1-\lambda_t)^2}{4} \widehat{\text{Tr}}_1 \otimes T_2$$

(136)

are positivity preserving on the states of two qubits. However, the Kossakowski matrix in the generator of the semigroup consisting of these maps is

$$K = \begin{pmatrix} 1 & 0 & 0 \\ 0 & 1 & 0 \\ 0 & 0 & 1 \end{pmatrix} \otimes \mathbb{1}_3 + \mathbb{1}_3 \otimes \begin{pmatrix} 1 & 0 & 0 \\ 0 & -1 & 0 \\ 0 & 0 & 1 \end{pmatrix} \tag{137}$$

and thus not positive definite. Therefore, Γ_t cannot be completely positive; its Choi matrix can better be studied by writing the totally symmetric projector $P_+^{(4)} \in M_{16}(\mathbf{C})$ as follows:

$$P_+^{(4)} = \frac{1}{4} \sum_{a,b;c,d=0}^{1} |ab\rangle\langle cd| \otimes |ab\rangle\langle cd| = (P_+^{(2)})_{13} \otimes (P_+^{(2)})_{24} \,, \tag{138}$$

with subindexes denoting to which qubits the 4×4 symmetric projectors $P_+^{(2)}$ refer to. Then, one computes

$$\Gamma_t \otimes \mathrm{id}_4[P_+^{(4)}] = \frac{1 + \lambda_t}{4} (P_+^{(2)})_{13} \otimes \left(2\lambda_t (P_+^{(2)})_{24} + V_{24} \right) \\ + \frac{1 - \lambda_t}{16} \left(4\lambda_t (P_+^{(2)})_{13} + (1 - \lambda_t)\mathbb{1}_{13} \right) \otimes V_{24} \,, \tag{139}$$

where V_{24} is the flip operator for the second and fourth qubits. Let $|\Psi_{11}\rangle$ be the anti-symmetric Bell state (9); then, $P^{(2)}|\Psi_{11}\rangle = 0$, $V|\Psi_{11}\rangle = -|\Psi_{11}\rangle$,

$$\Gamma_t \otimes \mathrm{id}_4[P_+^{(4)}](|\Psi_{11}\rangle)_{13} \otimes (|\Psi_{11}\rangle)_{24} = -\frac{(1 - \lambda_t)^2}{16}(|\Psi_{11}\rangle)_{13} \otimes (|\Psi_{11}\rangle)_{24} \,, \tag{140}$$

whence, as $\lambda_t = \exp(-2t)$, the Choi matrix of Γ_t is not positive for all $t > 0$.

2. Since all $\Gamma_t : M_4(\mathbf{C}) \mapsto M_4(\mathbf{C})$ are positive maps, a natural question is whether they are decomposable in the sense of Definition 4. Since $T \circ T = \mathrm{id}$ and $\widehat{\mathrm{Tr}} \circ T = \widehat{\mathrm{Tr}}$, where $\widehat{\mathrm{Tr}}$ is the trace operation $\mathrm{Tr}[X] = \mathrm{Tr}(X)\mathbb{1}$, and $T_1 \otimes T_2 = T_{12}$, expression (136) can be recast as

$$\Gamma_t = \lambda_t \frac{1 + \lambda_t}{2} \mathrm{id}_{12} + \frac{1 - \lambda_t^2}{4} \widehat{\mathrm{Tr}}_1 \\ + \frac{1 - \lambda_t}{2} \left(\lambda_t T_1 + \frac{1 - \lambda_t}{2} \widehat{\mathrm{Tr}}_1 \right) \circ T_{12} \,. \tag{141}$$

The first line is the sum of two completely positive maps and is thus completely positive; in order to check the state of the second line, we compute the Choi matrix of $\Lambda_t = \lambda_t T_1 + \frac{1 - \lambda_t}{2} \widehat{\mathrm{Tr}}_1$ as done in the previous example:

$$\Lambda_t \otimes \mathrm{id}_4[P_+^{(4)}] = \frac{1}{2} \left(\lambda_t V_{13} + \frac{1 - \lambda_t}{2} \mathbb{1}_{13} \right) \otimes (P_+^{(2)})_{24} \,. \tag{142}$$

Since the flip operator V has eigenvalues ± 1, the non-null eigenvalues of $\Lambda_t \otimes$ id$_4[P_+^{(4)}]$ are $(1+\lambda_t)/4$ and $(1-3\lambda_t)/4$; the latter is positive only if $t = 0$ and $t \geq$ $(\log 3)/2$ in which case Γ_t results are decomposable. When $0 < t < (\log 3)/2$, the Choi matrix has a negative eigenvalue and Λ_t is not completely positive; however, this does not mean that Γ_t is not decomposable as the decomposition into (141) is not unique. This question will be answered in the next section in connection with certain PPT-entangled states of two plus two qubits.

Exercise 5

1. *Consider a spin* $1/2$ *particle in a constant magnetic field, that is, a qubit with Hamiltonian* $H = \sum_{i=1}^{3} h_i \sigma_i$, $\boldsymbol{h} \in \mathbb{R}^3$. *By using the algebra of the Pauli matrices, show that, in terms of the Bloch vector, the time evolution equation* (103) *reads*

$$\frac{d\boldsymbol{r}}{dt} = 2\boldsymbol{h} \cdot \boldsymbol{r} \tag{143}$$

2. *Prove that the linear map* $\Lambda[X] = \sum_{i=1}^{3} (d_i/2)\Big(\sigma_i X \sigma_i - X\Big)$ *acts as follows*:

$$\mathbb{D}[\sigma_1] = -(d_2 + d_3)\,\sigma_1 \ , \quad \mathbb{D}[\sigma_2] = -(d_1 + d_3)\,\sigma_2 \ , \quad \mathbb{D}[\sigma_3] = -(d_1 + d_2)\,\sigma_3 \ . \tag{144}$$

3. *Prove that the projector onto the totally symmetric state* (14) *of two qubits can be written as*

$$P_+^{(2)} = \frac{1}{4}\Big(\mathbb{1}_2 \otimes \mathbb{1}_2 + \sigma_1 \otimes \sigma_1 - \sigma_2 \otimes \sigma_2 + \sigma_3 \otimes \sigma_3\Big) . \tag{145}$$

4. *Given a set of traceless matrices* $F_i \in M_d(\mathbb{C})$, $1 \leq i \leq d^2 - 1$ *plus* $F_{d^2} = \mathbb{1}_d/\sqrt{d}$ *such that* $\mathrm{Tr}\Big(F_j^\dagger F_i\Big) = \delta_{ij}$, *show that they constitute a Hilbert–Schmidt orthonormal basis in* $M_d(\mathbb{C})$, *namely that, for all* $X \in M_d(\mathbb{C})$,

$$X = \sum_{i=1}^{d^2} \mathrm{Tr}\,(F_i^\dagger X)\, F_i \ . \tag{146}$$

4.2 PPT-Entangled States

As already observed, when the dimensions of the parties involved are not as indicated in Corollary 4, there can exist bipartite-entangled states which remain positive under partial transposition.

As an example, consider a system consisting of two pairs of qubits described by the matrix algebra $M_{16}(\mathbb{C}) = M_4(\mathbb{C}) \otimes M_4(\mathbb{C})$. Consider the standard basis $|i\rangle$,

$i = 0, 1$ and the Pauli matrices σ_α, $\alpha = 0, 1, 2, 3$, with $\sigma_0 = \mathbb{1}_2$, the 2×2 identity matrix and $\sigma_3 |i\rangle = (-1)^i |i\rangle$. Set $\sigma_{\alpha\beta} = \sigma_\alpha \otimes \sigma_\beta$ and use the totally symmetric state $P_+^{(4)}$ to construct 16 orthonormal states (a basis in $\mathbf{C}^{16}$, see Exercise 4.2) and orthogonal one-dimensional projections

$$|\Psi_{\alpha\beta}\rangle = \mathbb{1}_4 \otimes \sigma_{\alpha\beta} |\Psi_+^{(4)}\rangle \,, \quad P_{\alpha\beta} = |\Psi_{\alpha\beta}\rangle\langle\Psi_{\alpha\beta}| = \mathbb{1}_4 \otimes \sigma_{\alpha\beta} \, P_+^{(4)} \, \mathbb{1}_4 \otimes \sigma_{\alpha\beta} \,. \quad (147)$$

An interesting class of states, called *lattice states* [5], is given by density matrices of the form

$$\rho_I = \frac{1}{\#(I)} \sum_{(\alpha,\beta)\in I} P_{\alpha\beta} \,, \quad (148)$$

namely, by equally weighted convex combinations of projections labeled by pairs (α, β) of indexes from a subset $I \subseteq L = \{(\alpha, \beta) : \alpha, \beta = 0, 1, 2, 3\}$ of cardinality $\#(I)$.

Lemma 6 *Let* $I = \{(0, 2), (1, 1), (2, 3), (3, 1), (3, 2), (3, 3)\}$; *then,*

$$\rho_I = \frac{1}{6}\left(P_{02} + P_{11} + P_{23} + P_{31} + P_{32} + P_{33}\right) \quad (149)$$

is a PPT lattice state.

Proof We shall use transposition over the first party; then, Example 1.1 one computes

$$T \otimes id_4[P_{\alpha\beta}] = \frac{1}{4}\left(\mathbb{1}_4 \otimes \sigma_{\alpha\beta}\right) V \left(\mathbb{1}_4 \otimes \sigma_{\alpha\beta}\right). \quad (150)$$

Furthermore, Exercises 4.5 and 4.3 yield the eigenvalue–eigenprojection relations

$$V|\Psi_{\gamma\delta}\rangle = V \, \mathbb{1}_4 \otimes \sigma_{\gamma\delta} \, V|\Psi_+\rangle = \mathbb{1}_4 \otimes (\sigma_{\gamma\delta})^T |\Psi_+\rangle = \varepsilon_\gamma \varepsilon_\delta |\Psi_{\gamma\delta}\rangle \,, \quad (151)$$

where $(\sigma_\alpha)^T = \varepsilon_\alpha \sigma_\alpha$ and $\varepsilon_2 = -1$, otherwise $= 1$. Then, inserting the resulting spectral decomposition of V, $V = \sum_{(\gamma,\delta)\in L} \varepsilon_\gamma \varepsilon_\delta \, P_{\gamma\delta}$, into (150), one obtains

$$T \otimes id_4[P_{\alpha\beta}] = \frac{1}{4} \sum_{(\gamma,\delta)\in L} \varepsilon_\gamma \varepsilon_\delta \, \mathbb{1}_4 \otimes \sigma_{\alpha\beta}\sigma_{\gamma\delta} \, P_+^{(4)} \, \mathbb{1}_4 \otimes \sigma_{\gamma\delta}\sigma_{\alpha\beta} \,. \quad (152)$$

Consider the algebraic relations between the Pauli matrices:

$$\sigma_0\sigma_\mu = \sigma_\mu \,, \quad \sigma_i\sigma_j = \delta_{ij} + i\sum_{k=1}^{3} \varepsilon_{ijk}\sigma_k \,, \quad i, j, k \neq 0 \,. \quad (153)$$

It proves convenient to recast them in the form $\sigma_\alpha \sigma_\gamma = \eta^\mu_{\alpha\gamma} \sigma_\mu$, where μ is uniquely determined by (α, β) through the Hermitian and unitary matrices

$$
\eta^0 = \mathbb{1}_4 \ , \ \eta^1 = \begin{pmatrix} 0 & 1 & 0 & 0 \\ 1 & 0 & 0 & 0 \\ 0 & 0 & 0 & i \\ 0 & 0 & -i & 0 \end{pmatrix} , \ \eta^2 = \begin{pmatrix} 0 & 0 & 1 & 0 \\ 0 & 0 & 0 & -i \\ 1 & 0 & 0 & 0 \\ 0 & i & 0 & 0 \end{pmatrix} , \ \eta^3 = \begin{pmatrix} 0 & 0 & 0 & 1 \\ 0 & 0 & i & 0 \\ 0 & -i & 0 & 0 \\ 1 & 0 & 0 & 0 \end{pmatrix} . \quad (154)
$$

Then, with $\varepsilon = \mathrm{diag}(1, 1, -1, 1)$, one writes (152) as

$$
\mathrm{T} \otimes \mathrm{id}_4[P_{\alpha\beta}] = \frac{1}{4} \sum_{(\mu,\nu)\in L} \left(\eta^\mu \varepsilon \eta^\mu\right)_{\alpha\alpha} \left(\eta^\nu \varepsilon \eta^\nu\right)_{\beta\beta} P_{\mu\nu} . \quad (155)
$$

The matrices $\eta^\mu \varepsilon \eta^\mu$ are diagonal: $\eta^0 \varepsilon \eta^0 = \varepsilon$, $\eta^1 \varepsilon \eta^1 = \mathrm{diag}(1, 1, 1, -1)$, $\eta^2 \varepsilon \eta^1 = \mathrm{diag}(-1, 1, 1, 1)$, and $\eta^3 \varepsilon \eta^1 = \mathrm{diag}(1, -1, 1, 1)$; by introducing the 4×4 real, symmetric matrix

$$
\tilde{\varepsilon} = \begin{pmatrix} 1 & 1 & -1 & 1 \\ 1 & 1 & 1 & -1 \\ -1 & 1 & 1 & 1 \\ 1 & -1 & 1 & 1 \end{pmatrix} \quad (156)
$$

one finally finds that any partial transposed lattice state is still diagonal with respect to the orthonormal basis $\left\{ |\Psi_{\alpha\beta}\rangle \right\}_{(\alpha,\beta)\in L}$:

$$
\mathrm{T} \otimes \mathrm{id}_4[\rho_I] = \frac{1}{4\#(I)} \sum_{(\mu,\nu)\in L} \left(\tilde{\varepsilon} X_I \tilde{\varepsilon}\right)_{\mu\nu} P_{\mu\nu} , \quad (157)
$$

with eigenvalues given by the entries of $(1/4\#(I))\tilde{\varepsilon} X_I \tilde{\varepsilon}$, where X_I is the characteristic matrix of the subset $I \subseteq L$:

$$
(X_I)_{\alpha\beta} = \begin{cases} 1 & (\alpha, \beta) \in I \\ 0 & \text{otherwise} \end{cases} . \quad (158)
$$

In the case of the subset I of this lemma,

$$
X_I = \begin{pmatrix} 0 & 0 & 1 & 0 \\ 0 & 1 & 0 & 0 \\ 0 & 0 & 0 & 1 \\ 0 & 1 & 1 & 1 \end{pmatrix} , \quad \tilde{\varepsilon} X_I \tilde{\varepsilon} = 4 \begin{pmatrix} 0 & 1 & 1 & 0 \\ 0 & 0 & 0 & 0 \\ 1 & 0 & 1 & 0 \\ 0 & 0 & 1 & 1 \end{pmatrix} , \quad (159)
$$

whence $\mathrm{T} \otimes \mathrm{id}_4[\rho_I] = \frac{1}{6}\left(P_{01} + P_{02} + P_{20} + P_{22} + P_{32} + P_{33}\right) \geq 0.$ $\qquad \square$

By using Lemma 5, we can now show that the PPT state ρ_I of above is entangled and, at the same time, that the positive map Γ_t of Examples 6.1 and 6.2 are not decomposable for $0 < t < (\log 3)/2$.

Proposition 4 *Consider the PPT state ρ_I in Lemma 6 and the maps $\Gamma_t = \gamma^{(1)} \otimes \gamma_t^{(2)}$ in (141). It turns out that, for $0 < t < (\log 3)/2$,*

$$\mathrm{Tr}\left(\rho_I \, \Gamma_t \otimes \mathrm{id}_4[P_+^{(4)}]\right) < 0 \,, \tag{160}$$

so that Γ_t is not decomposable and ρ_I entangled.

Proof Denote by S_μ the completely positive maps on $M_2(\mathbf{C})$ given by $S_\mu(X) = \sigma_\mu X \sigma_\mu$, where $\sigma_\mu, 0 \leq \mu \leq 4$, are the Pauli matrices and by $S_{\mu\nu}$ the 16 completely positive maps on $M_4(\mathbf{C})$ defined by $S_{\mu\nu}[X] = \sigma_{\mu\nu} X \sigma_{\mu\nu}$ where $\sigma_{\mu\nu} = \sigma_\mu \otimes \sigma_\nu$. Then one rewrites the maps $\gamma_t^{(1,2)}$ in Examples 5.1 and 5.2 and Example 6.1 as follows:

$$\gamma_t^{(1)} = \frac{1 + 3\lambda_t}{4} S_0 + \frac{1 - \lambda_t}{4} \sum_{i=1}^{3} S_i \tag{161}$$

$$\gamma_t^{(2)} = \frac{3 + \lambda_t}{4} S_0 + \frac{1 - \lambda_t}{4} \sum_{i=1}^{3} \varepsilon_i \, S_i \tag{162}$$

$$
\begin{aligned}
\Gamma_t = {} & \frac{(1 + 3\lambda_t)(3 + \lambda_t}{16} S_{00} + \frac{1 - \lambda_t^2}{8} \sum_{i=1}^{3} S_{i0} \\
& + \frac{(1 + 3\lambda_t)(1 - \lambda_t)}{16} \sum_{i=1}^{3} \varepsilon_i \, S_{0i} + \frac{(1 - \lambda_t)^2}{16} \sum_{i,j=1}^{3} \varepsilon_j S_{ij} \,.
\end{aligned}
\tag{163}
$$

Because of the symmetry of $P_+^{(4)}$, the projections in (147) also read

$$
\begin{aligned}
P_{\alpha\beta} &= \left(\mathbb{1}_4 \otimes \sigma_{\alpha\beta}\right) P_+^{(4)} \left(\mathbb{1}_4 \otimes \sigma_{\alpha\beta}\right) \\
&= \left(\sigma_{\alpha\beta} \otimes \mathbb{1}_4\right) P_+^{(4)} \left(\sigma_{\alpha\beta} \otimes \mathbb{1}_4\right) = S_{\alpha\beta} \otimes \mathrm{id}_4[P_+^{(4)}] \,.
\end{aligned}
\tag{164}
$$

Then,

$$
\begin{aligned}
\Gamma_t \otimes \mathrm{id}_4[P^{(4)}] = {} & \frac{(1 + 3\lambda_t)(3 + \lambda_t)}{16} P_{00} + \frac{1 - \lambda_t^2}{8} \sum_{i=1}^{3} P_{i0} \\
& + \frac{(1 + 3\lambda_t)(1 - \lambda_t)}{16} \sum_{i=1}^{3} \varepsilon_i \, P_{0i} + \frac{(1 - \lambda_t)^2}{16} \sum_{i,j=1}^{3} \varepsilon_j \, P_{ij}
\end{aligned}
\tag{165}
$$

and, because of the orthogonality of the $P_{\alpha\beta}$, one computes

$$
\mathrm{Tr}\left(\rho_I\, \Gamma_t \otimes \mathrm{id}_4[P_+^{(4)}]\right) = \frac{1-\lambda_t}{6\times 16}\left(-(1+3\lambda_t) + (1-\lambda_t)\sum_{i,j=1}^{3}\varepsilon_j\right)
$$
$$
= \frac{1-\lambda_t}{48}\left(1-3\lambda_t\right)
$$

(166)

which becomes negative for $1/3 < \lambda_t e^{-2t} < 1$. $\square$

References

1. Alicki, R., Fannes, M.: Quantum Dynamical Systems. Oxford University Press, Moscow (2001)
2. Alicki, R., Lendi, K.: *Quantum Dynamical Semigroups and Applications*. Lect. Notes Phys. **717**. Springer, Berlin (2007)
3. Bell, J.S.: Physics **1**, 195 (1965)
4. Benatti, F., Floreanini, R.: Int. J. Mod. Phys. B **19**, 19 (2005)
5. Benatti, F., Floreanini, R., Liguori, A.M.: J. Math. Phys. **48**, 052103 (2007)
6. Benatti, F., Floreanini, R., Piani, M.: Open Syst. Inform. Dyn. **11**, 325 (2004)
7. Benatti, F., Floreanini, R., Romano, R.: J. Phys. A **35**, L551 (2002)
8. Bengtsson, I., Życzkowski, K.: Geometry of Quantum States: An Introduction to Quantum Entanglement. Cambridge University Press, Cambridge (2007)
9. Bennett, C.H., Di Vincenzo, D.P., Smolin, J., et al.: Phys. Rev. A **54**, 3824 (1996)
10. Bratteli, O., Robinson, D.W.: Operator Algebras and Quantum Statistical Mechanics Equilibrium States. Models in Quantum Statistical Mechanics II, 2nd edn. Springer, Berlin (1997)
11. Breuer, H.-P., Petruccione, F.: The Theory of Open Quantum Systems. Oxford University Press, Oxford (2002)
12. Bruss, D.: J. Math. Phys. **43**, 4237 (2002)
13. Bruss, D., Leuchs, G.: Lectures on Quantum Information. Wiley, Berlin (2006)
14. Einstein, A., Podolsky, B., Rosen, N.: Phys. Rev. **47**, 77 (1935)
15. Gorini, V., Frigerio, A., Verri, M., et al.: Rep. Math. Phys. **13**, 149 (1978)
16. Hastings, M.B.: Nat. Phys. **5**, 255 (2009)
17. Horodecki, P.: Phys. Lett. A **232**, 333 (1997)
18. Horodecki, M., Horodecki, P.: Phys. Rev. A **59**, 4206 (1999)
19. Horodecki, M., Horodecki, P., Horodecki, R.: Phys. Lett. A **223**, 1 (1996)
20. Horodecki, R., Horodecki, P., Horodecki, M., et al.: Rev. Mod. Phys. **81**, 865 (2009)
21. Kossakowski, A.: Bull. Acad. Pol. Sci. **12**, 1021 (1972)
22. Kraus, K.: *States, Effects and Operations*. Lect. Notes Phys. **190**. Springer, Berlin (1983)
23. Lieb, E.H., Ruskai, M.B.: J. Math. Phys. **14**, 1938 (1973)
24. Lindblad, G.: Commun. Math. Phys. **48**, 119 (1976)
25. Nielsen, M.A., Chuang, I.L.: Quantum Computation and Quantum Information. Cambridge University Press, Cambridge (2000)
26. Nielsen, M.A., Petz, D.: Quantum Inf. Comp. **5**, 507 (2005)
27. Ohya, M., Petz, D.: Quantum Entropy and Its Use. Springer, Berlin (1993)
28. Peres, A.: Phys. Rev. Lett. **77**, 1413 (1996)
29. Ruskai, M.B.: Rep. Math. Phys. **60**, 1 (2007)

30. Schrödinger, E.: Proc. Camb. Philol. Soc. **31**, 446 (1936)
31. Stinespring, W.F.: Proc. Am. Math. Soc. **6**, 211 (1955)
32. Størmer, E.: Acta Math. **110**, 233 (1963)
33. Uhlmann, A.: Commun. Math. Phys. **54**, 21 (1977)
34. Vedral, V.: Rev. Mod. Phys. **74**, 197 (2002)
35. Wootters, W.K.: Phys. Rev. Lett. **80**, 2245 (1998)
36. Woronowicz, S.L.: Rep. Math. Phys. **10**, 165 (1976)

Field-Theoretical Methods

R. Alicki

1 Introduction

Field-theoretical methods are important tools in quantum information. Photons – the quanta of electromagnetic field – are the most frequently used carriers of quantum information. Laser pulses described by coherent states of electromagnetic field are the typical means of control for essentially all implementations of quantum information processing. On the other hand, systems of interacting fermions in solids provide promising implementations of quantum information processing. Last but not least quantum fields are natural models of reservoirs producing noise, always present in real systems. We restrict ourselves mainly to bosonic fields and begin with the simplest example of a single degree of freedom – a quantum harmonic oscillator. We describe it in terms of creation and annihilation operators, define evolution in the Heisenberg picture, introduce unitary Weyl operators and coherent states with their statistics and dynamics. Squeezed states as examples of the "nonclassical states" are briefly presented also. Then we discuss the case of infinite number of degrees of freedom quantizing a classical field in a finite volume. Particle number representation, Fock space, and local structure of quantum fields are discussed. The interpretation in terms of particles (photons, phonons, etc.) and the relation between spin and statistics are presented. The model of quantum bosonic fields coupled to classical sources is studied and the origin of coherent states is shown. Thermal states of non-interacting bosons and Bose–Einstein statistics are briefly discussed. The mechanism leading to Bose–Einstein condensation is also presented. The Gaussian states being generalizations of vacuum, squeezed vacuum, and thermal states are introduced and discussed in the context of entanglement. Finally, the second quantization of fermions is also outlined including Fermi–Dirac statistics and thermal states for the free Fermi gas.

R. Alicki (✉)
Institute of Theoretical Physics and Astrophysics, University of Gdańsk, Gdańsk, Poland,
fizra@univ.gda.pl

Alicki, R.: *Field-Theoretical Methods*. Lect. Notes Phys. **808**, 151–174 (2010)
DOI 10.1007/978-3-642-11914-9_5

2 The Quantum Harmonic Oscillator

A classical harmonic oscillator, described in terms of canonical position and momentum variables q and p, has Hamiltonian (we set mass $m = 1$ for simplicity, ω is an angular frequency)

$$H(q, p) = \frac{1}{2}(p^2 + \omega^2 q^2) \, . \tag{1}$$

The solution of the Hamilton's equations of motion

$$\frac{dq}{dt} = \frac{\partial}{\partial p} H(q, p) = p, \quad \frac{dp}{dt} = -\frac{\partial}{\partial q} H(q, p) = -\omega^2 q \tag{2}$$

can be written in a convenient form using a *complex amplitude* $z = \omega q + ip$

$$\frac{dz}{dt} = -i\omega z \, , \quad z(t) = z(0) \exp(-i\omega t) \, . \tag{3}$$

2.1 Heisenberg Canonical Quantization

The Heisenberg approach to quantization consists in replacing the classical variables q, p by non-commuting objects $\hat{q}$, $\hat{p}$ which can be represented by self-adjoint operators acting on a Hilbert space $\mathcal{H}$ ("infinite Hermitian matrices"). They should satisfy *canonical commutation relation* (CCR) in the form

$$[\hat{q}, \hat{p}] \equiv \hat{q}\hat{p} - \hat{p}\hat{q} = i\hbar \, . \tag{4}$$

The essential information concerning the physics of a quantum harmonic oscillator can be extracted from the purely algebraic CCR. Define a properly normalized *quantum complex amplitude* and its adjoint (Hermitian conjugation)

$$\hat{a} = \frac{1}{\sqrt{2\hbar\omega}}(\omega\hat{q} + i\hat{p}) \, , \quad \hat{a}^\dagger = \frac{1}{\sqrt{2\hbar\omega}}(\omega\hat{q} - i\hat{p}) \tag{5}$$

for which a CCR equivalent to (4) holds

$$[\hat{a}, \hat{a}^\dagger] = 1 \, . \tag{6}$$

Then the quantum Hamiltonian $\hat{H} = \frac{1}{2}(\hat{p}^2 + \omega^2\hat{q}^2)$ can be expressed in terms of $\hat{a}, \hat{a}^\dagger$ as

$$\hat{H} = \hbar\omega\left(\hat{a}^\dagger\hat{a} + \frac{1}{2}\right) \, . \tag{7}$$

In the following we omit the irrelevant constant $\hbar\omega/2$ and use as Hamiltonian for the harmonic oscillator the operator

$$\hat{H}_o = \hbar\omega\,\hat{a}^\dagger\hat{a}\ . \tag{8}$$

The self-adjoint, positive operator $\hat{n} = \hat{a}^\dagger\hat{a}$ possesses a complete set of normalized eigenvectors $|n\rangle$ satisfying

$$\hat{n}|n\rangle = n|n\rangle\ ,\ n = 0, 1, 2, \dots\ ,\ \langle n|n'\rangle = \delta_{nn'}\ . \tag{9}$$

Indeed, CCR (6) implies $\hat{n}\hat{a} = \hat{a}(\hat{n} - 1)$ $(\hat{n}\hat{a}^\dagger = \hat{a}^\dagger(\hat{n} + 1))$ and therefore the vector $\hat{a}|n\rangle$ $(\hat{a}^\dagger|n\rangle)$ is an (unnormalized) eigenvector of $\hat{n}$ corresponding to the eigenvalue $n - 1$ $(n + 1)$. Hence, the positivity of $\hat{n}$ and the relation $\langle n|\hat{a}^\dagger\hat{a}|n\rangle = n$ imply that $n = 0, 1, 2, \dots$ and the following formulas:

$$\hat{a}|n\rangle = \sqrt{n}|n - 1\rangle\ ,\quad \hat{a}^\dagger|n\rangle = \sqrt{n + 1}|n + 1\rangle\ ,\ n = 0, 1, 2, \dots\ , \tag{10}$$

which justify the names *annihilation*, *creation*, and *particle number* operators for $\hat{a}$, $\hat{a}^\dagger$, and $\hat{n}$, respectively. Obviously, $|n\rangle$ are also eigenstates of the Hamiltonian (8) corresponding to the eigenvalues $E_n = \hbar\omega n$. The ground state $|0\rangle$ of the quantum harmonic oscillator is called the *vacuum*.

For any classical observable $F(q, p) \equiv F(\alpha^*, \alpha) = \sum_{k,\ell} c_{k\ell}(\alpha^*)^k\alpha^\ell$, where

$$\alpha = \frac{1}{\sqrt{2\hbar\omega}}(\omega q + ip) \tag{11}$$

we can define its unique quantum counterpart by means of the *normal ordering*

$$\hat{F} = \sum_{k,\ell} c_{k\ell}(\hat{a}^\dagger)^k\hat{a}^\ell\ . \tag{12}$$

The time evolution in Heisenberg picture of $\hat{a}$ is given by

$$\hat{a}(t) \equiv \exp\left(\frac{i}{\hbar}\hat{H}_o t\right)\hat{a}\,\exp\left(-\frac{i}{\hbar}\hat{H}_o t\right) = e^{-i\omega t}\hat{a}\ . \tag{13}$$

This formula, completely analogous to the classical expression for the amplitude α (see (11) and (3)), can be obtained by differentiating both sides of (13) using $[\hat{H}_o, \hat{a}] = -i\,\hbar\omega\hat{a}$. Therefore all quantum observables written in terms of the normal ordering (12) evolve in Heisenberg picture in a similar way to their classical counterparts

$$\hat{F}(t) = \sum_{k,\ell} c_{k\ell}e^{i\omega t(k-\ell)}(\hat{a}^\dagger)^k\hat{a}^\ell\ . \tag{14}$$

2.2 The Weyl Unitary Operators

The position and momentum operators $\hat{q}$, $\hat{p}$ define a *quantum phase space (plane)* with a natural notion of translation operators called *displacement operators* or *Weyl unitaries*

$$\hat{W}(\alpha) = \exp(\alpha \hat{a}^\dagger - \alpha^* \hat{a}) . \tag{15}$$

Indeed, using CCR one can easily prove (step 1 of Exercise 1) that

$$\hat{W}(\alpha) \, \hat{a} \, \hat{W}(\alpha)^\dagger = \hat{a} - \alpha ; \tag{16}$$

namely, position and momentum are shifted in such a way that $\alpha = (\omega q + \mathrm{i} p)/\sqrt{2\hbar\omega}$. A useful operator identity (step 2 of Exercise 1),

$$\exp(\hat{A}) \, \exp(\hat{B}) = \exp\left(\frac{1}{2}[\hat{A}, \hat{B}]\right) \exp(\hat{A} + \hat{B}) , \tag{17}$$

is valid if

$$[\hat{A}, [\hat{A}, \hat{B}]] = [\hat{B}, [\hat{A}, \hat{B}]] = 0 \tag{18}$$

can be applied to prove the following group properties of the Weyl unitaries

$$\hat{W}(0) = \mathbb{1} , \quad \hat{W}(-\alpha) = \hat{W}(\alpha)^\dagger , \tag{19}$$

$$\hat{W}(\alpha) \hat{W}(\beta) = \exp(\mathrm{i} \operatorname{Im}\{\alpha\beta^*\}) \hat{W}(\alpha + \beta) \tag{20}$$

and the formal composition formulas

$$\begin{aligned}
\hat{W}(\alpha) &= \exp(-|\alpha|^2/2) \, \exp(\alpha \hat{a}^\dagger) \exp(-\alpha^* \hat{a}) \\
&= \exp(|\alpha|^2/2) \, \exp(-\alpha^* \hat{a}) \exp(\alpha \hat{a}^\dagger) .
\end{aligned} \tag{21}$$

2.3 Coherent States

Shifting the vacuum vector by Weyl unitaries, we obtain a family of *coherent vectors (states)* or *exponential vectors* defined as

$$|\alpha\rangle = \hat{W}(\alpha)|0\rangle , \quad \alpha \in \mathbf{C} . \tag{22}$$

Coherent vectors possess the following properties:

- They are linearly independent but not orthogonal eigenvectors of the (non-self-adjoint) operator $\hat{a}$:

$$\hat{a}|\alpha\rangle = \alpha|\alpha\rangle , \quad \langle\alpha|\beta\rangle = \exp(-|\alpha|^2/2 - |\beta|^2/2 + \alpha^*\beta) , \tag{23}$$

and form an over-complete set, i.e.,

$$\mathbb{1} = \frac{1}{\pi} \int_C d^2\alpha \, |\alpha\rangle\langle\alpha| . \tag{24}$$

- Their representation in terms of particle number eigenvectors reads

$$|\alpha\rangle = \exp(-|\alpha|^2/2) \sum_{n=0}^{\infty} \frac{\alpha^n}{\sqrt{n!}} |n\rangle , \tag{25}$$

which implies a Poissonian probability distribution of the particle number

$$p(n) = |\langle n|\alpha\rangle|^2 = \exp(-|\alpha|^2)\frac{|\alpha|^n}{n!} . \tag{26}$$

- The mean value of a normally ordered operator $\hat{F}$ (see (12)) reproduces the classical expression

$$\langle\alpha|\hat{F}|\alpha\rangle = F(\alpha^*, \alpha) = \sum_{k,\ell} c_{k\ell}(\alpha^*)^k\alpha^\ell , \tag{27}$$

and a coherent state evolves into a coherent one following the classical trajectory,

$$\exp\left(-\frac{i}{\hbar}\hat{H}_o t\right) |\alpha\rangle = |\alpha(t)\rangle , \quad \alpha(t) = e^{-i\omega t}\alpha . \tag{28}$$

- The manifest Schrödinger *position representation* in terms of the Hilbert space $\mathcal{H} = L^2(\mathbf{R})$ with position and momentum operators given by

$$(\hat{q}\psi)(x) = x\psi(x) , \quad (\hat{p}\psi)(x) = -i\hbar\frac{d}{dx}\psi(x) \tag{29}$$

leads to the following wave function for the coherent state $|\alpha\rangle$:

$$\phi_\alpha(x) = (\pi\hbar)^{-1/4} \exp\left(\frac{i}{\hbar}px\right) \exp\left(-\frac{\omega}{2\hbar}(x-q)^2\right) . \tag{30}$$

The Fourier transform of $\phi_\alpha(x)$ gives the *momentum representation* of the coherent state

$$\tilde{\phi}_\alpha(v) = (\pi\hbar)^{-1/4} \exp\left(\frac{i}{\hbar}q\,(p-v)\right) \exp\left(-\frac{(v-p)^2}{2\omega\hbar}\right). \qquad (31)$$

- Computing the corresponding Gaussian probability distributions $|\phi_\alpha(x)|^2$ and $|\tilde{\phi}_\alpha(v)|^2$ one can check that for coherent states the Heisenberg uncertainty relation $\Delta\hat{q}\Delta\hat{p} \geq \hbar/2$ becomes an equality in a symmetric way with respect to position and momentum, i.e., $\Delta\hat{q} = \sqrt{\hbar/2\omega}$, $\Delta\hat{p} = \sqrt{\hbar\omega/2}$. Hence, the coherent vector provides the best quantum analogue of the classical state localized at the phase-space point (q, p).

Proof Using (10) we obtain

$$(\hat{a}^\dagger)^n|0\rangle = \sqrt{n!}|n\rangle \qquad (32)$$

and then from (21) it follows that

$$|\alpha\rangle = \hat{W}(\alpha)|0\rangle = \exp(-|\alpha|^2/2)\exp(\alpha\hat{a}^\dagger)|0\rangle$$
$$= \exp(-|\alpha|^2/2)\sum_{n=0}^\infty \frac{\alpha^n}{\sqrt{n!}}|n\rangle. \qquad (33)$$

A straightforward computation based on formulas (25) and (10) yields (26), (23), and (24). To prove (24) we use the parametrization $\alpha = Re^{i\Theta}$, $d^2\alpha = R\,d R\,d\Theta$ and first integrate over Θ. Formula (14) follows directly from the eigenvector condition (23) and the classical behavior of evolution (28) from (13). To prove (30) we write condition (23) in the position representation (29) and obtain a linear differential equation

$$\frac{d}{dx}\psi(x) = \frac{1}{\hbar}\left(i\,p - \omega(x-q)\right)\psi(x) \qquad (34)$$

with normalized solution (30).

2.4 Glauber–Sudarshan Representation, Squeezed Coherent Vectors

A mixed state of the quantum harmonic oscillator is given by a density matrix $\hat{\rho}$. One can prove that any $\hat{\rho}$ can be written in terms of coherent vectors as

$$\hat{\rho} = \int_C d^2\alpha\,P(\alpha)|\alpha\rangle\langle\alpha|, \qquad (35)$$

where $P(\alpha)$ is a function or a singular distribution (e.g., for a coherent state $\hat{\rho} = |\alpha_0\rangle\langle\alpha_0|$, $P(\alpha) = \delta^{(2)}(\alpha - \alpha_0)$) called Glauber–Sudarshan P-function.

It is normalized (i.e., $\int_C d^2\alpha\, P(\alpha) = 1$) but not necessarily positive. Quantum (mixed) states described by non-positive $P(\alpha)$ are called *non-classical* in quantum optics. For example, eigenstates of $\hat{n}$, called often *Fock states*, $\hat{\rho} = |n\rangle\langle n|$, are non-classical. The name *Gaussian wave packets* is used for pure states given by Gaussian wave functions in position (momentum) representation or equivalently eigenstates of $\hat{q} + \xi\hat{p}$, $\xi \in \mathbf{C}$. It implies that the Heisenberg uncertainty relation becomes an equality. The Gaussian wave functions are non-classical if $\Delta\hat{q} < \sqrt{\hbar/2\omega}$ or $\Delta\hat{p} < \sqrt{\hbar\omega/2}$. The notion of Gaussian wave function is equivalent to the notion of *coherent squeezed vector* $|\alpha, \xi\rangle$ which is defined by

$$|\alpha, \xi\rangle = \hat{W}(\alpha)\hat{S}(\xi)|0\rangle\,, \quad \hat{S}(\xi) = \exp\left[\frac{1}{2}\left(\xi^*\hat{a}^2 - \xi(\hat{a}^\dagger)^2\right)\right]\,, \quad \alpha, \xi \in \mathbf{C}\,. \quad (36)$$

The state $|0, \xi\rangle$ is called *squeezed vacuum*. It is not difficult to show (step 4 of Exercise 1) that the following relation holds:

$$\hat{S}(\xi)^\dagger \hat{a}\hat{S}(\xi) = \hat{a}\cosh r - \hat{a}^\dagger e^{i2\theta}\sinh r\,, \quad (37)$$

where $\xi = re^{i2\theta}$.

A clear geometrical picture of $|\alpha, \xi\rangle$ is given in terms of the *quadrature operators* $\hat{X}, \hat{Y}$ which are dimensionless "position" and "momentum" observables

$$\hat{X} = \frac{1}{\sqrt{2}}\left(\hat{a} + \hat{a}^\dagger\right)\,, \quad \hat{Y} = \frac{1}{i\sqrt{2}}\left(\hat{a} - \hat{a}^\dagger\right) \quad (38)$$

satisfying $[\hat{X}, \hat{Y}] = i$. Using directly definitions (36), (38), and equation (37) one obtains (see Exercises step 5 of 1) the mean values

$$\langle\hat{X}\rangle = \sqrt{2}\mathrm{Re}\,\alpha\,, \quad \langle\hat{Y}\rangle = \sqrt{2}\mathrm{Im}\,\alpha \quad (39)$$

and dispersions

$$\sigma^2(\hat{X}) \equiv \langle\hat{X}^2\rangle - \langle\hat{X}\rangle^2 = \frac{1}{2}e^{-2r}\,, \quad \sigma^2(\hat{Y}) \equiv \langle\hat{Y}^2\rangle - \langle\hat{Y}\rangle^2 = \frac{1}{2}e^{2r}\,, \quad (40)$$

where the shorthand notation $\langle\hat{A}\rangle$ means $\langle\alpha, \xi|\hat{A}|\alpha, \xi\rangle$. For a large r the fluctuations of $\hat{X}$ are highly reduced ("squeezed") at the expense of increased fluctuations of $\hat{Y}$, what explains the name of the vectors $|\alpha, \xi\rangle$.

The harmonic oscillator's dynamics preserves the form of squeezed coherent vectors (step 6 of Exercise 1):

$$\exp\left(-\frac{i}{\hbar}\hat{H}_o t\right)|\alpha, \xi\rangle = |e^{-i\omega t}\alpha\,, e^{-i2\omega t}\xi\rangle\,. \quad (41)$$

Remark 1

1. In quantum optics a quantum harmonic oscillator provides a good model for a single mode of radiation confined in an optical cavity (q and p variables can be seen as the amplitudes of the magnetic and electric fields, respectively). If this cavity is a part of a laser working in the regime of *laser action*, then the coherent states with their Poisson statistics perfectly describe the states of light. On the other hand, below the threshold of laser action the thermal states of light appear (see Sect. 4.3).
2. There has been a whole variety of successful demonstrations of coherent and squeezed states. Besides the experiments with light fields using lasers and nonlinear optics the coherent and squeezed states have also been realized using trapped ions, phonon states in crystal lattices, or atom ensembles.

Exercise 1

1. *Prove formula (16).*
   Hint: Compute the derivative of the operator-valued function

$$\hat{A}(t) = \hat{W}(t\alpha)\,\hat{a}\,\hat{W}(t\alpha)^{\dagger} \tag{42}$$

   and integrate the obtained differential equation up to $t = 1$.
2. *Prove formula (17).*
   Hint: Compute the derivative of the operator-valued function

$$\hat{G}(t) = \exp(t\hat{A})\,\exp(t\hat{B})\,\exp(-t(\hat{A}+\hat{B})) \tag{43}$$

   and integrate the obtained differential equation up to $t = 1$.
3. *Using (17) and (18) prove (20) and (21).*
4. *Prove relation (37) .*
   Hint: Compute $\hat{b}_z = \hat{S}(\xi)\,\hat{a}\,\hat{S}(\xi)^{\dagger}$ using the operator identity

$$e^{\hat{A}}\,\hat{B}\,e^{-\hat{A}} = \hat{B} + \frac{1}{1!}\left[\hat{A},\,\hat{B}\right] + \frac{1}{2!}\left[\hat{A},\,\left[\hat{A},\,\hat{B}\right]\right] + \frac{1}{3!}\left[\hat{A},\,\left[\hat{A},\,\left[\hat{A},\,\hat{B}\right]\right]\right] + \cdots . \tag{44}$$

5. *Prove formulas (39) and (40).*
6. *Prove (41) using the following identity valid for any function F and any operators $\hat{A}$, $\hat{B}$:*

$$e^{\hat{A}}\,F(\hat{B})\,e^{-\hat{A}} = F(e^{\hat{A}}\,\hat{B}\,e^{-\hat{A}}) . \tag{45}$$

3 Quantum Bosonic Fields

The theory of quantum bosonic field allows for two interpretations. The first treats it as a quantization of the macroscopic classical field theory like classical electrodynamics or the theory of acoustic waves in solids. The second interprets it as a theory of many-body systems consisting of quantum particles, each of which described by the suitable quantum wave equation ("second quantization"). In the latter case Maxwell equations are treated as the Schrödinger equation for a single photon.

3.1 Quantization of Classical Fields

Classical field theory is a convenient formalism describing macroscopic properties of either fundamental interactions (electromagnetic, gravitational, etc.) or providing simplified continuous models of many-body systems (acoustic waves). By a classical field we mean a (generally multi-component) function $\phi(\mathbf{x}; t)$ satisfying some linear wave equation which we do not specify here.

For mathematical simplicity we consider a real scalar field confined in a finite region Ω of the one-, two-, or three-dimensional space with specified boundary conditions. The particular (complex-valued) solutions of the wave equation with periodic time dependence can be written as

$$u_k(\mathbf{x}, t) = e^{-i\omega(\mathbf{k})t} u_k(\mathbf{x}) , \tag{46}$$

where k represents a certain multi-index labeling the *modes* u_k. Modes satisfy orthogonality conditions

$$\int_\Omega d\mathbf{x}\, u_k^*(\mathbf{x}) u_\ell(\mathbf{x}) = c_k \delta_{k\ell} \tag{47}$$

with yet unspecified normalization constants c_k.

According to the superposition principle, any solution of the wave equation can be written in terms of modes and complex amplitudes α_k

$$f(\mathbf{x}, t) = \sum_k \left(\alpha_k(t) u_k(\mathbf{x}) + \alpha_k^*(t) u_k^*(\mathbf{x}) \right) , \tag{48}$$

where

$$\alpha_k(t) = \alpha_k e^{-i\omega(\mathbf{k})t} . \tag{49}$$

The linearity of wave equations suggests a quadratic dependence of the corresponding energy. Indeed, for all examples the energy of the field $f(\mathbf{x}, t)$ is given by a quadratic form

$$\mathcal{E}(f) = \int_\Omega d^r\mathbf{x}\, f(\mathbf{x}, t) \hat{D} f(\mathbf{x}, t) , \tag{50}$$

where $\hat{D}$ is typically a differential operator whose form depends on the wave equation. Moreover, the modes u_k are eigenfunctions of $\hat{D}$, i.e.,

$$\hat{D}u_k = \lambda_k u_k . \tag{51}$$

Choosing the normalization constants as

$$c_k = \sqrt{\frac{\hbar\omega_k}{2\lambda_k}} , \tag{52}$$

we can write energy (50) in terms of the complex amplitudes as

$$\mathcal{E}(f) = \sum_k \hbar\omega_k \alpha_k^* \alpha_k . \tag{53}$$

If we replace the complex amplitudes α_k and α_k^* by a set of independent annihilation and creation operators $\hat{a}_k$ and $\hat{a}_k^\dagger$ satisfying a general form of CCR

$$[\hat{a}_k, \hat{a}_\ell^\dagger] = \delta_{k\ell} , \quad [\hat{a}_k, \hat{a}_\ell] = [\hat{a}_k^\dagger, \hat{a}_k^\dagger] = 0, \tag{54}$$

we obtain the quantum Hamiltonian for the field in the bounded region

$$\hat{H}_F = \sum_k \hbar\omega_k \hat{a}_k^\dagger \hat{a}_k \tag{55}$$

and the oscillations of the quantum amplitudes in the Heisenberg picture

$$\hat{a}_k(t) = \exp\left(\frac{i}{\hbar}\hat{H}_F t\right)\hat{a}_k \exp\left(-\frac{i}{\hbar}\hat{H}_F t\right) = e^{-i\omega_k t}\hat{a}_k . \tag{56}$$

Expressions (55) and (56), when compared with (7) and (13), show that the quantum field in a finite space region is equivalent to a (possibly infinite) family of independent quantum harmonic oscillators. The quantum analogue of the classical field is now the operator-valued function

$$\hat{f}(\mathbf{x}, t) = \sum_k \left(\hat{a}_k e^{-i\omega(\mathbf{k})t} u_k(\mathbf{x}) + \hat{a}_k^\dagger e^{i\omega(\mathbf{k})t} u_k^*(\mathbf{x})\right) . \tag{57}$$

Remark 2

1. We have used a notation suitable for a scalar field f. For a multi-component field we can treat $\mathbf{x}$ as a combined continuous–discrete variable $\mathbf{x} = (x, \mu)$ with $x \in \mathbf{R}^\ell$ and $\mu = 1, 2, \ldots, M$. In this case the integral $\int d\mathbf{x}$ means $\sum_\mu \int d^\ell x$.
2. The quantization procedure which was presented here can be made more rigorous using the canonical formalism of classical field theory [2, 6].

3.2 Fock Space

The Hilbert space for the quantized field can be formally treated as the Hilbert space of a family of harmonic oscillators, i.e., a tensor product of single-oscillator Hilbert spaces $\bigotimes_k \mathcal{H}_k$. For an infinite number of modes this is a subtle mathematical problem and it is convenient to construct explicitly the appropriate Hilbert space, called *Fock space* and denoted by $\mathcal{F}$.

The Fock space is by definition spanned by a countable family of vectors which form an orthonormal basis of $\mathcal{F}$ denoted by $|\{n_k\}\rangle \equiv |n_{k_1}, n_{k_2}, \ldots, n_{k_m}\rangle$, where $\{n_k\}$ is an arbitrary sequence of numbers $n_k = 0, 1, 2, \ldots$ labeled by the indices $\{k\}$ and with only a finite number of non-zero elements explicitly denoted by $n_{k_1}, n_{k_2}, \ldots, n_{k_m}$. We interpret such a vector as a quantum state of the field for which we observe n_k *particles* (excitations) corresponding to the mode u_k. In this picture, called *particle number representation*, the modes u_k span the *single-particle Hilbert space* $\mathcal{H}_1$ and the particles are obviously indistinguishable. The state $|0\rangle$ with all $n_k = 0$ is again called the vacuum. The formula

$$|\{n_k\}\rangle \equiv |n_{k_1}, n_{k_2}, \ldots, n_{k_m}\rangle = \frac{\left(\hat{a}_{k_1}^\dagger\right)^{n_{k_1}}}{\sqrt{n_{k_1}!}} \frac{\left(\hat{a}_{k_2}^\dagger\right)^{n_{k_2}}}{\sqrt{n_{k_2}!}} \cdots \frac{\left(\hat{a}_{k_m}^\dagger\right)^{n_{k_m}}}{\sqrt{n_{k_m}!}} |0\rangle , \tag{58}$$

motivated by (32), can be used as a consistent definition of the creation operators $\hat{a}_k^\dagger$ which increase by one the number of particles in the state (mode) u_k. Computing the adjoint operator $\hat{a}_k$, one can easily obtain its action on the basis vectors

$$\hat{a}_k | \ldots, n_k, \ldots \rangle = \sqrt{n_k} | \ldots, n_k - 1, \ldots \rangle . \tag{59}$$

It is also not difficult to check that such operators satisfy CCR (54) and that the states $|\{n_k\}\rangle$ are joint eigenvectors for the particle number operators $\hat{n}_k = \hat{a}_k^\dagger \hat{a}_k$

$$\hat{n}_k |\{n_k\}\rangle = n_k |\{n_k\}\rangle . \tag{60}$$

For any $N = 0, 1, 2, \ldots$ we define a subspace of $\mathcal{F}$, denoted by $\mathcal{H}_N$, spanned by the vectors $|\{n_k\}\rangle$ satisfying $\sum_k n_k = N$. Then we can decompose the Fock space into a direct sum

$$\mathcal{F} = \bigoplus_{N=0}^{\infty} \mathcal{H}_N . \tag{61}$$

The subspace $\mathcal{H}_0$ is a one-dimensional *ray* generated by the vacuum $|0\rangle$. The single-particle Hilbert space $\mathcal{H}_1 \equiv \mathcal{H}$ is a complex manifold spanned by the vectors $\hat{a}_k^\dagger |0\rangle$ and can be identified with a complex Hilbert space containing the linear combinations of the properly normalized modes (47) and (52)

$$\phi(\mathbf{x}) = \sum_k \phi_k e_k(\mathbf{x}), \quad e_k(\mathbf{x}) = \sqrt{\frac{2\lambda_k}{\hbar \omega_k}} u_k(\mathbf{x}), \tag{62}$$

with a scalar product satisfying

$$\langle \phi | \psi \rangle = \int_\Omega \mathbf{dx}\, \phi^*(\mathbf{x}) \psi(\mathbf{x}) = \sum_k \phi_k^* \psi_k . \tag{63}$$

The N-particle Hilbert space $\mathcal{H}_N$ is spanned by the vectors which can be identified with the formal products $u_{k_1} \circ u_{k_2} \circ \cdots \circ u_{k_N}$ (the same mode can appear many times and the order of multiplication is irrelevant!). Therefore, the structure of $\mathcal{H}_N$ is identical to that of the N-fold *symmetric tensor product* and we can write

$$\mathcal{H}_N = \frac{1}{N!} \sum_\pi S_\pi \left(\mathcal{H}^{\otimes N} \right) . \tag{64}$$

The operator S_π corresponds to a permutation of particles and acts on the product state as follows:

$$S_\pi \phi_1(\mathbf{x}_1) \phi_2(\mathbf{x}_2) \cdots \phi_N(\mathbf{x}_N) = \phi_{\pi(1)}(\mathbf{x}_1) \phi_{\pi(2)}(\mathbf{x}_2) \cdots \phi_{\pi(N)}(\mathbf{x}_N), \tag{65}$$

and the sum in (64) is taken over all permutations of the set $\{1, 2, ..., N\}$.

The Fock space $\mathcal{F}$ can be considered as an "exponential" of the underlying single-particle Hilbert space $\mathcal{H}$

$$\mathcal{F}(\mathcal{H}) = \bigoplus_{N=0}^\infty \frac{1}{N!} \sum_\pi S_\pi \left(\mathcal{H}^{\otimes N} \right) . \tag{66}$$

Remark 3 We use the name "field" for both the true time-dependent fields (48) and the single-particle wave functions (62). However, the complexification of the former is isomorphic to the later and even isometric if the proper scalar products are defined.

3.3 Local Structure of Quantum Fields

It is not difficult to check by comparing orthonormal bases of the Hilbert spaces appearing on both sides that the following relation holds:

$$\mathcal{F}(\mathcal{H} \oplus \mathcal{K}) = \mathcal{F}(\mathcal{H}) \otimes \mathcal{F}(\mathcal{K}) . \tag{67}$$

The single-particle Hilbert space $\mathcal{H}$ possesses a natural local structure due to the $\mathbf{x}$-dependence of its elements (complex wave functions) $\phi(\mathbf{x})$ (see (62)). If we decompose a space region Ω into disjoint subsets $\Omega = \Omega_1 \cup \Omega_2 \cup \cdots \cup \Omega_k$ any wave function can be written as an orthogonal sum of wave functions localized in the Ω_j and this generates a corresponding decomposition of the Hilbert space:

$$\phi(\mathbf{x}) = \phi_1(\mathbf{x}) \oplus \phi_2(\mathbf{x}) \oplus \cdots \oplus \phi_k(\mathbf{x}) \Longrightarrow \mathcal{H} = \mathcal{K}_1 \oplus \mathcal{K}_2 \oplus \cdots \oplus \mathcal{K}_k . \qquad (68)$$

Then from (67) we obtain the following decomposition of the Fock space:

$$\mathcal{F}(\mathcal{H}) = \mathcal{F}(\mathcal{K}_1) \otimes \mathcal{F}(\mathcal{K}_2) \otimes \cdots \otimes \mathcal{F}(\mathcal{K}_k) . \qquad (69)$$

Physically, the above decomposition means that the quantum field localized in a certain subset of Ω is a *physical subsystem* of the total system which is localized in Ω.

The local structure of the above construction can be expressed in terms of observables. For a wave function (62) (often called *test function*) we can define the *smeared annihilation and creation operators*

$$\hat{a}(\phi) = \sum_k \phi_k^* \hat{a}_k , \quad \hat{a}^\dagger(\phi) = \sum_k \phi_k \hat{a}_k^\dagger . \qquad (70)$$

They satisfy another form of CCR

$$[\hat{a}(\phi), \hat{a}^\dagger(\psi)] = \langle \phi | \psi \rangle , \quad [\hat{a}(\phi), \hat{a}(\psi)] = [\hat{a}^\dagger(\phi), \hat{a}^\dagger(\psi)] = 0 . \qquad (71)$$

Taking products, sums (and proper limits) of operators $\hat{a}(\phi)$ and $\hat{a}^\dagger(\psi)$ with different test functions we obtain all physically relevant observables for the quantum field. In particular, if we restrict ourselves to test functions with support in a given subset $\Lambda \subset \Omega$ we generate all *local observables* associated with the space region Λ. Due to (71) two observables localized in disjoint subsets of Ω commute.

The important class of operators on the Fock space describing additive quantum observables is defined as follows. For a single-particle observable which can be represented as

$$\hat{A} = \sum_j |\psi_j\rangle\langle\phi_j| \qquad (72)$$

we define its *second quantized* counterpart acting on the Fock space

$$\hat{A}_\mathrm{F} = \sum_j \hat{a}^\dagger(\psi_j)\hat{a}(\phi_j) . \qquad (73)$$

A simple calculation of the action of $\hat{A}_\mathrm{F}$ on the basic vectors $|\{n_k\}\rangle$ using definition (70) shows that the N-particle Hilbert subspaces are invariant under $\hat{A}_\mathrm{F}$ and that

$$\hat{A}_\mathrm{F}|_{\mathcal{H}_N} = \sum_{k=1}^n \mathbb{1} \otimes \cdots \otimes \underset{k}{\hat{A}} \otimes \cdots \otimes \mathbb{1} \qquad (74)$$

acts as a sum of single-particle operators. Important single-particle observables are the Hamiltonian $\hat{H}_F$, the particle number at the mode $\hat{n}_k$, and the total particle number $\hat{N} = \sum_k \hat{n}_k$ which are second quantizations of the operators $\hat{H}_o = \sum_k \hbar \omega_k |e_k\rangle\langle e_k|$, $\hat{P}_k = |e_k\rangle\langle e_k|$, and $\mathbb{1}$, respectively.

By continuity arguments, one can extend the second quantization rule (73) to more general single-particle operators than (72). For example, if $\hat{D}$ is a differential operator acting on $\mathcal{H}$ we can write

$$\hat{D}_F = \int_\Omega d\mathbf{x} \, \hat{\psi}^\dagger(\mathbf{x}) \hat{D} \hat{\psi}(\mathbf{x}), \tag{75}$$

where $\hat{\psi}(\mathbf{x})$ and $\hat{\psi}^\dagger(\mathbf{x})$ are local annihilation and creation fields defined as

$$\hat{\psi}(\mathbf{x}) = \sum_k e_k(\mathbf{x}) \hat{a}_k, \quad \hat{\psi}^\dagger(\mathbf{x}) = \sum_k e_k^*(\mathbf{x}) \hat{a}_k^\dagger, \tag{76}$$

satisfying yet another version of CCR

$$[\hat{\psi}(\mathbf{x}), \hat{\psi}^\dagger(\mathbf{y})] = \delta(\mathbf{x} - \mathbf{y}), \quad [\hat{\psi}(\mathbf{x}), \hat{\psi}(\mathbf{y})] = [\hat{\psi}^\dagger(\mathbf{x}), \hat{\psi}^\dagger(\mathbf{y})] = 0. \tag{77}$$

3.4 Photons, Phonons, Spin, and Statistics

The most important example of a quantized field is an electromagnetic field in vacuum. As classical field equations we choose Maxwell's equations for the electric and magnetic fields $\mathbf{E}(\mathbf{x}, t)$, $\mathbf{B}(\mathbf{x}, t)$ which can be written in a compact form using a complex vector field $\mathbf{Z}(\mathbf{x}, t) = \mathbf{E}(\mathbf{x}, t) + i\mathbf{B}(\mathbf{x}, t)$

$$\nabla \cdot \mathbf{Z} = 0, \quad \frac{\partial \mathbf{Z}}{\partial t} = -ic\nabla \times \mathbf{Z}. \tag{78}$$

Introducing a box L^3 and periodic boundary conditions, we can find the solutions of (78) in terms of plane waves

$$\mathbf{Z}(\mathbf{x}, t) = \mathbf{Z}_0 \exp\{i(\mathbf{k} \cdot \mathbf{x} - \omega(\mathbf{k}) t)\}, \quad \mathbf{k} \cdot \mathbf{Z}_0 = 0, \quad ic\mathbf{k} \times \mathbf{Z}_0 = \omega(\mathbf{k})\mathbf{Z}_0. \tag{79}$$

Here $\mathbf{k}$ takes on the discrete values $\mathbf{k} = (\frac{2\pi m_1}{L}, \frac{2\pi m_2}{L}, \frac{2\pi m_3}{L})$, $m_\ell = 0, \pm1, \pm2, \ldots$, and we have the dispersion law $\omega(\mathbf{k}) = c|\mathbf{k}|$. The electric field, the magnetic field, and the wave vector $\mathbf{k}$ are mutually orthogonal; hence we have two possible polarizations of the plane wave denoted by an index $\lambda = \pm1$. The energy is given by the quadratic form

$$\mathcal{E} = \frac{1}{8\pi} \int d\mathbf{x} \, (\mathbf{E}^2 + \mathbf{B}^2) = \frac{1}{8\pi} \int d\mathbf{x} \, \overline{\mathbf{Z}} \cdot \mathbf{Z} \tag{80}$$

and therefore we can apply the quantization procedure of Sect. 2.1 by introducing annihilation and creation operators $\hat{a}_{\mathbf{k},\lambda}$, $\hat{a}^\dagger_{\mathbf{k},\lambda}$ for plane waves with given wave vectors and polarizations. Hence we obtain the following parametrization of the quantized electric field:

$$\hat{\mathbf{E}}(\mathbf{x}) = i \sum_{\mathbf{k},\lambda} \left(\frac{2\pi\hbar c|\mathbf{k}|}{L^3}\right)^{1/2} \mathbf{e}_{\mathbf{k},\lambda} \left(e^{i\mathbf{k}\cdot\mathbf{x}} \hat{a}_{\mathbf{k},\lambda} - e^{-i\mathbf{k}\cdot\mathbf{x}} \hat{a}^\dagger_{\mathbf{k},\lambda}\right). \tag{81}$$

To quantize the oscillations of interacting atoms, ions, or molecules in a solid state we associate with any normal oscillation mode described by the wave vector $\mathbf{k}$ and the index α corresponding to the possible types of oscillations (three polarizations of *acoustic modes*, different branches of *optical modes*) annihilation and creation operators $\hat{a}_{\mathbf{k},\alpha}$ and $\hat{a}^\dagger_{\mathbf{k},\alpha}$ and a dispersion law $\omega(\mathbf{k}, \alpha)$. In contrast to electromagnetic waves, the number of modes is finite and proportional to the volume. The (quasi) particles corresponding to this picture are called *phonons*.

The multi-particle structure of the Fock space (61) suggests its direct application to the description of many-body systems consisting of particles which are not associated with macroscopic classical fields in an obvious way. However, it follows from relativistic quantum field theory that only particles with an integer spin $S = 0, 1, 2, \ldots$ called *bosons* can be described by the formalism of above (e.g., photons, pions, W,Z-bosons). Particles with a spin $S = 1/2, 3/2, \ldots$ called *fermions* (e.g., electrons, protons, neutrons,) must satisfy *Pauli's exclusion principle* which implies that the possible eigenvalues of the particle number operators $\hat{n}_k$ should be equal to 0 or 1. This can be achieved with a second quantization procedure involving *canonical anti-commutation relations* instead of CCR (see Sect. 5). The relation between spin and statistics is valid not only for elementary particles but also for composed ones like nuclei or atoms with the spin replaced by the total quantum angular momentum with respect to the particle's mass center (e.g., He^4 atoms are bosons, while He^3 are fermions). However, in this case *super-selection rules* forbidding superpositions of states with different total particle numbers must be taken into account.

Exercise 2

1. *Prove formula (67).*
   Hint: Choose bases in $\mathcal{H}$ and $\mathcal{K}$ and construct the corresponding bases of $\mathcal{F}(\mathcal{H})$, $\mathcal{F}(\mathcal{K})$, and $\mathcal{F}(\mathcal{H}) \otimes \mathcal{F}(\mathcal{K})$.
2. *Using (76) prove (77).*
3. *Using (79) and (81) compute the quantized magnetic field $\hat{\mathbf{B}}(\mathbf{x})$. Check that the quantized formula for the energy $\frac{1}{8\pi} \int (\hat{\mathbf{E}}^2 + \hat{\mathbf{B}}^2) d\mathbf{x}$ reproduces the Hamiltonian $\hat{H}_F = \sum_{\mathbf{k},\lambda} \hbar\omega(\mathbf{k}) \hat{a}^\dagger_{\mathbf{k},\lambda} \hat{a}_{\mathbf{k},\lambda}$ up to an irrelevant constant.*

4 Coherent and Thermal States for Bosons

We discuss now two types of states for the bosonic field which can be created either by the interaction with classical sources (coherent states) or by a weak coupling of the system to the external heat bath (thermal states).

4.1 Weyl Unitaries and Coherent States

As a quantum bosonic field is equivalent to a set of independent quantum harmonic oscillators, all mathematical tools used in the case of a single oscillator can be easily generalized. In particular, the Weyl unitaries, now defined in terms of smeared operators $\hat{a}(\phi)$ and $\hat{a}^\dagger(\phi)$ as

$$\hat{W}(\phi) = \exp\left(\hat{a}^\dagger(\phi) - \hat{a}(\phi)\right), \tag{82}$$

satisfy the group properties

$$\hat{W}(0) = \mathbb{1}, \quad \hat{W}(-\phi) = \hat{W}(\phi)^\dagger, \tag{83}$$

$$\hat{W}(\phi)\,\hat{W}(\psi) = \exp\left(i\mathrm{Im}\langle\phi|\psi\rangle\right)\hat{W}(\phi + \psi) \tag{84}$$

and the formal composition formulas

$$\begin{aligned}
\hat{W}(\phi) &= \exp\left(-\frac{1}{2}\langle\phi|\phi\rangle\right)\exp\left(\hat{a}^\dagger(\phi)\right)\exp\left(-\hat{a}(\phi)\right)\\
&= \exp\left(\frac{1}{2}\langle\phi|\phi\rangle\right)\exp\left(\hat{a}(\phi)\right)\exp\left(-\hat{a}^\dagger(\phi)\right).
\end{aligned} \tag{85}$$

Similar to (16) Weyl unitaries generate shifts of the field

$$\hat{W}(\phi)\,\hat{a}(\psi)\,\hat{W}(\phi)^\dagger = \hat{a}(\psi) - \phi \tag{86}$$

and a family of *coherent vectors* or *exponential vectors* labeled by the elements ϕ of the single-particle Hilbert space (fields)

$$|[\phi]\rangle = \hat{W}(\phi)\,|0\rangle, \quad \phi \in \mathcal{H}. \tag{87}$$

We use the notation $|[\phi]\rangle$ to distinguish coherent states from the elements ϕ of $\mathcal{H}$ written in Dirac notation as $|\phi\rangle$.

4.1.1 Properties of Coherent Vectors

1. Representation in terms of N-particle states

$$|[\phi]\rangle = \exp\left(-\frac{1}{2}\langle\phi|\phi\rangle\right)\sum_{N=0}^{\infty}\frac{1}{\sqrt{N!}}\underbrace{\phi\otimes\phi\otimes\cdots\otimes\phi}_{N}. \tag{88}$$

2. Eigenvector condition

$$\hat{a}(\psi)\,|[\phi]\rangle = \langle\psi|\phi\rangle\,|[\phi]\rangle\,. \tag{89}$$

3. Scalar product

$$\langle[\phi]|[\psi]\rangle = \exp\left\{-\frac{1}{2}\Big(\langle\phi|\phi\rangle + \langle\psi|\psi\rangle - 2\langle\psi|\phi\rangle\Big)\right\}\,. \tag{90}$$

4. The mean value of a normally ordered local operator $\hat{F} = F\Big(\hat{\psi}^\dagger(\mathbf{x});\hat{\psi}(\mathbf{x})\Big)$ (see (12)) reproduces the classical expression

$$\langle[\phi]|\hat{F}|[\phi]\rangle = F\Big(\phi^*(\mathbf{x}),\phi(\mathbf{x})\Big)\,. \tag{91}$$

5. Coherent states evolve into coherent ones and follow classical solutions, i.e.,

$$\exp\left(-\frac{i}{\hbar}\hat{H}_F\,t\right)|[\phi]\rangle = |[\phi_t]\rangle\,,\quad \phi_t = \exp\left(-\frac{i}{\hbar}\hat{H}_o\,t\right)\phi\,. \tag{92}$$

Coherent states in quantum electrodynamics perfectly describe coherent radiation emitted, for example, by macroscopic antennae in the case of radio-waves or masers and lasers in microwave and optical regimes, respectively.

4.2 Bosonic Fields with Classical Sources

In the previous sections we have discussed the free dynamics of the quantum field governed by the Hamiltonian $\hat{H}_F$ (55). The simplest model describing the interaction of the field with other physical systems involves a classical external source $\chi(\mathbf{x},t)$ which can be treated as an element of $\mathcal{H}$ denoted by $|\chi_t\rangle$. The Hamiltonian is given by

$$\hat{H}_F^\chi(t) = \hat{H}_F + i\hbar\left(\hat{a}(\chi_t) - \hat{a}^\dagger(\chi_t)\right)\,. \tag{93}$$

Our aim is to compute the time evolution of the coherent state under the dynamics governed by (93). We first compute the form of the unitary propagator in the interaction picture (**T** denotes time-ordering operation)

$$\hat{V}(t) = \exp\left\{\frac{i}{\hbar}\hat{H}_F\,t\right\}\mathbf{T}\exp\left\{-\frac{i}{\hbar}\int_0^t \hat{H}_F^\chi(s)\,ds\right\}\,, \tag{94}$$

where the time ordering means that in the series expansion of the time-ordered exponential the kth power has the form

$$\int_0^t ds_1 \int_0^{s_1} ds_2 \cdots \int_0^{s_{k-1}} ds_k\, \hat{H}_F^\chi(s_1)\,\hat{H}_F^\chi(s_2)\cdots\hat{H}_F^\chi(s_k)\,. \tag{95}$$

The unitary operator $\hat{V}(t)$ satisfies the following differential equation:

$$\frac{d}{dt}\hat{V}(t) = \exp\left\{\frac{i}{\hbar}\hat{H}_F t\right\}\left(\hat{a}(\chi_t) - \hat{a}^\dagger(\chi_t)\right)\exp\left\{\frac{i}{\hbar}\hat{H}_F t\right\}\hat{V}(t)$$
$$= \left(\hat{a}(\tilde{\chi}_t) - \hat{a}^\dagger(\tilde{\chi}_t)\right)\hat{V}(t), \tag{96}$$

where

$$\tilde{\chi}_t = \exp\left\{-\frac{i}{\hbar}\hat{H}_o t\right\}\chi_t. \tag{97}$$

Equation (96) can be solved in terms of the ordered exponential

$$\hat{V}(t) = \lim_{N\to\infty}\exp\left((\hat{a}(\tilde{\chi}_{t_N}) - \hat{a}^\dagger(\tilde{\chi}_{t_N}))\Delta t\right)$$
$$\cdots\exp\left((\hat{a}(\tilde{\chi}_{t_1}) - \hat{a}^\dagger(\tilde{\chi}_{t_1}))\Delta t\right)\exp\left((\hat{a}(\tilde{\chi}_0) - \hat{a}^\dagger(\tilde{\chi}_0))\Delta t\right), \tag{98}$$

where $\Delta t = t/N$ and $t_k = t(k/N)$. The right-hand side of (98) is a product of Weyl unitaries (82) and hence by (84) also a Weyl operator (the phase factor disappears in the limit $N \to \infty$)

$$\hat{V}(t) = \hat{W}\left(\int_0^t \tilde{\chi}_s \, ds\right). \tag{99}$$

Now we can compute the evolution of an initially coherent state $|[\psi_0]\rangle$

$$\mathbf{T}\exp\left\{-\frac{i}{\hbar}\int_0^t \hat{H}_F^\chi(s)\,ds\right\}|[\psi_0]\rangle = \exp\left\{-\frac{i}{\hbar}\hat{H}_F t\right\}\hat{V}(t)\,|[\psi_0]\rangle$$
$$= \exp\left\{i\mathrm{Im}\int_0^t \langle\tilde{\chi}_s|\psi_0\rangle\right\}|[\psi_t]\rangle, \tag{100}$$

where

$$\psi_t = \exp\left\{-\frac{i}{\hbar}\hat{H}_o t\right\}\psi_0 + \int_0^t \exp\left\{-\frac{i}{\hbar}\hat{H}_o(t-s)\right\}\chi_s\,ds. \tag{101}$$

It follows that a coherent state evolves into another coherent state (with a certain phase factor) which is determined by a field ψ_t satisfying a suitable equation with respect to the external source χ_t

$$\frac{d}{dt}\psi_t = -\frac{i}{\hbar}\hat{H}_o\psi_t + \chi_t. \tag{102}$$

In particular, by a proper choice of the classical source we can generate an arbitrary coherent state from the vacuum, i.e., the coherent state with $\psi_0 = 0$.

Remark 4 All the formulas and proofs of above, involving smeared operators $\hat{a}(\phi)$ and $\hat{a}^\dagger(\phi)$, are valid in the infinite volume limit, i.e., for test functions defined on the whole $\mathbf{R}^\ell$.

4.3 Thermal States of Non-interacting Bosons

From the general Boltzmann principle: *At thermal equilibrium the probability of finding a physical system in a micro-state with energy E is proportional to* $\exp(-E/k_B T)$; it follows that the thermal state at the temperature T of a quantum system with Hamiltonian $\hat{H}$ is described by the density matrix (*canonical ensemble*)

$$\hat{\rho}_\beta = Z^{-1}(\beta) \exp(-\beta \hat{H}) \,, \quad \beta = \frac{1}{k_B T} \,, \quad Z(\beta) = \mathrm{Tr}\exp(-\beta \hat{H}) \,. \quad (103)$$

For a single harmonic oscillator with $\hat{H}_o = \hbar\omega\hat{n}$, we can easily compute the normalization constant (*partition sum*) $Z(\beta)$ and the mean particle number $\bar{n}$

$$Z(\beta) = \mathrm{Tr}\exp(-\beta \hat{H}_o) = \sum_{n=0}^{\infty} e^{-\beta\hbar\omega n} = \frac{1}{1 - e^{-\beta\hbar\omega}} \,, \quad (104)$$

$$\bar{n} = -\frac{1}{\hbar\omega} \frac{\partial}{\partial\beta} \log Z(\beta) = \frac{1}{e^{\beta\hbar\omega} - 1} \,. \quad (105)$$

For non-interacting bosons the Hamiltonian $\hat{H}_F = \sum_k \hbar\omega_k \hat{n}_k$ should be modified to

$$\hat{H}_F^\mu = \hat{H}_F - \mu\hat{N} = \sum_k (\varepsilon_k - \mu)\hat{n}_k \,, \quad \varepsilon_k = \hbar\omega_k, \quad (106)$$

where the *chemical potential* μ is a new parameter which determines the average number of particles in the system (for massless particles we put $\mu = 0$). The thermal equilibrium state is given by a density matrix (*grand canonical ensemble*)

$$\hat{\rho}_{\beta,\mu} = Z^{-1}(\beta, \mu) \exp(-\beta \hat{H}_F^\mu) \,, \quad Z(\beta, \mu) = \mathrm{Tr}\exp(-\beta \hat{H}_F^\mu), \quad (107)$$

which is equivalent to a tensor product of properly parametrized single harmonic oscillator density matrices. Hence, the average particle number in the single-particle state (mode) e_k is given by the *Bose–Einstein statistics*

$$\bar{n}_k = \frac{1}{e^{\beta(\varepsilon_k - \mu)} - 1} \,, \quad (108)$$

and the von Neumann entropy reads

$$S(\hat{\rho}_{\beta,\mu}) = -k_B \text{Tr}(\hat{\rho}_{\beta,\mu} \log \hat{\rho}_{\beta,\mu})$$
$$= k_B \sum_k \left((1 + \bar{n}_k) \log(1 + \bar{n}_k) - \bar{n}_k \log \bar{n}_k\right) . \tag{109}$$

4.4 Bose–Einstein Condensation

For a large number of bosons the mean number of particles in the equilibrium state
(see (108))

$$N \equiv \sum_k \bar{n}_k = \sum_k \frac{1}{e^{\beta(\varepsilon_k - \mu)} - 1} \tag{110}$$

can be identified with an actual number of particles in the sample. The chemical potential $\mu \leq 0$ depends on N and β. For a fixed N the character of its
β-dependence can be extracted from the identity $0 = dN = (\partial N/\partial \mu)d\mu + (\partial N/\partial \beta)d\beta$. By a direct computation the following inequality follows

$$\frac{d\mu}{d\beta} \geq 0 \quad \text{or} \quad \frac{d\mu}{dT} \leq 0 . \tag{111}$$

Another useful inequality reads

$$\text{for } \mu = 0 \text{ and } \varepsilon_k > 0 , \quad \frac{d\bar{n}_k}{dT} > 0 . \tag{112}$$

Many models of bosonic systems, including a non-relativistic gas in a three-
dimensional confining potential, exhibit a critical temperature T_c for which $\mu(T_c)$,
according to (111), approaches zero when $T \searrow T_c$.[1] Below this temperature the
chemical potential remains equal to zero and therefore for $T < T_c$ according to (112)

$$N' = \sum_{k \neq 0} \frac{1}{e^{\varepsilon_k/kT} - 1} < N . \tag{113}$$

Notice that in the sum above the ground state (typically non-degenerate) with $\varepsilon_0 \simeq 0$ is omitted because the ground state occupation number $N_0 = 1/(e^{\varepsilon_0/kT} - 1)$
becomes singular and can be determined only from the relation $N_0 + N' = N$. The
fraction of bosons occupying a ground state below the critical temperature form
a *Bose–Einstein condensate* and this phenomenon is a unique example of *phase
transition* for noninteracting particle systems. The fact that a macroscopic number
of particles can occupy a single quantum state leads to many collective/coherent

---

[1] The presented arguments become sharp in the thermodynamic limit.

phenomena which are intensively studied in the last decade, both experimentally and theoretically, in the context of *ultracold gases* [10].

4.5 Gaussian States and Entanglement

The vacuum, squeezed vacuum, and the thermal state for a bosonic field are examples of a more general family of (zero-mean) *Gaussian states*. We restrict ourselves to a finite mode case with $\{\hat{a}_k, \hat{a}_k^\dagger, k = 1, 2, ..., K\}$ and introduce quadrature operators $\hat{X}_{2k-1} = \frac{1}{\sqrt{2}}\left(\hat{a}_k + \hat{a}_k^\dagger\right)$, $\hat{X}_{2k} = \frac{1}{i\sqrt{2}}\left(\hat{a}_k - \hat{a}_k^\dagger\right)$ satisfying commutation relations

$$[\hat{X}_j, \hat{X}_k] = i J_{jk}, \tag{114}$$

where $\mathbf{J} = [J_{jk}]$ is the standard symplectic matrix.

Define a new parametrization of the Weyl unitaries

$$\hat{W}(\mathbf{x}) = \exp\left\{i \sum_{j=1}^{2K} x_j \hat{X}_j\right\}, \quad \mathbf{x} = (x_1, x_2, ..., x_{2K}), \quad x_j \in \mathbf{R}. \tag{115}$$

A state given by a density matrix $\hat{\rho}_G$ is called Gaussian if

$$\mathrm{Tr}\left(\hat{\rho}_G \hat{W}(\mathbf{x})\right) = \exp\left\{-\frac{1}{2}\mathbf{x}^T \mathbf{G}\mathbf{x}\right\}. \tag{116}$$

Here $\mathbf{G}$ is a real symmetric correlation matrix given by

$$\mathbf{G} = [G_{mn}], \quad G_{mn} = \mathrm{Tr}\left(\hat{\rho}_G(\hat{X}_m \hat{X}_n + \hat{X}_n \hat{X}_m)\right), \tag{117}$$

satisfying a necessary and sufficient condition

$$\mathbf{G} + i\mathbf{J} \geq 0. \tag{118}$$

Consider a bipartite system which consists of two oscillators and a Gaussian state with a correlation matrix $\mathbf{G}_{11} = \mathbf{A}, \mathbf{G}_{22} = \mathbf{B}, \mathbf{G}_{12} = \mathbf{C}$. The partial transposition used to detect bipartite-entangled states of two finite-level systems (see chapter "Bipartite Quantum Entanglement", Sect. 3) is realized by a time reversal anti-unitary map $(\hat{X}_1, \hat{X}_2) \to (\hat{X}_1, -\hat{X}_2)$. Despite the infinite dimensionality of the two oscillators' Hilbert space, it turns out that transposition is an exhaustive entanglement witness for bipartite Gaussian states. Indeed, the separability criterion can be expressed by the single inequality [1]

$$\det\mathbf{A}\,\det\mathbf{B} + \left(\frac{1}{4} + |\det\mathbf{C}|\right)^2 - \det(\mathbf{AJCJBJC}^T\mathbf{J}) \geq \frac{1}{4}(\det\mathbf{A} + \det\mathbf{B}). \tag{119}$$

Exercise 3

1. *Using* (17), (18), *and* (71) *prove formulas* (83), (84), (85), *and* (86) *and* (88), (89), (90), (91), *and* (92).
2. *Check that* (101) *is a solution of* (102).
3. *Compute the von Neumann entropy for the thermal state of a single harmonic oscillator in terms of* $\bar{n}$.
4. *Prove inequalities* (111) *and* (112).

5 Second Quantization of Fermions

In contrast to bosons, due to the Pauli's exclusion principle, fermions cannot macro-scopically occupy a single quantum state and therefore a wave function of a fermion has no classical meaning of a measurable macroscopic field. However, the description of many fermion systems in terms of the second quantization is very similar to the bosonic one. Choosing again a certain collection of modes $\{u_k\}$, being solutions of a single-fermion Schrödinger equation, we can define the corresponding anni-hilation and creation operators $\hat{c}_k, \hat{c}_k^\dagger$. The new particle number operators $\hat{n}_k$ must have only 0, 1 eigenvalues that can be achieved imposing the following canonical anti-commutation relations (CAR):

$$\{\hat{c}_k, \hat{c}_\ell^\dagger\} = \delta_{k\ell} \,, \quad \{\hat{c}_k, \hat{c}_\ell\} = \{\hat{c}_k^\dagger, \hat{c}_\ell^\dagger\} = 0, \tag{120}$$

where $\{\hat{A}, \hat{B}\} \equiv \hat{A}\hat{B} + \hat{B}\hat{A}$. Therefore (step 1 of Exercise 2), the particle number operators satisfy the relations

$$\hat{n}_k^2 = \hat{n}_k \,, \quad [\hat{n}_k, \hat{n}_\ell] = 0 \,. \tag{121}$$

5.1 Fock Space and Observables

The minimal Hilbert space, called fermionic Fock space and denoted by $\mathcal{F}^a$, which supports the structure of operators (120, 121) is spanned by the Fock vectors being the joint eigenvectors of all $\hat{n}_k$

$$\hat{n}_k|\{n_k\}\rangle = n_k|\{n_k\}\rangle \,, \quad n_k = 0, 1 \,. \tag{122}$$

Similar to (61), for any $N = 0, 1, 2, \dots$ we define a subspace of $\mathcal{F}^a$, denoted by $\mathcal{H}_N^a$, spanned by the vectors $|\{n_k\}\rangle$ satisfying $\sum_k n_k = N$. Then we can decompose the Fock space into a direct sum

$$\mathcal{F}^a = \bigoplus_{N=0}^{\infty} \mathcal{H}_N^a, \tag{123}$$

while now $\mathcal{H}_N^a$ is an *antisymmetric tensor product* spanned by the totally anti-symmetrized formal products $u_{k_1} \circ u_{k_2} \circ \cdots \circ u_{k_N}$ (now the same mode can appear only once!). The subspace $\mathcal{H}_0^a$ is a one-dimensional ray generated by the vacuum $|0\rangle$ and the single-particle Hilbert space $\mathcal{H}_1^a \equiv \mathcal{H}$ is spanned by the modes u_k. The local structure is described by the relation identical to (67). The formalism of local quantum fields, smeared fields, extension to infinite spaces, etc., is similar to the bosonic case presented in Sect. 3.3.

Operators for many fermion systems are generated by all $\hat{c}_k, \hat{c}_k^\dagger$ and form the *CAR algebra*. However, due to the fact that the single-fermion wave functions are not measurable, only the observables which are *even* possess physical meaning. Important additive (single-particle) observables are the free Hamiltonian and the total number of particles

$$\hat{H}_F = \sum_k \varepsilon_k \hat{c}_k^\dagger \hat{c}_k \ , \quad \hat{N} = \sum_k \hat{c}_k^\dagger \hat{c}_k \ . \tag{124}$$

From (122) it follows that in the case of finite number of modes K the Fock space $\mathcal{F}^a$ is 2^K-dimensional and, moreover, the CAR algebra coincides with the whole algebra of $2^K \times 2^K$ complex matrices, i.e., is isomorphic to the algebra of K qubits. The manifest form of this isomorphism is given in terms of the *Jordan–Wigner transformation*

$$\hat{c}_k \equiv \underbrace{\sigma_z \otimes \cdots \otimes \sigma_z}_{k-1} \otimes \sigma_- \otimes \mathbb{1} \otimes \cdots \otimes \mathbb{1} \ , \quad k = 1, 2, \ldots, K, \tag{125}$$

with $\sigma_- = (\sigma_x - i\sigma_y)/2$ (see step 4 of Exercise 2). Here, σ_x, σ_y, and σ_z denote the Pauli matrices.

5.2 Thermal States of Non-interacting Fermions

Following Sect. 4.3 we can apply the grand canonical ensemble (107) to non-interacting fermions with the Hamiltonian $\hat{H}_F$ given by (124). Computing the partition sum in a Fock basis (122), one obtains

$$Z(\beta, \mu) = \sum_{\{n_k\}} e^{-\beta \sum_k \left(n_k (\varepsilon_k - \mu) \right)} = \prod_k \left(1 + e^{-\beta(\varepsilon_k - \mu)} \right) . \tag{126}$$

This implies the following expression for the average particle number in the single-particle state (mode) (*Fermi–Dirac statistics*):

$$\bar{n}_k = \frac{1}{e^{\beta(\varepsilon_k - \mu)} + 1} \ , \tag{127}$$

whence the corresponding von Neumann entropy (compare (108) and (109)) reads

$$S(\hat{\rho}_{\beta,\mu}) = -k_B \sum_k \left(\bar{n}_k \log \bar{n}_k + (1 - \bar{n}_k) \log(1 - \bar{n}_k) \right) \tag{128}$$

and gives a clear interpretation in terms of *particle* and *hole* occupation numbers $\bar{n}_k$ and $1 - \bar{n}_k$, respectively.

Exercise 4

1. *Prove relations* (121) *using* (120).
2. *Prove formula* (126) *for the fermionic partition sum.*
3. *Derive the Fermi–Dirac statistics from* (126).
4. *Check CAR relations* (120) *for the operators defined by* (125).

6 Further Reading

Section 2: The basic properties and applications of coherent vectors, squeezed states, and Glauber representation are presented in [4] in the context of quantum optics. More mathematical approach with important generalizations can be found in [9].

Section 3: For introduction to quantum field theory see [7, 6]; the more advanced material, relevant for quantum electrodynamics, can be found in [5].

Section 4: Coherent states for electromagnetic field and their origin are discussed in [5]. For a mathematically rigorous theory of thermal states including systems in the thermodynamic limit see [3]. A simple approach to Bose–Einstein condensation can be found in [8]. Application of Gaussian states in quantum information is studied in [1].

Section 5: Basic information about fermionic systems can be found in [2]; for more advanced material see [3].

References

1. Adesso, G., Illuminati, F.: Bipartite and multipartite entanglement of Gaussian states. In: Cerf, J., Leuchs, G., Polzik, E.S. (eds.) Quantum Information with Continuous Variables of Atoms and Light, pp. 1–22. World Scientific, Singapore (2007)
2. Alicki, R., Fannes, M.: Quantum Dynamical Systems. Oxford University Press, Moscow (2001)
3. Bratteli, O., Robinson, D.W.: Operator Algebras and Quantum Statistical Mechanics Equilibrium States. Models in Quantum Statistical Mechanics II, 2nd edn. Springer, Berlin (1997)
4. Carmichael, H.J.: Statistical Methods in Quantum Optics 1. Springer, Berlin (1999)
5. Cohen-Tannoudji, C., Dupont-Roc, J., Grynberg, G.: Photons and Atoms. Introduction to Quantum Electrodynamics. Wiley, New York (1989)
6. Greiner, W., Reinhardt, J.: Field Quantization. Springer, Berlin (1996)
7. Haken, H.: Quantum Field Theory of Solids. North-Holland, Amsterdam (1976)
8. Huang, K.: Introduction to Statistical Physics. Taylor and Francis, London (2001)
9. Perelomov, A.: Generalized Coherent States and Their Applications. Springer, Berlin (1986)
10. Sokol, P., Griffin, A., Snoke, D.W., et al.: Bose Einstein Condensation. Cambridge University Press, Cambridge (1995)

Quantum Entropy and Information

Nilanjana Datta

1 Introduction

As seen in chapter "Classical Information Theory", classical information theory is
the mathematical theory of information-processing tasks such as storage and transmission of information. It was born out of a seminal paper by Claude Shannon in
1948.

Two fundamental tasks in classical information theory are the storage and transmission of information. For efficient use of the resources available for storage of
information, it is essential to compress data.[1] This is known as *source coding* and
involves encoding the messages emitted by an information source, by exploiting
redundancies which are typically present in them. As regards transmission of information, the biggest hurdle that one faces is the presence of noise in a communications channel, which leads to a distortion of messages transmitted through it. In
order to counteract its effect, one needs to add redundancy to the messages before
transmitting them. This is known as *channel coding*. The idea central to both source
and channel coding is to encode messages in a manner which allows them to be
later decoded with an arbitrarily low probability of error, while optimizing the rate,
i.e., the ratio between the sizes of the message and its corresponding codeword
(see Sect. 3 for details). Note that there is a duality between the problems of data
compression and data transmission. In the first case one exploits the redundancy
present in the messages to compress the data, whereas in the second case one adds
redundancy in a controlled manner to combat errors introduced by the noise in the
channel.

Classical information theory relies on the assumption that the states of the physical system, in which information is encoded, can be perfectly distinguished, perfectly copied, and measured with arbitrary precision. For most practical applications

N. Datta (✉)
Statistical Laboratory, DPMMS, University of Cambridge, Cambridge, UK,
n.datta@statslab.cam.ac.uk

[1] In this chapter, the words *data, information, signals* and *outputs of an information source* will be
used interchangeably.

Datta, N.: *Quantum Entropy and Information*. Lect. Notes Phys. **808**, 175–214 (2010)
DOI 10.1007/978-3-642-11914-9_6　　　　　　　© Springer-Verlag Berlin Heidelberg 2010

of this theory, the laws of classical physics provide a sufficiently close approximation to the system's behavior and hence the above assumptions are indeed satisfied. However, in order to investigate the fundamental limits which are imposed by the basic laws of physics on our ability to encode, decode, process, and transmit information, we need to go beyond the ideas of classical physics.

Quantum information theory generalizes classical information theory to systems which are governed by the laws of quantum mechanics. It deals with how the quantum mechanical properties of physical systems can be exploited to obtain the limits of efficient storage and transmission of information. The underlying quantum mechanics leads to important differences between the two theories, at times yielding distinctively new features, which have no classical analogues [8, 10].

In this chapter, we will address some key aspects of the problems of data compression and data transmission in quantum information theory. The main topics that will be covered are Schumacher compression, quantum channels, the Holevo bound, and the transmission of classical information through a quantum channel.

2 Preliminaries

In quantum information processing systems, information is stored in the quantum states of a physical system. In the following, we will only consider systems with finite-dimensional Hilbert spaces. The fundamental unit of quantum information is called the "quantum bit" or *qubit*. A qubit is a vector in a 2-dimensional Hilbert space (the *single qubit space*), a notion already encountered in chapter "Quantum Probability and Quantum Information Theory", Sect. 1.2. In analogy with the classical bit,[2] we call the elements of an orthonormal basis in the 2-dimensional Hilbert space $|0\rangle$ and $|1\rangle$. Intuitively, the states $|0\rangle$ and $|1\rangle$ are analogous to the values 0 and 1 that a (classical) bit can take. However, there is an important difference between a qubit and a bit. Superpositions of the states $|0\rangle$ and $|1\rangle$ of the form

$$|\psi\rangle = a|0\rangle + b|1\rangle, \tag{1}$$

where $a, b \in \mathbf{C}$, with $|a|^2 + |b|^2 = 1$ can also exist.

There are various physical realizations of a qubit (see chapters "Photonic Realization of Quantum Information Protocols" and "Physical Realizations of Quantum Information"), e.g., an electronic or nuclear spin, or a polarized photon. Basis states $|0\rangle$ and $|1\rangle$ of a single qubit space correspond to a fixed pair of reliably distinguishable states of the qubit, e.g., horizontal and vertical polarizations of the photon or the spin-up $(|\uparrow\rangle)$ and spin-down $(|\downarrow\rangle)$ states of an electron along a particular axis. In the case of the polarized photon, superpositions of these basis states correspond to other polarizations, e.g., the state

---

[2] A bit is the basic indivisible unit of classical information which takes one of two possible values – 0 and 1.

$$\frac{1}{\sqrt{2}}\left(|0\rangle + i\,|1\rangle\right) \tag{2}$$

corresponds to right circular polarization. Such a state is not orthogonal to the basis states $|0\rangle$ and $|1\rangle$ and hence cannot be reliably distinguished from them, even in principle.[3] With regard to any measurement which distinguishes the states $|0\rangle$ and $|1\rangle$, it behaves like $|0\rangle$ with probability $1/2$ and like $|1\rangle$ with probability $1/2$.

In the real world, however, there are no perfectly isolated systems. Real systems suffer from unwanted interactions with their environments and are hence *open*. In quantum information processing systems, these interactions manifest themselves as noise, which damages the information that the system encodes, thus leading to errors. This noise process, known as *decoherence*, plays a vital role in quantum information theory and, in order to understand it, we need to study the properties of states of *open quantum systems*.

An open system can be considered to be a subsystem of a larger closed system, in which the interactions with the environment are incorporated (see chapter "Bipartite Quantum Entanglement", Sect. 4). The most general description of the state of a quantum system, which also holds for such composite systems, is provided by the density matrix formalism. A density matrix (or density operator) ρ is a positive semi-definite, linear operator of unit trace, acting on the Hilbert space $\mathcal{H}$ of the system: $\rho \geq 0$; $\mathrm{Tr}\,\rho = 1$. The density matrix formalism also provides a description of a system whose precise state is not known (see chapter "Hilbert Space Methods for Quantum Mechanics", Sect. 2.1).

Consider a system described by a statistical mixture of state vectors $|\psi_1\rangle$, $|\psi_2\rangle$, $\ldots, |\psi_n\rangle$, with corresponding probabilities $p_1, p_2 \ldots p_n$, $\sum_i p_i = 1$. The state of the system is, therefore, characterized by an ensemble $\{p_i, |\psi_i\rangle\}_{i=1}^n$. The density matrix, ρ, of such a system is defined as follows:

$$\rho := \sum_{i=1}^{n} p_i |\psi_i\rangle\langle\psi_i|. \tag{3}$$

The state vectors $|\psi_i\rangle$ are normalized but they need not be mutually orthogonal. Strictly speaking, the operator ρ is the density operator and the matrix representing it is the density matrix. Note that different ensembles may give rise to the same density matrix.

A quantum system whose state vector is known precisely is said to be in a *pure state*, and in this case, there is only one term in the sum (3) and the density matrix is a one-dimensional projector; e.g., if the system is known to be in a state $|\psi_2\rangle$ then $p_2 = 1$, $p_i = 0\,\forall i \neq 2$, and $\rho = |\psi_2\rangle\langle\psi_2|$. If there is more than one term in the sum (3) then the density matrix ρ is said to be *mixed*. Equivalently, a density matrix ρ is pure if $\mathrm{Tr}\,(\rho^2) = 1$ and mixed if $\mathrm{Tr}\,(\rho^2) < 1$. Henceforth, we shall use the word

---

[3] Non-orthogonal states of a quantum mechanical system cannot be reliably distinguished by any measurement (see, e.g., [8]).

state to refer to a density matrix, whereas the word *pure state* shall be used to refer to both the state vector $|\psi\rangle$ and the corresponding density matrix $\rho = |\psi\rangle\langle\psi|$. The system is said to be in a mixed state if its density matrix is mixed. Pure states are extremal points of the convex set of all states (see step 1 of Exercise 1 of chapter "Bipartite Quantum Entanglement").

The expectation value of any observable A in a state ρ is given by a positive, linear, normalized functional ϕ (see chapter "Hilbert Space Methods for Quantum Mechanics", Sects. 1.5 and 2.1):

$$\phi(A) \equiv \langle A \rangle = \mathrm{Tr}\,(A\rho), \tag{4}$$

where $A \in \mathcal{B}(\mathcal{H})$ and $A = A^\dagger$. Here $\mathcal{B}(\mathcal{H})$ denotes the algebra of all operators acting in $\mathcal{H}$ and $A^\dagger$ denotes the Hermitian conjugate of the operator A. As seen in chapter "Hilbert Space Methods for Quantum Mechanics", Sect. 2.3, when one considers a composite (bipartite) system AB, its Hilbert space is $\mathcal{H}_{AB} = \mathcal{H}_A \otimes \mathcal{H}_B$, where $\mathcal{H}_A, \mathcal{H}_B$ are the Hilbert spaces of the two parts (see chapter "Hilbert Space Methods for Quantum Mechanics", Sect. 1.3). If the system is in the state ρ_{AB} then the state of the subsystem A is given by the operator

$$\rho_A := \mathrm{Tr}_B \rho_{AB}, \tag{5}$$

which acts in $\mathcal{H}_A$. Here Tr_B denotes a trace over the Hilbert space $\mathcal{H}_B$ alone. It is referred to as the *partial trace* over the Hilbert space of the system B. Moreover, if T_A is an operator in $B(\mathcal{H}_A)$, and $\mathbb{1}_B$ is the identity operator in $B(\mathcal{H}_B)$, then (see step 1 of Exercise 1)

$$\mathrm{Tr}\,((T_A \otimes \mathbb{1}_B)\rho_{AB}) = \mathrm{Tr}_A(\rho_A T_A)\,. \tag{6}$$

The operator ρ_A is referred to as the *reduced density matrix* of the subsystem A. Similarly, the reduced density matrix of the subsystem B is given by $\rho_B := \mathrm{Tr}_A \rho_{AB}$. Reduced density matrices satisfy $\rho_A \geq 0$, $\mathrm{Tr}\,\rho_A = 1$.

The most general description of the dynamics of an open quantum system is provided by the quantum operations formalism. A *quantum operation* (or super-operator) Φ is a linear, completely positive, trace-preserving map (CPT) which takes density matrices to density matrices (see chapter "Hilbert Space Methods for Quantum Mechanics", Sect. 2.4 and chapter "Quantum Probability and Quantum Information Theory", Sect. 5):

$$\Phi : \rho \mapsto \rho', \quad \rho \in B(\mathcal{H}), \ \rho' \in B(\mathcal{H}'), \ \rho, \rho' \geq 0, \ \mathrm{Tr}\,\rho = 1 = \mathrm{Tr}\,\rho'. \tag{7}$$

In general a quantum operation captures the dynamical change to the state of a system, which occurs as the result of some physical process: ρ is the initial state before the process and $\Phi(\rho)$ is the final state after the process occurs. There are various physical processes which are of relevance in quantum information theory, e.g., time evolution of the state of an open system, compression of data from a quantum information source and transmission of quantum information through a noisy quantum channel. The latter is the quantum analogue of the transmission of information through a classical communications channel. In view of this analogy, a quantum operation is also referred to as a *quantum channel*. The channel acts on an input state ρ, yielding the output state $\Phi(\rho) = \rho'$. In general ρ and ρ' may be in different Hilbert spaces $\mathcal{H}$ and $\mathcal{H}'$, respectively. In this case, $\mathcal{H}$ and $\mathcal{H}'$ are referred to as the input and output Hilbert spaces of the channel Φ.

Exercise 1

1. *Consider a bipartite system AB with Hilbert space $\mathcal{H}_{AB} = \mathcal{H}_A \otimes \mathcal{H}_B$. Let the system be in a state ρ_{AB} and let ρ_A and ρ_B denote the reduced density matrices of its two parts. Prove that if $T = T_A \otimes \mathbb{1}_B$ is an operator acting on $\mathcal{H}_{AB}$ then*

$$\mathrm{Tr}\,(\rho_{AB}T) = \mathrm{Tr}\,_A(\rho_A T_A). \tag{8}$$

2. *Show that the following three operators,*

$$E_1 = \frac{\sqrt{2}}{1+\sqrt{2}}|1\rangle\langle 1|, \quad E_2 = \frac{\sqrt{2}}{2+2\sqrt{2}}(|0\rangle - |1\rangle)(\langle 0| - \langle 1|), \tag{9}$$

and $E_3 = \mathbb{1} - E_1 - E_2$, form a POVM (see [8], chapter "Hilbert Space Methods for Quantum Mechanics", Sect. 1.5 and chapter "Quantum Probability and Quantum Information Theory", Sect. 5). Namely, their sum is the identity so that $\rho \mapsto \sum_{i=1}^{3} \sqrt{E_i}\,\rho\,\sqrt{E_i}$ is a completely positive trace-preserving map. Suppose Alice gives Bob a state prepared in one of the two states $|\psi_1\rangle = |0\rangle$ or $|\psi_2\rangle = (|0\rangle + |1\rangle)/\sqrt{2}$. Show that if Bob does a measurement characterized by these POVM elements on the state he receives, he never makes an error of misidentification. Discuss the possible outcomes.

3 Rudiments of Classical Information Theory

For the benefit of the reader, some of the topics of chapter "Classical Information Theory"[4] are briefly reviewed here. The issues of storage and transmission of *classical* information were first addressed by Claude Shannon in 1948. He laid the foundation of classical information theory by answering the following key questions:

(Q1) *What is the limit to which information can be reliably compressed, that is, compressed in a manner such that it can be recovered later with arbitrarily low probability of error?*

---

[4] The symbol for the Shannon entropy and related quantities used in this chapter is H instead of h.

This question is of relevance because there is often a physical limit to the amount of space available for the storage of data, e.g., in the memory of a mobile phone.

(Q2) *What is the amount of information that can be reliably transmitted through a communications channel ?*

The relevance of this question arises from the fact that one of the biggest hurdles that one faces in the transmission of information is the presence of noise in communication channels, e.g., a conversation over a crackling telephone line or a spacecraft sending photos from a distant planet. The effect of this noise is to distort the information transmitted through it.

The answer to **(Q1)** is given in *Shannon's noiseless channel coding theorem* [14]. It states that the limit to which information from a source can be compressed is given by the *Shannon entropy* (see chapter "Classical Information Theory", Sect. 1.2) – a quantity characteristic of the source.

The outputs (or signals) of a classical information source are sequences of *letters* (or *symbols*) chosen from a finite set J – the *source alphabet*, according to a given probability distribution. Some examples of source alphabets are (i) binary alphabet: $J = \{0, 1\}$; (ii) telegraph English: J consists of the 26 letters of the English alphabet and a space; (iii) written English: J consists of the 26 letters of the English alphabet in upper and lower cases and punctuation marks.

The simplest example of a source is a *memoryless source*. It is characterized by a probability distribution $\{p(u)\}$, and each use of the source results in a *letter* $u \in J$ being emitted with probability $p(u)$. A signal emitted from the source does not depend on the previously emitted signals. A memoryless source can therefore be modeled by a sequence of independent, identically distributed (i.i.d.) random variables $U_1, U_2, \ldots, U_n$ with common probability mass function $p(u) = P(U_k = u)$, $u \in J$, for all $1 \leq k \leq n$. The signals of the source are sequences of letters $(u_1, u_2, \ldots, u_n) \in J^n$ taken by these variables and

$$p(u_1, u_2, \ldots, u_n) \equiv P(U_1 = u_1, \ldots, U_n = u_n) = \prod_{i=1}^{n} p(u_i). \qquad (10)$$

A memoryless information source is also referred to as an *i.i.d. information source*.

The Shannon entropy of the source is a function of the source probability distribution and is given by

$$H(\{p(u)\}) = -\sum_{u \in J} p(u) \log p(u) . \qquad (11)$$

Here and henceforth, logarithms indicated by log are taken to base 2. This is natural in the context of coding with binary alphabets. We use the convention $0 \log 0 = 0$, which is justified by continuity $\left(\lim_{x \to 0} x \log x = 0\right)$. Hence, if an event $\{U = u\}$ has zero probability $\left(p(u) = 0\right)$ then it does not contribute to the entropy. The Shannon entropy of a single random variable $H(X)$ with probability mass function $p(x)$, $x \in J$ is given by

$$H(\{p(x)\}) = -\sum_{x \in J} p(x) \log p(x). \tag{12}$$

It is a measure of the uncertainty of the random variable X. It also quantifies how much information we gain on average when we learn the value of X.

Example 1 (Binary Entropy) Let X be a random variable which takes the value 1 with probability p, and the value 0 with probability $(1 - p)$. Then

$$H(X) = -p \log p - (1 - p) \log(1 - p) =: H(p). \tag{13}$$

This is called *binary entropy* and we denote it by the symbol $H(p)$. In particular, $H(p) = 1$ bit when $p = 1/2$.

Data compression is possible because an information source typically produces some outputs more frequently than others. In other words, there is frequently redundancy in the information. During data compression one exploits the redundancy in the data to form the most compressed version possible. There are different ways of doing this. One way is called *variable length coding*. This is done by encoding the signals in a way such that the outputs which occur more frequently are assigned shorter descriptions (i.e., fewer bits) than the less frequent ones. For example, a good compression for a source of English text could be achieved by using fewer bits to represent the letter e than the letter z. This is because e occurs much more frequently than z in the English text. Another way is called *fixed length coding*. This is done by identifying a set of signals which have a high probability of occurrence (*typical* signals), and assigning unique fixed length binary strings to each of these signals. All other signals (which are *atypical*) are assigned a single binary string of the same length as those assigned to the typical ones. However, in either case of coding, there is a limit to which data from a classical information source can be reliably compressed. This is given by Shannon's noiseless channel coding theorem. This theorem tells us that a classical i.i.d. source described by the probabilities $\{p(u)\}$ can be compressed so that, on an average, each use of the source can be represented using $H(\{p(u)\})$ bits of information. Moreover, it tells us that if the source is compressed any further, i.e., if fewer than $H(\{p(u)\})$ bits are used to represent it, then there is a high probability of error in retrieving the original information from the compressed signals. Hence the Shannon entropy quantifies the optimal rate of compression that can be achieved.

The answer to (**Q2**) is given in *Shannon's noisy channel coding theorem* [14]. To combat the effects of noise in a communications channel, the message to be sent is suitably encoded by the sender and the resulting codewords (which are, e.g., binary sequences) are then transmitted through the channel. The set of codewords constitute a *classical error-correcting code*. The idea behind the encoding is to introduce redundancy in the message so that upon decoding the received message, the receiver can retrieve the original message with a low probability of error, even if part of the message is lost or corrupted due to noise. The amount of redundancy which needs to be added to the original message depends on how much noise is present in the channel.

A *discrete*[5] channel is defined by an input alphabet J_X, an output alphabet J_Y, and a set of conditional distributions

$$p(\underline{y}^{(N)}|\underline{x}^{(N)}). \tag{14}$$

The input to the channel is the sequence $\underline{x}^{(N)} := (x_1, \ldots, x_N)$, with x_i, $i = 1, \ldots, N$, being letters from J_X. The output of the channel is the word $\underline{y}^{(N)} := (y_1, \ldots, y_N) \in J_Y^N$. The conditional probability of receiving the word $\underline{y}^{(N)}$ at the output of the channel, given that the codeword $\underline{x}^{(N)} \in J_X^N$ was sent, is denoted by $p(\underline{y}^{(N)}|\underline{x}^{(N)})$.

We suppose that these conditional distributions are known to both sender and receiver. Each of the possible input sequences $\underline{x}^{(N)}$ induces a probability distribution on the output sequences $\underline{y}^{(N)}$. The correspondence between input and output sequences is not one- to- one. Two different input sequences may give rise to the same output sequence. However, we will see later that it is possible to choose a subset of input sequences so that with high probability, there is only one highly likely input that could have caused the particular output. The receiver at the output end of the channel can then reconstruct the input sequences at the output with negligible probability of error.

A simple class of discrete channels are the so-called *memoryless channels*. A channel is said to be memoryless if the probability distribution of the output depends only on the input at that time and is independent of previous channel inputs or outputs.

Definition 1 A memoryless channel is one in which

$$p(\underline{y}^{(N)}|\underline{x}^{(N)}) = \prod_{i=1}^{N} p(y_i|x_i) \tag{15}$$

for $\underline{y}^{(N)} = (y_1, \ldots, y_N)$ and $\underline{x}^{(N)} = (x_1, \ldots, x_N)$. Here $p(y_i|x_i)$ is a letter-to-letter channel probability, i.e., the conditional probability to obtain the letter y_i as the output, given that the letter x_i has been sent through the channel. A memoryless channel is completely characterized by the probabilities $p(y_i|x_i)$, $i = 1, \ldots, N$.

To send a binary code word of length N through a discrete, memoryless channel, the latter is used N times, a single bit being transmitted on each use. The process of

---

[5] A channel is not always discrete. An example of a channel which is not discrete is a *Gaussian channel*. This channel has a *continuous* alphabet. It consists of the set of all real numbers. This is a time-discrete channel with output $Y_i = X_i + Z_i$ at time i, where X_i is the input and Z_i is the noise. The Z_is are assumed to be IID random variables which have a Gaussian distribution of mean 0 and a given variance N (say).

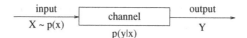

Fig. 1 Classical channel

information transmission through such a channel can be schematically represented as in Fig. 1.

Here X is the random variable at the input of the channel and Y is the corresponding induced random variable at the output of the channel. The channel is characterized by the conditional probabilities $p(y|x)$ where x, y are the values taken by X and Y, respectively.

Transmission of information through a communications channel involves a sender and a receiver. In quantum information theory the sender is popularly called Alice and the receiver is called Bob. We shall adopt these personalities in the classical case too. Information transmission involves the following steps. Suppose Alice wants to send a message $M \in \mathcal{M}$ to Bob. Let the number of messages be given by

$$|\mathcal{M}| = 2^{NR}. \tag{16}$$

She encodes her message M using the map

$$\mathcal{C}^N : \mathcal{M} \to J_X^N, \tag{17}$$

where $J_X = \{0, 1\}$ in general. This encoding assigns to each of Alice's possible messages an input sequence, or *code word* $\underline{x}^{(N)} = (x_1, \dots, x_N)$, $x_i \in J_X \, \forall \, i = 1, \dots, N$. Each code word is sent through N uses of the channel to Bob, who decodes the channel output using a decoding map

$$\mathcal{D}^N : J_Y^N \to \mathcal{M}, \tag{18}$$

which assigns a message in the set $\mathcal{M}$ to each possible output of N uses of the channel. For a given pair of encoding and decoding maps, the probability of error is defined to be the maximum probability, over all messages $M \in \mathcal{M}$, that the decoded output of the channel is not equal to the message M:

$$p\left(\mathcal{C}^N, \mathcal{D}^N\right) := \max_{M \in \mathcal{M}} \text{Prob}\left(\mathcal{D}^N(Y) \neq M | X = \mathcal{C}^N(M)\right). \tag{19}$$

The rate R of the encoding–decoding scheme, or equivalently, of the corresponding code, is given by the ratio of the size of the message to the size of the corresponding code word (sizes being measured in bits):

$$R = \frac{\log |\mathcal{M}|}{N}. \tag{20}$$

It is hence equal to the number of bits of message transmitted per use of the channel. A rate R is said to be *achievable* if there exists a sequence of encoding–decoding pairs $(\mathcal{C}^N, \mathcal{D}^N)$ such that $p(\mathcal{C}^N, \mathcal{D}^N) \to 0$ as $N \to \infty$.

It is natural to ask whether for any arbitrary channel it is always possible to find a code which ensures that arbitrarily reliable transmission of information through it can be achieved asymptotically, i.e., in the limit $N \to \infty$, where N denotes the length of the code words sent through the channel.

Shannon proved that this is indeed possible[6] as long as there is some correlation between the input and the output of the channel. *Shannon's noisy channel coding theorem* tells us that the optimal rate of a noisy channel, i.e., the maximum amount of information (measured in bits) that can be transmitted reliably, per use of the channel, is given by a quantity called the *channel capacity*. It also gives a prescription for the calculation of capacities of memoryless channels.

Theorem 1 (Shannon's noisy channel coding theorem) *The channel capacity of a discrete memoryless channel is given by*

$$C = \max_{p(x)} H(X : Y). \tag{21}$$

In the previous expression $H(X : Y)$ is the mutual information (see chapter "Classical Information Theory", Sect. 1.2) of the random variables X and Y:

$$H(X : Y) = H(X) + H(Y) - H(X, Y), \tag{22}$$

while $H(X)$, $H(Y)$ are the Shannon entropies of the random variables X, Y, and $H(X, Y)$ is the joint entropy[7] of X and Y. The maximum in (21) is taken over all possible input distributions $p(x)$. Using properties of $H(X)$, $H(Y)$, and $H(X, Y)$ (see e.g. [1]), it is easy to prove that (i) $C \geq 0$; (ii) $C \leq \log |J_X|$, and (iii) $C \leq \log |J_Y|$.

Exercise 2 *Suppose there are two independent, discrete, memoryless channels, with capacities C_1 and C_2 bits/s, respectively. Consider the compound channel formed by using these two channels in parallel in the following sense: at every second a symbol is transmitted through channel 1 (from its input alphabet) and a symbol through channel 2 (from its input alphabet); each channel thus emits one symbol each second. Prove that the capacity C of the compound channel is given by $C = C_1 + C_2$.*

---

[6] More precisely, he showed that this is possible for *discrete memoryless* channels.

[7] If X and Y have a joint probability distribution $\{p(x, y)\}$ then

$$H(X, Y) := -\sum_{x,y} p(x, y) \log p(x, y). \tag{23}$$

4 Quantum Entropy

Shannon entropy plays a pivotal role in classical information theory. It quantifies the minimal physical resources needed to store information emitted by a classical information source. It provides a limit to which data can be compressed reliably. In quantum information theory the quantity analogous to the Shannon entropy is called the von Neumann entropy. It has been introduced in chapter "Hilbert Space Methods for Quantum Mechanics", Sect. 3.1 and further elaborated upon in chapter "Bipartite Quantum Entanglement", Sect. 2.2; in this section we will study other of its properties which are useful for the analysis of quantum informational tasks.

In a later section we shall see that the von Neumann entropy quantifies the incompressible information content (or data compression limit) for a memoryless (or i.i.d.) quantum source (to be defined below) – just as the Shannon entropy quantifies the information content of a classical i.i.d. information source.

In analogy with the classical entropies (see chapter "Classical Information Theory", Sect. 1.2), we define quantum joint and conditional entropies for composite quantum systems, and mutual information of two subsystems.

The *quantum joint entropy* $S(A, B)$ of a composite system with two components A and B is defined as

$$S(A, B) = -\mathrm{Tr}\,(\rho_{AB} \log \rho_{AB}), \qquad (24)$$

where ρ_{AB} is the density matrix of the composite system AB.

The *quantum conditional entropy* $S(A|B)$ is defined as

$$S(A|B) = S(A, B) - S(B). \qquad (25)$$

The *quantum mutual information* of two subsystems A and B of a composite system AB is defined as

$$
\begin{aligned}
S(A : B) &= S(A) + S(B) - S(A, B), \\
&= S(A) - S(A|B) = S(B) - S(B|A). \qquad (26)
\end{aligned}
$$

For classical random variables X and Y, $H(X) \leq H(X, Y)$. This is intuitively obvious since the uncertainty of the random variable X cannot be more than the uncertainty of the pair of random variables X and Y. However, this intuition is not valid in the quantum case. For a bipartite quantum system AB, $S(A)$ can indeed exceed $S(A, B)$ as seen in chapter "Bipartite Quantum Entanglement", Sect. 2.1, in the case of a maximally entangled state of two d-dimensional systems: $S(A, B) = 0$ since the joint state is a pure state; however, $\rho_A = \mathrm{Tr}\,_B \rho_{AB} = \mathbb{1}_A/d$, and hence $S(A) = \log d$. In this case, therefore the conditional entropy $S(B|A) = S(A, B) - S(A)$ is *negative*. Notice that Corollary 2 in chapter "Bipartite Quantum Entanglement", Sect. 2.1, expresses a direct relationship between entanglement and negative conditional entropy: *A pure state* $|\Psi_{AB}\rangle$ *of a bipartite system* AB *is entangled if and only if* $S(B|A) < 0$.

A number of important consequences follow from the fact that the von Neumann entropy is strongly subadditive [7], namely that, for any state ρ_{ABC} of a tripartite system,

$$S(\rho_{ABC}) + S(\rho_B) \le S(\rho_{AB} + S(\rho_{BC}). \tag{27}$$

1. *Conditioning reduces entropy* When you condition on two systems B and C, the entropy is less than when you condition on system B alone.

$$S(A|BC) \le S(A|B). \tag{28}$$

Proof

$$\begin{aligned}
S(A|BC) &= S(A, B, C) - S(B, C) \\
&\le S(A, B) + S(B, C) - S(B) - S(B, C) \quad \text{by (27),} \\
&= S(A|B). \tag{29}
\end{aligned}$$

In the above, $S(A, B, C)$ denotes the von Neumann entropy of the composite system ABC.   □

2. *Discarding quantum systems never increases mutual information*

$$S(A : B) \le S(A : BC). \tag{30}$$

Proof

$$\begin{aligned}
S(A : BC) &= S(A) + S(B, C) - S(A, B, C) \\
&\ge S(A) + S(B, C) - S(A, B) - S(B, C) + S(B) \quad \text{by (27),} \\
&= S(A : B). \tag{31}
\end{aligned}$$

□

3. *Quantum operations never increase mutual information* Let AB denote a composite system. Let Φ denote a CPT map acting on the state of the subsystem B alone. Let $A'B'$ denote the composite system after this action. Then

$$S(A' : B') \le S(A : B). \tag{32}$$

Proof As mentioned in Sect. 2, the action of Φ on the state of the subsystem B can be considered to result from the action of a unitary operator U_{BC} on the system B and an ancilla C [16]. The ancilla is assumed to be initially in a pure state $|0_C\rangle$. Let C' denote the ancilla after the action U_{BC}. Hence,

$$\rho_{A'B'C'} = (\mathbb{1}_A \otimes U_{BC})(\rho_{AB} \otimes |0_C\rangle\langle 0_C|)(\mathbb{1}_A \otimes U_{BC}^\dagger), \tag{33}$$

where $\mathbb{1}_A$ denotes the identity operator acting on the Hilbert space $\mathcal{H}_A$ of the system A. Taking the partial trace over the Hilbert space of the ancilla yields

$$\rho_{A'B'} = (\text{id}_A \otimes \Phi_B)(\rho_{AB}) \equiv \text{Tr}_{C'}(\rho_{A'B'C'}), \tag{34}$$

where id $_A$ denotes the identity map on $B(\mathcal{H}_A)$. Note that

$$S(A : B) = S(A : BC), \tag{35}$$

since C is initially uncorrelated with A and B. It is easy to verify that

$$S(A : BC) = S(A' : B'C'). \tag{36}$$

Moreover, it follows from (30) that

$$S(A' : B'C') \geq S(A' : B'). \tag{37}$$

The desired inequality (32) follows from (35), (36), and (37). $\qquad\qquad\square$

5 Data Compression in Quantum Information Theory

A *quantum information source* is defined by a set of pure states $|\Psi_k\rangle$, acting on a given Hilbert space $\mathcal{H}$, and a set of corresponding probabilities $\{p_k\}$. Here $|\Psi_k\rangle$ is the pure state of a quantum mechanical system and p_k is the probability that the system is in the pure state $|\Psi_k\rangle$. From the information theoretic point of view, we interpret the $|\Psi_k\rangle$ as signals (or output) of the source, and the p_k's as the probabilities with which the signals are produced. Hence, we can equivalently characterize a quantum information source by $\{\rho, \mathcal{H}\}$ where ρ is a density matrix $\rho = \sum_k p_k|\Psi_k\rangle\langle\Psi_k|$, and $\mathcal{H}$ the Hilbert space on which it acts. Note that the pure states $|\Psi_k\rangle$ need not be mutually orthogonal.

To study data compression, we consider a sequence of density matrices ρ_n with $n \to \infty$ acting on Hilbert spaces denoted by $\mathcal{H}_n$, of increasing dimensions N_n, given by

$$\rho^{(n)} = \sum_k p_k^{(n)} |\Psi_k^{(n)}\rangle\langle\Psi_k^{(n)}|, \tag{38}$$

with $p_k^{(n)} \geq 0$ and $\sum_k p_k^{(n)} = 1$. The state vectors $|\Psi_k^{(n)}\rangle$ need not be mutually orthogonal. We interpret the $|\Psi_k^{(n)}\rangle$ as signal states and $p_k^{(n)}$ as their probabilities of occurrence, with $\dim \mathcal{H}_n = N_n$, and N_n increasing with n. The optimal rate of data compression, in the limit $n \to \infty$, is defined through Eqs. (41) and (42).

To compress data from such a source one encodes each signal state $|\Psi_k^{(n)}\rangle$ by a state $\widetilde{\rho}_k^{(n)} \in B(\widetilde{\mathcal{H}}_n)$, where $\dim\widetilde{\mathcal{H}}_n = d_c(n) < N_n$. Thus, a compression scheme is a map

$$C^{(n)} : |\Psi_k^{(n)}\rangle\langle\Psi_k^{(n)}| \mapsto \widetilde{\rho}_k^{(n)} \in B(\widetilde{\mathcal{H}}_n). \tag{39}$$

A corresponding decompression scheme is a map:

$$\mathcal{D}^{(n)} : B(\widetilde{\mathcal{H}}_n) \mapsto B(\mathcal{H}_n). \tag{40}$$

Both $\mathcal{C}^{(n)}$ and $\mathcal{D}^{(n)}$ must be CPT maps.

In classical information theory, data compression corresponds to a reduction in the number of bits required to store information emitted by a classical information source. In the quantum case, the idea of data compression is analogous, with bits being replaced by qubits, and a classical information source being replaced by a quantum information source. In other words, the quantity that one compresses in the quantum case is the *dimension of the Hilbert space* $\widetilde{\mathcal{H}}_n$. The goal is, therefore, to make the dimension $d_c(n)$ as small as possible (subject to the condition that the information carried in the signal states can be retrieved with high accuracy upon decompression).

The *rate of compression* is defined as

$$R_n := \frac{\log(\dim\widetilde{\mathcal{H}}_n)}{\log(\dim\mathcal{H}_n)} = \frac{\log d_c(n)}{\log N_n}. \tag{41}$$

It is natural to consider the original Hilbert space, $\mathcal{H}_n$, to be the Hilbert space of n qubits (the n-qubit space). In this case $N_n = 2^n$ and hence $\log N_n = n$. As in the case of classical data compression, we are interested in finding the optimal *limiting rate of data compression*, which in this case is given by

$$R_\infty := \lim_{n\to\infty} R_n \equiv \lim_{n\to\infty} \frac{\log d_c(n)}{n}. \tag{42}$$

Unlike classical signals, quantum signal states are not completely distinguishable. This is because they are, in general, not mutually orthogonal. As a result, perfectly reconstructing a quantum signal state from its compressed version is often an impossible task and therefore too stringent a requirement for the reliability of a compression–decompression scheme. Instead, a reasonable requirement is that a state, which is nearly indistinguishable from the original signal state, can be reconstructed from the compressed state $\widetilde{\rho}_k^{(n)}$. A measure of indistinguishability useful for this purpose is the *ensemble average fidelity* defined as follows (see chapter "Hilbert Space Methods for Quantum Mechanics", Sect. 3.3).

$$F_n := \sum_k p_k^{(n)} \langle \Psi_k^{(n)} | \mathcal{D}^{(n)}(\widetilde{\rho}_k^{(n)}) | \Psi_k^{(n)} \rangle. \tag{43}$$

This fidelity satisfies $0 \leq F_n \leq 1$ and $F_n = 1$ if and only if $\mathcal{D}^{(n)}\left(\widetilde{\rho}_k^{(n)}\right) = |\Psi_k^{(n)}\rangle\langle\Psi_k^{(n)}|$ for all k. A compression–decompression scheme is said to be *reliable* if $F_n \to 1$ as $n \to \infty$.

The key idea behind data compression is the fact that some signal states have a higher probability of occurrence than others (these states playing a role analogous to the typical sequences of classical information theory). These signal states span a subspace, of the original Hilbert Space of the source, which is referred to as the *typical subspace*.

5.1 Schumacher's Theorem for Memoryless Quantum Sources

The notion of a typical subspace was introduced by Ohya and Petz [9]. It was first used in the context of quantum information theory by Schumacher, in his seminal paper [11]. He considered the simplest class of quantum information sources, namely quantum *memoryless* or IID sources. For such a source the density matrix $\rho^{(n)}$, defined through (38), acts on a tensor product Hilbert space $\mathcal{H}_n = \mathcal{H}^{\otimes n}$ and is itself given by a tensor product

$$\rho^{(n)} = \pi^{\otimes n}. \tag{44}$$

Here $\mathcal{H}$ is a fixed Hilbert space (representing an elementary quantum subsystem) and π is a density matrix acting on $\mathcal{H}$; e.g., $\mathcal{H}$ can be a single qubit space, in which case dim $\mathcal{H} = 2$, $\mathcal{H}_n$ is the n-qubit space and π is the density matrix of a single qubit. If the spectral decomposition of π is given by

$$\pi = \sum_{i=1}^{d} q_i |\phi_i\rangle\langle\phi_i|, \tag{45}$$

where $d = $ dim $\mathcal{H}$, then the eigenvalues and eigenvectors of $\rho^{(n)}$ are, respectively, given by

$$\lambda_{\underline{j}}^{(n)} = q_{j_1} q_{j_2} \ldots q_{j_n}, \tag{46}$$

and

$$|\psi_{\underline{j}}^{(n)}\rangle = |\phi_{j_1}\rangle \otimes |\phi_{j_2}\rangle \otimes \cdots \otimes |\phi_{j_n}\rangle. \tag{47}$$

Thus we can write the spectral decomposition of the density matrix $\rho^{(n)}$ of an i.i.d. source as

$$\rho^{(n)} = \sum_{\underline{j}} \lambda_{\underline{j}}^{(n)} |\psi_{\underline{j}}^{(n)}\rangle\langle\psi_{\underline{j}}^{(n)}|, \tag{48}$$

where the sum is over all possible sequences $\underline{j} := (j_1 \ldots j_n)$, with each j_i taking d values. Hence we see that the eigenvalues $\rho^{(n)}$ are labeled by a classical sequence of indices $\underline{j} = j_1 \ldots j_n$.

The von Neumann entropy of such a source is given by

$$S(\rho^{(n)}) \equiv S(\pi^{\otimes n}) = nS(\pi) = nH(\{q_i\}). \tag{49}$$

Let $T_{\varepsilon}^{(n)}$ be the classical typical subset of indices $(j_1 \ldots j_n)$ for which

$$\left| -\frac{1}{n} \log(q_{j_1} \ldots q_{j_n}) - S(\pi) \right| \le \varepsilon, \tag{50}$$

as in the theorem of typical sequences (see chapter "Classical Information Theory", Sect. 2.2). Defining $\mathcal{T}_{\varepsilon}^{(n)}$ as the space spanned by the eigenvectors $|\psi_{\underline{j}}^{(n)}\rangle$ with $\underline{j} \in T_{\varepsilon}^{(n)}$ then yields the quantum analogue of the theorem of typical sequences, the typical subspace theorem (given below). We refer to $\mathcal{T}_{\varepsilon}^{(n)}$ as the *typical subspace* (or more precisely, the ε-typical subspace).

Theorem 2 (Typical subspace theorem) *Fix $\varepsilon > 0$. Then for any $\delta > 0 \; \exists \; n_0(\delta) > 0$ such that $\forall n \ge n_0(\delta)$ and $\rho^{(n)} = \pi^{\otimes n}$, the following are true:*
(a) $\mathrm{Tr}\left(P_{\varepsilon}^{(n)} \rho^{(n)}\right) > 1 - \delta$ *and   (b)* $(1-\delta) \, 2^{n(S(\pi)-\varepsilon)} \le \dim \mathcal{T}_{\varepsilon}^{(n)} \le 2^{n(S(\pi)+\varepsilon)}$,
where $P_{\varepsilon}^{(n)}$ is the orthogonal projection onto the subspace $\mathcal{T}_{\varepsilon}^{(n)}$.

Note that $\mathrm{Tr}\,(P_{\varepsilon}^{(n)} \rho^{(n)})$ gives the probability of the typical subspace $\mathcal{T}_{\varepsilon}^{(n)}$.

As $\mathrm{Tr}\,(P_{\varepsilon}^{(n)} \rho^{(n)})$ approaches unity for n sufficiently large, $\mathcal{T}_{\varepsilon}^{(n)}$ carries almost all the weight of $\rho^{(n)}$. Let $\mathcal{T}_{\varepsilon}^{(n)\perp}$ denote the orthocomplement of the typical subspace, i.e., for any pair of vectors $|\psi\rangle \in \mathcal{T}_{\varepsilon}^{(n)}$ and $|\phi\rangle \in \mathcal{T}_{\varepsilon}^{(n)\perp}$, $\langle\phi|\psi\rangle = 0$. It follows from the above theorem that the probability of a signal state belonging to $\mathcal{T}_{\varepsilon}^{(n)\perp}$ can be made arbitrarily small for n sufficiently large.

Let $P_{\varepsilon}^{(n)}$ denote the orthogonal projection onto the typical subspace $\mathcal{T}_{\varepsilon}^{(n)}$. The encoding (compression) of the signal states $|\Psi_k^{(n)}\rangle$ of (38) is done in the following manner. $\mathcal{C}^{(n)} : |\Psi_k^{(n)}\rangle\langle\Psi_k^{(n)}| \mapsto \widetilde{\rho}_k^{(n)}$ where

$$\widetilde{\rho}_k^{(n)} := \alpha_k^2 |\widetilde{\Psi}_k^{(n)}\rangle\langle\widetilde{\Psi}_k^{(n)}| + \beta_k^2 |\Phi_0\rangle\langle\Phi_0|. \tag{51}$$

Here

$$|\widetilde{\Psi}_k^{(n)}\rangle := \frac{P_{\varepsilon}^{(n)}|\Psi_k^{(n)}\rangle}{||P_{\varepsilon}^{(n)}|\Psi_k^{(n)}\rangle||} \; ; \; \alpha_k := ||P_{\varepsilon}^{(n)}|\Psi_k^{(n)}\rangle|| \; ; \; \beta_k = ||(\mathbb{1} - P_{\varepsilon}^{(n)})|\Psi_k^{(n)}\rangle||, \tag{52}$$

and $|\Phi_0\rangle$ is any fixed state in $\mathcal{T}_{\varepsilon}^{(n)}$.

Obviously $\widetilde{\rho}_k^{(n)} \in B(\mathcal{T}_{\varepsilon}^{(n)})$, and hence the typical subspace $\mathcal{T}_{\varepsilon}^{(n)}$ plays the role of the compressed space. The decompression $\mathcal{D}^{(n)}(\widetilde{\rho}_k^{(n)})$ is defined as the extension of $\widetilde{\rho}_k^{(n)}$ on $\mathcal{T}_{\varepsilon}^{(n)}$ to $\mathcal{H}_n$:

$$\mathcal{D}^{(n)}\left(\widetilde{\rho}_{\underline{k}}^{(n)}\right) = \widetilde{\rho}_{\underline{k}}^{(n)} \oplus 0. \tag{53}$$

The fidelity of this compression–decompression scheme satisfies

$$
\begin{aligned}
F_n &= \sum_k p_k^{(n)} \langle \Psi_k^{(n)} | \tilde{\rho}_k^{(n)} | \Psi_k^{(n)} \rangle \\
&= \sum_k p_k^{(n)} \left[\alpha_k^2 |\langle \Psi_k^{(n)} | \tilde{\Psi}_k^{(n)} \rangle|^2 + \beta_k^2 |\langle \Psi_k^{(n)} | \Phi_0 \rangle|^2 \right] \\
&\geq \sum_k p_k^{(n)} \alpha_k^2 |\langle \Psi_k^{(n)} | \tilde{\Psi}_k^{(n)} \rangle|^2 = \sum_k p_k^{(n)} \alpha_k^4 \\
&\geq \sum_k p_k^{(n)} (2\alpha_k^2 - 1) = 2 \sum_k p_k^{(n)} \alpha_k^2 - 1.
\end{aligned}
\tag{54}
$$

Using the typical subspace theorem, Schumacher [11] proved the following analogue of Shannon's noiseless channel coding theorem for memoryless quantum information sources:

Theorem 3 (Schumacher's quantum coding theorem) *Let* $\{\rho_n, \mathcal{H}_n\}$ *be an i.i.d. quantum source:* $\rho_n = \pi^{\otimes n}$ *and* $\mathcal{H}_n = \mathcal{H}^{\otimes n}$. *If* $R > S(\pi)$ *then there exists a reliable compression scheme of rate* R. *If* $R < S(\pi)$ *then any compression scheme of rate* R *is unreliable.*

Proof (i) $R > S(\pi)$: Choose $\varepsilon > 0$ such that $R \geq S(\pi) + \varepsilon$.

For a given $\delta > 0$, choose the typical subspace as above and choose n large enough so that (a) and (b) of the Typical Subspace Theorem hold. We have that

$$
\dim \mathcal{T}_\varepsilon^{(n)} \leq 2^{n(S(\pi)+\varepsilon)} \leq 2^{nR}.
\tag{55}
$$

Hence, $\mathcal{T}_\varepsilon^{(n)}$ is a subspace of the compressed space $\tilde{\mathcal{H}}_n$. In this case one can show that the compression–decompression scheme discussed previously is indeed reliable. To see this, note that

$$
\alpha_k^2 = \langle \Psi_k^{(n)} | P_\varepsilon^{(n)} | \Psi_k^{(n)} \rangle,
\tag{56}
$$

and hence

$$
\text{right-hand side of (54)} = 2 \sum_k p_k^{(n)} \langle \Psi_k^{(n)} | P_\varepsilon^{(n)} | \Psi_k^{(n)} \rangle - 1
$$

$$
= 2 \text{Tr} \left(P_\varepsilon^{(n)} \rho^{(n)} \right) - 1.
\tag{57}
$$

From the statement (a) of the typical subspace theorem, it follows that the right-hand side of (54) $> 1 - 2\delta$. However, δ can be made arbitrarily small for sufficiently large n, and this implies that there exists a reliable compression scheme of rate R whenever $R > S(\pi)$.

(ii) Suppose $R < S(\pi)$. Choose $\varepsilon > 0$ such that $R < S(\pi) - \varepsilon$. Let the compression map be $C^{(n)}$. We may assume that the compressed space $\tilde{\mathcal{H}}_n$ is a subspace of $\mathcal{H}_n$ with $\dim \widetilde{\mathcal{H}}_n = 2^{nR}$. We denote the projection onto $\widetilde{\mathcal{H}}_n$ as $\widetilde{P}_n$ and let $\tilde{\rho}_k^{(n)} = \mathcal{C}^{(n)}\left(|\Psi_k^{(n)}\rangle \langle \Psi_k^{(n)}|\right)$.

$$
\begin{aligned}
F_n &= \sum_k p_k^{(n)} \langle \Psi_k^{(n)} | \mathcal{D}^{(n)}(\tilde{\rho}_k^{(n)}) | \Psi_k^{(n)}\rangle, \\
&= \sum_k p_k^{(n)} \langle \Psi_k^{(n)} | P_\varepsilon^{(n)} \mathcal{D}^{(n)}(\tilde{\rho}_k^{(n)}) P_\varepsilon^{(n)} | \Psi_k^{(n)}\rangle \\
&\quad + \sum_k p_k^{(n)} \langle \Psi_k^{(n)} | (\mathbb{1} - P_\varepsilon^{(n)}) \mathcal{D}^{(n)}(\tilde{\rho}_k^{(n)})(\mathbb{1} - P_\varepsilon^{(n)}) | \Psi_k^{(n)}\rangle \\
&\quad + \sum_k p_k^{(n)} \langle \Psi_k^{(n)} | (\mathbb{1} - P_\varepsilon^{(n)}) \mathcal{D}^{(n)}(\tilde{\rho}_k^{(n)}) P_\varepsilon^{(n)} | \Psi_k^{(n)}\rangle \\
&\quad + \sum_k p_k^{(n)} \langle \Psi_k^{(n)} | P_\varepsilon^{(n)} \mathcal{D}^{(n)}(\tilde{\rho}_k^{(n)})(\mathbb{1} - P_\varepsilon^{(n)}) | \Psi_k^{(n)}\rangle, \\
&:= (I) + (II) + (III) + (IV),
\end{aligned}
\tag{58}
$$

where $P_\varepsilon^{(n)}$ is the orthogonal projection onto the typical subspace $T_\varepsilon^{(n)}$ of $\rho^{(n)}$.

Since $\tilde{\rho}_k^{(n)}$ is concentrated on $\widetilde{\mathcal{H}}_n$, we have $\tilde{\rho}_k^{(n)} \le \widetilde{P}_n$ and hence $\mathcal{D}^{(n)}(\tilde{\rho}_k^{(n)}) \le \mathcal{D}^{(n)}(\widetilde{P}_n)$, for any decompression map $\mathcal{D}^{(n)}$. Inserting this into the first term on the right-hand side of (58) we get

$$
\begin{aligned}
(I) &\le \sum_k p_k^{(n)} \langle \Psi_k^{(n)} | P_\varepsilon^{(n)} \mathcal{D}^{(n)}(\widetilde{P}_{(n)}) P_\varepsilon^{(n)} | \Psi_k^{(n)}\rangle, \\
&= \mathrm{Tr}\,(\rho^{(n)} P_\varepsilon^{(n)} \mathcal{D}^{(n)}(\widetilde{P}_{(n)}) P_\varepsilon^{(n)}), \\
&= \sum_{\underline{j} \in T_\varepsilon^{(n)}} \lambda_{\underline{j}}^{(n)} \langle \psi_{\underline{j}}^{(n)} | \mathcal{D}^{(n)}(\widetilde{P}_n) | \psi_{\underline{j}}^{(n)}\rangle.
\end{aligned}
\tag{59}
$$

In the above sum over $\underline{j} \in T_\varepsilon^{(n)}$ we have $\lambda_{\underline{j}}^{(n)} \le 2^{-n(S(\pi)-\varepsilon)}$. It can therefore be bounded as follows:

$$
\begin{aligned}
\sum_{\underline{j} \in T_\varepsilon^{(n)}} \lambda_{\underline{j}}^{(n)} \langle \psi_{\underline{j}}^{(n)} | \mathcal{D}^{(n)}(\widetilde{P}_n) | \psi_{\underline{j}}^{(n)}\rangle &\le 2^{-n(S(\pi)-\varepsilon)} \sum_{\underline{j}} \langle \psi_{\underline{j}}^{(n)} | \mathcal{D}^{(n)}(\widetilde{P}_n) | \psi_{\underline{j}}^{(n)}\rangle, \\
&= 2^{-n(S(\pi)-\varepsilon)} \mathrm{Tr}\left(\mathcal{D}^{(n)}(\widetilde{P}_n)\right), \\
&= 2^{-n(S(\pi)-\varepsilon)} 2^{nR},
\end{aligned}
\tag{60}
$$

since $\mathcal{D}^{(n)}$ is a CPT map and $\dim \widetilde{\mathcal{H}}_n = 2^{nR}$. Since $R < S(\pi) - \varepsilon$, it is clear that the right-hand side of (60) tends to 0 as $n \to \infty$.

In the second term (B) on the right-hand side of (58) we use the fact that $\mathcal{D}^{(n)}(\tilde{\rho}_k^{(n)}) \le \mathbb{1}$ (since $\mathcal{D}^{(n)}(\tilde{\rho}_k^{(n)})$ is a density matrix) to get

$$(II) \leq \sum_k p_k^{(n)} \langle \Psi_k^{(n)} | (\mathbb{1} - P_\varepsilon^{(n)}) | \Psi_k^{(n)} \rangle,$$

$$= \mathrm{Tr}\,(\rho^{(n)}(\mathbb{1} - P_\varepsilon^{(n)})) = \sum_{k \notin T_\varepsilon^{(n)}} \lambda_k^{(n)} \tag{61}$$

By the typical subspace theorem, the right-hand side of (61) tends to 0 as $n \to \infty$.

The third and fourth terms on the right-hand side of (58) are conjugates of each other. Using the Cauchy–Schwarz inequality, we obtain

$$(III) \leq \left(\sum_k p_k^{(n)} \langle \Psi_k^{(n)} | \mathcal{D}^{(n)}(\tilde{\rho}_k^{(n)})^{1/2} P_\varepsilon^{(n)} \mathcal{D}^{(n)}(\tilde{\rho}_k^{(n)})^{1/2} | \Psi_k^{(n)} \rangle \right)^{1/2}$$

$$\times \left(\sum_k p_k^{(n)} \langle \Psi_k^{(n)} | (\mathbb{1} - P_\varepsilon^{(n)}) \mathcal{D}^{(n)}(\tilde{\rho}_k^{(n)})(\mathbb{1} - P_\varepsilon^{(n)}) | \Psi_k^{(n)} \rangle \right)^{1/2}$$

$$\leq \left(\sum_k p_k^{(n)} \langle \Psi_k^{(n)} | \mathcal{D}^{(n)}(\tilde{\rho}_k^{(n)})^{1/2} \mathcal{D}^{(n)}(\tilde{\rho}_k^{(n)})^{1/2} | \Psi_k^{(n)} \rangle \right)^{1/2}$$

$$\times \left(\sum_k p_k^{(n)} \langle \Psi_k^{(n)} | (\mathbb{1} - P_\varepsilon^{(n)}) \mathcal{D}^{(n)}(\tilde{\rho}_k^{(n)})(\mathbb{1} - P_\varepsilon^{(n)}) | \Psi_k^{(n)} \rangle \right)^{1/2}$$

$$\leq \left(\sum_k p_k^{(n)} \langle \Psi_k^{(n)} | \Psi_k^{(n)} \rangle \right)^{1/2} \times \left(\sum_k p_k^{(n)} \langle \Psi_k^{(n)} | (\mathbb{1} - P_\varepsilon^{(n)}) | \Psi_k^{(n)} \rangle \right)^{1/2} \tag{62}$$

where we have used the inequality $\mathcal{D}^{(n)}(\tilde{\rho}_k^{(n)}) \leq \mathbb{1}$. Hence,

$$(III) \leq \left(\mathrm{Tr}\,(\mathbb{1} - P_\varepsilon^{(n)})\rho^{(n)} \right)^{1/2} = \left(\sum_{k \notin T_\varepsilon^{(n)}} \lambda_k^{(n)} \right)^{1/2}, \tag{63}$$

which tends to 0 as $n \to \infty$ by the Typical Subspace Theorem. Since the fourth term on the right-hand side of (58) is a conjugate of the third term, it also tends to zero. Hence we see that for $R < S(\pi)$ the fidelity F_n tends to 0 as $n \to \infty$. Therefore there is no reliable compression–decompression scheme in this case. $\qquad\square$

6 Quantum Channels

In classical information theory, Shannon's noisy channel coding theorem asserts that a classical communications channel has a well-defined information carrying capacity and it provides a formula for calculating it. However, Shannon's theorem is not applicable to communications channels which incorporate intrinsically quantum effects. In this section we shall discuss the transmission of information through noisy quantum channels.

As mentioned earlier, the quantum analogue of a classical stochastic communications channel is a linear, completely positive, trace-preserving (CPT) map. It is referred to as a *quantum channel* and we denote it as Φ. The channel acts on an input state ρ, yielding the output state $\Phi(\rho)$ (see Fig. 2). The noise in the channel distorts the message sent through it and hence $\Phi(\rho) \neq \rho$ in general. An example of a quantum communications channel is an optical fibre. The input to the channel is a photon in some quantum state. The latter suffers from the effects of noise in the optical fibre as it passes through it and consequently emerges from it in a transformed quantum state.

Fig. 2 A quantum channel

A quantum channel Φ is said to be *memoryless* if using the channel n times successively is described by the map

$$\Phi^{\otimes n} = \Phi \otimes \Phi \otimes \ldots \otimes \Phi \quad (n \text{ times}). \tag{64}$$

In this case, the action of each use of the channel is identical and it is independent for different uses. If Φ acts on density matrices in $B(\mathcal{H})$, then $\Phi^{\otimes n}$ acts on density matrices in $B(\mathcal{H}^{\otimes n})$. For example, consider $\mathcal{H}$ to be the single qubit space. Then the input to $\Phi^{\otimes n}$ is an n-qubit state (say $\rho^{(n)}$). In other words, the n qubits are sent through the channel one by one and the resultant output state is given by $\Phi^{\otimes n}(\rho^{(n)})$ (see Fig. 3).

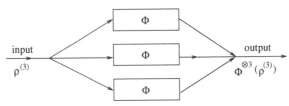

Fig. 3 Information transmission through a product channel $\Phi^{\otimes 3}$. Here $\rho^{(3)} \in B(\mathcal{H}^{\otimes 3})$

As in the classical case, noise present in a quantum channel distorts the information transmitted through it, thus introduce errors in the received message. In order to study particular examples of noisy quantum channels, we need a way to characterize these errors. In the following section it will be shown that these errors can be characterized by the actions of the four matrices: the 2×2 identity matrix σ_0, and the three Pauli matrices σ_x, σ_y, and σ_z.

$$\sigma_0 = \begin{pmatrix} 1 & 0 \\ 0 & 1 \end{pmatrix}, \quad \sigma_x = \begin{pmatrix} 0 & 1 \\ 1 & 0 \end{pmatrix}, \quad \sigma_y = \begin{pmatrix} 0 & -i \\ i & 0 \end{pmatrix}, \quad \sigma_z = \begin{pmatrix} 1 & 0 \\ 0 & -1 \end{pmatrix}. \tag{65}$$

The action of these operators on the basis vectors $|0\rangle$ and $|1\rangle$ of the single qubit space $\mathcal{H}$ are

$$\sigma_0|0\rangle = |0\rangle \, , \; \sigma_0|1\rangle = |1\rangle \, ; \quad \sigma_x|0\rangle = |1\rangle \, , \; \sigma_x|1\rangle = |0\rangle$$
$$\sigma_z|0\rangle = |0\rangle \, , \; \sigma_z|1\rangle = -|1\rangle \, ; \quad \sigma_y|0\rangle = \mathrm{i}\,|1\rangle \, , \; \sigma_y|1\rangle = -\mathrm{i}\,|0\rangle \, . \tag{66}$$

6.1 Characterization of Quantum Errors

Decoherence, which results in errors, arises from the interaction of a system with its environment. A simple characterization of errors can therefore be obtained by considering the interaction of a system with its environment. Consider a single qubit, which is initially in a pure state, and which interacts with its environment in an arbitrary manner. Without loss of generality we can assume that the initial state of the environment is a pure state (which we denote as $|0\rangle_E$). This is because if the environment is in a mixed state, we can always *purify* it by adding an extra fictitious system. The latter does not affect the dynamics of the original single qubit (which we refer to as the principal system).

The evolution of the qubit and its environment can be described by a unitary transformation U such that

$$U|0\rangle \otimes |0\rangle_E = |0\rangle \otimes |e_{00}\rangle_E + |1\rangle \otimes |e_{01}\rangle_E \, , \; U|1\rangle \otimes |0\rangle_E = |0\rangle \otimes |e_{10}\rangle_E + |1\rangle \otimes |e_{11}\rangle_E \, . \tag{67}$$

Here $|e_{ij}\rangle_E$, $i, j \in \{0, 1\}$ denote the four (corresponding) states of the environment. They *need not* be normalized or mutually orthogonal.

Under U, an arbitrary state $|\psi\rangle = a|0\rangle + b|1\rangle$ of the single qubit evolves into $U\left(\left(a|0\rangle + b|1\rangle\right) \otimes |0\rangle_E\right) = |\Psi\rangle$; explicitly,

$$
\begin{aligned}
|\Psi\rangle &= a\left(|0\rangle \otimes |e_{00}\rangle_E + |1\rangle \otimes |e_{01}\rangle_E\right) + b\left(|0\rangle \otimes |e_{10}\rangle_E + |1\rangle \otimes |e_{11}\rangle_E\right) \\
&= \left(a|0\rangle + b|1\rangle\right) \otimes \frac{|e_{00}\rangle_E + |e_{11}\rangle_E}{2} + \left(a|0\rangle - b|1\rangle\right) \otimes \frac{|e_{00}\rangle_E - |e_{11}\rangle_E}{2} \\
&\quad + \left(a|1\rangle + b|0\rangle\right) \otimes \frac{|e_{01}\rangle_E + |e_{10}\rangle_E}{2} + \left(a|1\rangle - b|0\rangle\right) \otimes \frac{|e_{01}\rangle_E - |e_{10}\rangle_E}{2} \, .
\end{aligned}
\tag{68}
$$

Now,

$$
\begin{aligned}
a\,|0\rangle + b\,|1\rangle &= \sigma_0\left(a\,|0\rangle + b\,|1\rangle\right) = \sigma_0\,|\psi\rangle \, , \\
a\,|0\rangle - b\,|1\rangle &= \sigma_z\left(a\,|0\rangle + b\,|1\rangle\right) = \sigma_z\,|\psi\rangle \, , \\
a\,|1\rangle + b\,|0\rangle &= \sigma_x\left(a\,|0\rangle + b\,|1\rangle\right) = \sigma_x\,|\psi\rangle \, , \\
a\,|1\rangle - b\,|0\rangle &= -\mathrm{i}\,\sigma_y\left(a\,|0\rangle + b\,|1\rangle\right) = -\mathrm{i}\,\sigma_y\,|\psi\rangle \, ,
\end{aligned}
\tag{69}
$$

where σ_0 is the 2×2 identity operator and σ_x, σ_y, and σ_z are the Pauli matrices. Hence,

$$U\left(|\psi\rangle \otimes |0\rangle_E\right) = \sigma_0 |\psi\rangle \otimes |e_0\rangle_E + \sigma_z |\psi\rangle \otimes |e_z\rangle_E$$
$$+ \sigma_x |\psi\rangle \otimes |e_x\rangle_E + \sigma_y |\psi\rangle \otimes |e_y\rangle_E , \qquad (70)$$

where the definition of the states $|e_i\rangle_E, i \in \{0, x, y, z\}$ follow from (68) and (69).

The effect of U on the qubit can, therefore, be expressed in terms of the unitary operators $\sigma_0, \sigma_x, \sigma_y$, and σ_z (acting on the single qubit space). Heuristically, we may interpret this by saying that one of four things might happen:
nothing (σ_0), a bit flip (σ_x), a phase flip (σ_z), or a combined flip ($\sigma_y = i\sigma_x\sigma_z$).

However, this classification should not be taken literally because unless the states $\{|e_0\rangle_E, |e_x\rangle_E, |e_y\rangle_E, |e_z\rangle_E\}$ are mutually orthogonal, there is no conceivable measurement that could perfectly distinguish between the four alternatives. This is because quantum mechanics allows only mutually orthogonal states to be distinguished perfectly.

Now consider the unitary transformation U of n qubits interacting with their environment. The effect of U on the n-qubit system can be expressed in terms of 4^n operators belonging to the set

$$\{\sigma_0, \sigma_x, \sigma_y, \sigma_z\}^{\otimes n}. \qquad (71)$$

In other words, each such operator can be expressed as a tensor product of n single-qubit operators, with each operator in the string chosen from the set $\{\sigma_0, \sigma_x, \sigma_y, \sigma_z\}$. Let $|\Psi\rangle$ be the initial state of the n-qubit system and assume that the environment is in the pure state $|0\rangle_E$. Then the action of an arbitrary unitary operator U on the composite system consisting of the n-qubit system and the environment can be expanded as

$$U|\Psi\rangle \otimes |0\rangle_E = \sum_a \mathbf{E}_a|\Psi\rangle \otimes |e_a\rangle_E, \qquad (72)$$

where the index a ranges over 4^n values. This is because each operator $\mathbf{E}_a$ is a tensor product of n operators, each of the latter being one of the four Pauli matrices $\sigma_0, \sigma_x, \sigma_y$, or σ_z.

The states $\{|e_a\rangle_E\}$ are the corresponding states of the environment – which are *not* assumed to be normalized or mutually orthogonal. An important feature of this expansion is that each $\mathbf{E}_a$ is a unitary operator.

In obtaining a characterization of quantum errors acting on messages carried by n qubits, we assume that errors are

- *locally independent,* i.e., errors on different qubits (or gates) are not correlated.
- *sequentially independent,* i.e., subsequent errors on the same qubit (or in the same gate) are not correlated.

These assumptions allow us to express an error operator acting on n qubits as a tensor product of error operators on individual qubits. Consequently, an equivalent characterization of error operators acting in the n-qubit Hilbert space $\mathcal{H}_n := \mathcal{H}^{\otimes n}$ is given by

$$\mathbf{E}_{\underline{a}} = \bigotimes_{1 \le j \le n} W_{\alpha_j}^{(j)}, \tag{73}$$

where $\underline{a} = (\alpha_1, \ldots, \alpha_n)$ (with $\alpha_j \in \{I, X, Y, Z\}$) being a *word* from a four-letter alphabet:

$$W_{\alpha_j} = \sigma_0, \ \sigma_x \ \sigma_y, \ \sigma_z,$$
$$\text{if} \quad \alpha_j = I, X, Y, Z \quad \text{respectively}, \tag{74}$$

and $W_{\alpha_j}^{(j)}$ is the corresponding 2×2 matrix in the jth position in the tensor product $\mathbf{E}_{\underline{a}}$. Hence, $\mathbf{E}_{\underline{a}}$ is a $2^n \times 2^n$ matrix. It is called a *Pauli operator*. From (73) it follows that on each qubit there is either no error or an error corresponding to σ_x, σ_y or σ_z (heuristically).

Let us now study the action of a Pauli operator $\mathbf{E}_{\underline{a}}$ on an n-qubit state. For this, it suffices to consider the action of $\mathbf{E}_{\underline{a}}$ on basis states $|\underline{x}\rangle$ of $\mathcal{H}_n$, where $\underline{x} \in \{0, 1\}^n$.

If $\underline{a}$ has non-trivial ($\ne I$) letters in positions $j_1, \ldots, j_r$ then $\mathbf{E}_{\underline{a}}$ can be alternatively expressed as the usual product of r $2^n \times 2^n$ matrices:

$$\mathbf{E}_{\underline{a}} = \mathbf{E}_{\alpha_{j_1}}^{(j_1)} \cdots \mathbf{E}_{\alpha_{j_r}}^{(j_r)}. \tag{75}$$

The matrices $\mathbf{E}_X^{(j)}$, $\mathbf{E}_Y^{(j)}$, and $\mathbf{E}_Z^{(j)}$ are defined through their action on the basis vectors $|\underline{x}\rangle$ in $\mathcal{H}_n$:

$$\mathbf{E}_X^{(j)} |\underline{x}\rangle = |\underline{x} + \underline{e}_j\rangle, \tag{76}$$

where $\underline{e}_j = (0, \ldots, 1_j, \ldots, 0)$ is a binary string of length n with a 1 occurring only in the jth position. (Here $\underline{x} + \underline{e}_j$ stands for mod2 addition.) Alternatively, we can write

$$\mathbf{E}_X^{(j)} |\underline{x}\rangle = |\underline{x}'\rangle, \quad \text{with} \ x_k' = x_k \ \text{for} \ k \ne j \ \text{and} \ x_j' = x_j + 1. \tag{77}$$

Note that all additions are modulo 2. Further,

$$\mathbf{E}_Y^{(j)} |\underline{x}\rangle = i(-1)^{x_j} |\underline{x} + \underline{e}_j\rangle,$$
$$\mathbf{E}_Z^{(j)} |\underline{x}\rangle = (-1)^{x_j} |\underline{x}\rangle. \tag{78}$$

Hence, $\mathbf{E}_X^{(j)}$, $\mathbf{E}_Y^{(j)}$, and $\mathbf{E}_Z^{(j)}$ act non-trivially only on the jth tensor factor of $|\underline{x}\rangle$, producing results which depend only on the values of x_j (and not on the values of the other x_k's). It is easy to see from the above definitions that $\mathbf{E}_{\underline{a}}$ is Hermitian and unitary: $\mathbf{E}_{\underline{a}}^{\dagger} = \mathbf{E}_{\underline{a}}^{-1} = \mathbf{E}_{\underline{a}}$. In a conventional language, $\alpha_j \in \{X, Y, Z\}$ denotes a "local" error at the jth position. Such an error affects the jth qubit only. Hence, the action of $\mathbf{E}_{\underline{a}}$ is determined by the "local" errors contained in the word $\underline{a} = (\alpha_1, \alpha_2, \ldots \alpha_n)$.

6.2 Examples of Single Qubit Channels

Let us consider two simple examples of single qubit channels, i.e., quantum channels acting on single qubits.

Bit flip channel: this channel flips the qubit sent through it with probability p and leaves it invariant with probability $(1 - p)$. If ρ is the input state to the channel then the output is

$$\Phi(\rho) = p\sigma_x \rho \sigma_x + (1 - p)\rho. \tag{79}$$

The corresponding Kraus operators are $A_1 = \sqrt{1 - p}\mathbb{1}$, $A_2 = \sqrt{p}\sigma_x$. This is the quantum analogue of the binary symmetric channel (see, e.g., [1]).

Depolarizing channel: this channel leaves the input qubit intact with probability $(1 - p)$, and it results in the occurrence of the following errors with probability $p/3$ each: bit flip (σ_x), phase flip (σ_z), and combined flip (σ_y).

$$\Phi\rho = (1 - p)\rho + \frac{p}{3}\left(\sigma_x \rho \sigma_x + \sigma_y \rho \sigma_y + \sigma_z \rho \sigma_z\right). \tag{80}$$

There are four Kraus operators:

$$A_1 = \sqrt{1 - p}\,\mathbb{1} \quad ; \quad A_2 = \sqrt{\frac{p}{3}}\,\sigma_x \quad ; \quad A_3 = \sqrt{\frac{p}{3}}\,\sigma_y \quad ; \quad A_4 = \sqrt{\frac{p}{3}}\,\sigma_z. \tag{81}$$

Alternatively, a depolarizing channel can be considered to leave a qubit unaffected with a certain probability, say $(1 - q)$, and to replace its state with the *completely mixed state* $\mathbb{1}/2$ with probability q. In other words the error completely randomizes the state with probability q:

$$\Phi(\rho) = (1 - q)\rho + q\frac{\mathbb{1}}{2}. \tag{82}$$

Note that $\rho + \sigma_x \rho \sigma_x + \sigma_y \rho \sigma_y + \sigma_z \rho \sigma_z = 2\mathbb{1}$ for any state ρ, hence

$$\Phi(\rho) = (1 - q)\rho + \frac{q}{4}\rho + \frac{q}{4}[\sigma_x \rho \sigma_x + \sigma_y \rho \sigma_y + \sigma_z \rho \sigma_z],$$

$$= (1 - \frac{3}{4}q)\rho + \frac{q}{4}\sum_{\alpha=x,y,z} \sigma_\alpha \rho \sigma_\alpha. \tag{83}$$

Comparing (83) with (80) we find that $q = (4/3)p$. The depolarizing channel can be generalized to quantum systems of dimension $d > 2$:

$$\Phi(\rho) = p\frac{\mathbb{1}}{d} + (1 - p)\rho, \tag{84}$$

since $\mathbb{1}/d$ represents the completely mixed state in a Hilbert space of d-dimensions.

Amplitude damping channel: this channel describes energy dissipation. It provides a simple model of the decay of the excited state of a 2-level atom due to spontaneous emission of a photon. If the system is in the excited state $|1\rangle$, then it has a probability p of decaying to its ground state $|0\rangle$, emitting a photon in the process. The Kraus operators of the channel are given by

$$A_1 = \begin{pmatrix} 1 & 0 \\ 0 & \sqrt{1-p} \end{pmatrix}, \quad A_2 = \begin{pmatrix} 0 & \sqrt{p} \\ 0 & 0 \end{pmatrix}. \tag{85}$$

Their actions on the states $|0\rangle$ and $|1\rangle$ of the atom are as follows:

$$\begin{aligned} A_1|0\rangle &= |0\rangle \; ; A_1|1\rangle = \sqrt{1-p}|1\rangle, \\ A_2|0\rangle &= 0 \; ; A_2|1\rangle = \sqrt{p}|0\rangle. \end{aligned} \tag{86}$$

The operator A_2 describes the decay of the atom from its excited state to its ground state, whereas the operator A_1 describes how the state evolves if there is no decay and spontaneous emission. Under the action of this channel, the density matrix ρ of the atom undergoes the following transformation:

$$\rho \equiv \begin{pmatrix} \rho_{00} & \rho_{01} \\ \rho_{10} & \rho_{11} \end{pmatrix} \longmapsto \begin{pmatrix} \rho_{00} + p\rho_{11} & \sqrt{1-p}\,\rho_{01} \\ \sqrt{1-p}\,\rho_{10} & (1-p)\rho_{11} \end{pmatrix}. \tag{87}$$

Consider applying the channel n times in succession. Then

$$\lim_{n\to\infty} \Phi(\rho) = \begin{pmatrix} \rho_{00} + \rho_{11} & 0 \\ 0 & 0 \end{pmatrix} = |0\rangle\langle 0|. \tag{88}$$

The final state of the atom is its ground state. This is intuitively obvious, since an atom in its excited state necessarily decays to its ground state. The amplitude damping channel is an example of a quantum channel which takes a *mixed initial state* $\rho = \sum_{i,j=0}^{1} \rho_{ij}|i\rangle\langle j|$ to a *pure final state* $(\rho_{00} + \rho_{11})|0\rangle\langle 0|$ (asymptotically).

7 Accessible Information and the Holevo Bound

7.1 Introduction

Suppose Alice has a classical information source which emits symbols $x \in J = \{1, 2, \ldots, M\}$ with corresponding probabilities $p(x)$. The source can be characterized by a classical random variable X with probability mass function $p(x) = P(X = x)$, $x \in J$. Alice wishes to communicate the symbols emitted by the source to Bob. To do this, she encodes the symbol x into a suitable quantum state ρ_x (i.e., in general, a mixed state density matrix) of some physical system. She then sends this state to Bob through a noiseless quantum channel. Bob does a measurement (POVM) on the state ρ_x and tries to guess the symbol x. Let Y denote the classical random variable corresponding to the outcome of Bob's POVM.

(**Q**) How much information can Bob gain about X through the measurement (i.e., from Y) ?

This is given by the *mutual information* $H(X : Y)$ (defined by (22)) of the random variables X and Y, since $H(X : Y)$ is a measure of how much information X and Y have in common.

To get the maximum information about X, Bob would need to choose a measurement which maximizes $H(X : Y)$. The maximum information amount of information about X that Bob can gain through any possible measurement is referred to as his *accessible information*. It can be viewed as the amount of classical information that can be stored and recovered from a quantum system. It is denoted by the symbol I_{acc} and defined as follows:

$$I_{\mathrm{acc}} = \max \ H(X : Y), \tag{89}$$

where the maximum is taken over all possible measurement schemes.

In the following section we obtain an upper bound to I_{acc} by first considering a simple example.

7.2 An Example

Suppose a classical source emits each of the signals $x \in \{1, 2, 3\}$ with probability $1/3$. For each x, Alice does the encoding $x \mapsto \rho_x$, with $\rho_x = |\psi_x\rangle\langle\psi_x|$, where

$$
\begin{aligned}
|\psi_1\rangle &= |0\rangle = (1, \ 0)^T, \\
|\psi_2\rangle &= -\frac{1}{2}|0\rangle + \frac{\sqrt{3}}{2}|1\rangle = \left(-\frac{1}{2}, \ \frac{\sqrt{3}}{2}\right)^T, \\
|\psi_3\rangle &= -\frac{1}{2}|0\rangle - \frac{\sqrt{3}}{2}|1\rangle = \left(-\frac{1}{2}, \ -\frac{\sqrt{3}}{2}\right)^T.
\end{aligned}
\tag{90}
$$

Note that the above signal states are not mutually orthogonal:

$$\langle \psi_x | \psi_y \rangle = -\frac{1}{2}\delta_{xy}, \quad x, y = 1, 2, 3 . \tag{91}$$

Note that the density matrix ρ corresponding to the ensemble of states $\{p_x, \rho_x\}_{x=1,2,3}$, where $p_x = 1/3$ for each $x \in \{1, 2, 3\}$, is given by the completely mixed state :

$$\rho := \sum_{x=1}^{3} p_x \, \rho_x = \frac{1}{3}\sum_{x=1}^{3} \rho_x = \mathbb{1}/2. \tag{92}$$

Alice sends her signal states to Bob through a noiseless quantum channel. Using the symmetry in Alice's signal states, it can be proved that the optimal measurement that Bob can do is characterized by POVM elements:

$$E_x = \frac{2}{3}(\mathbb{1} - |\psi_x\rangle\langle\psi_x|), \quad x = 1, 2, 3 . \tag{93}$$

If Bob receives the pure state $\rho_x = |\psi_x\rangle\langle\psi_x|$, then the probability that his measurement yields the outcome x is given by

$$P(Y = x | X = x) = \mathrm{Tr}\,(E_x \rho_x) = 0 . \tag{94}$$

The probability of an outcome $y \in \{1, 2, 3\}$, where $y \neq x$, is given by

$$\begin{aligned}
p(y|x) = P(Y = y | X = x) &= \mathrm{Tr}\,(E_y \rho_x), \\
&= \frac{2}{3}(1 - |\langle\psi_x|\psi_y\rangle|^2), \\
&= \frac{2}{3}(1 - \frac{1}{4}) = \frac{1}{2} \quad \text{for all } y \neq x .
\end{aligned} \tag{95}$$

A measurement outcome x, therefore, implies that Alice definitely did not send the state ρ_x. However, from this outcome, Bob can only conclude that the state that Alice sent could have been either one of the other two, with equal probability ($p = 1/2$).

Moreover, we find that $H(X) = \log 3$, and $H(X|Y) = 1$. Hence the mutual information is given by

$$H(X : Y) = H(X) - H(X|Y) = \log 3 - 1 < 1 . \tag{96}$$

On the other hand we have that $S(\rho) = \log 2 = 1$, since $\rho = \mathbb{1}/2$. Hence, we see that

$$\max H(X : Y) < S(\rho) , \tag{97}$$

and therefore the accessible information I_{acc} is bounded above by the von Neumann entropy $S(\rho)$. However, the actual upper bound for the accessible information can be made stronger, and is given by the well-known *Holevo bound* [5].

Theorem 4 (Holevo bound) *Suppose Alice has a classical source, characterized by a random variable X, which takes values $x \in J = \{1, 2, \ldots, M\}$ with probabilities $p(x)$. She encodes the symbol x into a quantum state ρ_x and sends it to Bob through a noiseless quantum channel. Bob does a measurement on it, described by a finite set of POVM elements $\{E_y\}$. Let Y be the classical random variable corresponding to the outcome of the measurement. Then for any such measurement that Bob may do, the mutual information $H(X : Y)$ satisfies the following upper bound:*

$$H(X : Y) \leq \chi , \tag{98}$$

where

$$\chi := S(\rho) - \sum_{x \in J} p_x S(\rho_x) , \tag{99}$$

and $\rho = \sum_{x \in J} p_x \rho_x$. The equality in (98) is achieved if all the ρ_x's commute (in which case they are simultaneously diagonalizable) and the measurement is performed in the simultaneous eigenbasis of all the ρ_x's.

The Holevo bound is thus an upper bound to the accessible information. The quantity χ is called the *Holevo χ quantity* or *Holevo information*.

Remark 1 The *Holevo χ quantity* not only depends on the state ρ but also on its "preparation", i.e., on the ensemble $\mathcal{E} := \{p_x, \rho_x\}$. To emphasize this we will write χ as $\chi(\mathcal{E})$. In particular $\chi(\mathcal{E})$ reduces to the von Neumann entropy $S(\rho)$ for an ensemble $\mathcal{E}$ of *pure states*. This is because $S(\rho_x) = 0$ if ρ_x is pure and hence in this case the second term on the right-hand side of (99) vanishes.

Proof The Holevo bound follows easily from the strong subadditivity (SSA) of the von Neumann entropy (see chapter "Hilbert Space Methods for Quantum Mechanics", Sect. 3.1). Since SSA is a property involving three systems, we first need to identify three systems to which we could apply the SSA. For this purpose we consider the *enlarged Hilbert Space* representation, obtained by embedding the classical random variable X into a dummy quantum system A. The latter can be viewed as a quantum register which keeps a record of the classical symbol x that Alice wants to communicate to Bob, and by definition its Hilbert space $\mathcal{H}_A$ has an orthonormal basis whose elements are labeled by the symbols x, i.e.,

$$\{|x\rangle : x \in J\} . \tag{100}$$

Let Q be a quantum system (e.g., a photon) in whose states, ρ_x, Alice encodes her message x. Let B be a quantum system representing Bob's measuring device. The latter is considered to be initially in some fixed pure state $|0_B\rangle$. Hence, the initial state of the composite system AQB, with Hilbert space $\mathcal{H}_A \otimes \mathcal{H}_Q \otimes \mathcal{H}_B$, is given by the density matrix

$$\rho_{AQB} = \sum_{x \in J} p_x |x\rangle\langle x| \otimes \rho_x \otimes |0_B\rangle\langle 0_B| . \qquad (101)$$

Bob's measurement is described by POVM elements $\{E_y\}$, which act on the Hilbert space of the quantum system Q. The corresponding outcome y of the measurement is recorded in the state of the measuring device. Let $A'Q'B'$ denote the composite system after this measurement. Hence, as a result of the measurement, the state ρ_{AQB} of the composite system gets transformed to

$$\rho_{A'Q'B'} = \sum_{x,y \in J} p_x |x\rangle\langle x| \otimes \sqrt{E_y}\rho_x\sqrt{E_y} \otimes |y\rangle\langle y| . \qquad (102)$$

Note the following

1. The quantum mutual entropies satisfy

$$S(A : Q) = S(A : QB), \qquad (103)$$

since B is initially uncorrelated with A and Q.

2.

$$S(A : QB) \geq S(A' : Q'B'), \qquad (104)$$

since a quantum operation (in this case Bob's POVM) acting on Q cannot increase the mutual information. This follows from the property (32) of Sect. 4.

3.

$$S(A' : Q'B') \geq S(A' : B'), \qquad (105)$$

since discarding quantum systems never increases mutual information. This follows from the property (30) of Sect. 4.

Putting these together, we obtain

$$S(A' : B') \leq S(A : Q). \qquad (106)$$

The inequality (106) is actually the Holevo bound. To see this, let us evaluate the terms on either side of the inequality. Let us start with the term $S(A : Q)$ appearing on the right hand side of (106). Note that

$$\rho_{AQ} = \text{Tr}_B \rho_{AQB} = \sum_{x \in J} p_x |x\rangle\langle x| \otimes \rho_x . \qquad (107)$$

Hence,

$$\rho_A = \text{Tr}_Q \rho_{AQ} = \sum_{x \in J} p_x |x\rangle\langle x| \,, \quad \rho_Q = \text{Tr}_A \rho_{AQ} = \sum_{x \in J} p_x \rho_x \equiv \rho. \qquad (108)$$

Therefore, $S(A) \equiv S(\rho_A) = H(\{p_x\})$ (the Shannon entropy corresponding to the probability distribution $\{p_x\}$) and $S(Q) \equiv S(\rho_Q)$. Moreover,

$$S(A, Q) \equiv S(\rho_{AQ}) = H(\{p_x\}) + \sum_x p_x S(\rho_x). \qquad (109)$$

This follows from step 3 of Exercise 3 in Sect. 2.2 of chapter "Bipartite Quantum Entanglement". This is because (i) the density matrices $|x\rangle\langle x| \otimes \rho_x$ have orthogonal supports for different x's, since the vectors $|x\rangle$ are mutually orthogonal, and (ii) $S(|x\rangle\langle x| \otimes \rho_x) = S(\rho_x)$. Hence,

$$S(A : Q) = S(A) + S(Q) - S(A, Q),$$
$$= S(\rho) - \sum_x p_x S(\rho_x) \equiv \chi. \qquad (110)$$

Let us now inspect the term $S(A' : B')$ appearing on the left-hand side of (106): Note that

$$\rho_{A'Q'B'} = \sum_{x,y \in J} p_x |x\rangle\langle x| \otimes \sqrt{E_y} \rho_x \sqrt{E_y} \otimes |y\rangle\langle y|, \qquad (111)$$

Therefore,

$$\rho_{A'B'} = \text{Tr}_{Q'} \rho_{AQ} = \sum_{x,y \in J} p_x |x\rangle\langle x| \text{Tr} \left(\sqrt{E_y} \rho_x \sqrt{E_y} \right) \otimes |y\rangle\langle y|. \qquad (112)$$

By the cyclicity of the trace (see chapter "Hilbert Space Methods for Quantum Mechanics", Sect. 1.5),

$$\text{Tr} \left(\sqrt{E_y} \rho_x \sqrt{E_y} \right) = \text{Tr} \left(E_y \rho_x \right) = p(y|x) \,. \qquad (113)$$

Moreover, using the relation $p(x, y) = p(x)p(y|x)$, we obtain

$$\rho_{A'B'} = \sum_{x,y \in J} p(x, y) |xy\rangle\langle xy|. \qquad (114)$$

Note that the pure states $\{|xy\rangle\}$ form an orthonormal basis of $\mathcal{H}_A \otimes \mathcal{H}_B$. Hence (114) is the spectral decomposition of $\rho_{A'B'}$. This implies that

$$S(A' : B') = H(X : Y). \qquad (115)$$

Substituting (110) and (115) in (106) we obtain

$$H(X : Y) \leq \chi. \tag{116}$$

This completes the proof of the Holevo bound. □

7.3 Properties of the Holevo χ Quantity

The Holevo χ quantity can be considered to be a generalization of the von Neumann entropy $S(\rho)$, which reduces to $S(\rho)$ for an ensemble of pure states.

It is *non-negative*: $\chi(\mathcal{E}) \geq 0$. This follows easily from the concavity of the von Neumann entropy:

$$S(\rho) = S\left(\sum_x p_x \rho_x\right) \geq \sum_x p_x S(\rho_x). \tag{117}$$

It can be expressed in terms of the relative entropy as follows. For an ensemble $\mathcal{E} = \{p_x, \rho_x\}$, and $\rho := \sum_x p_x \rho_x$,

$$\chi(\mathcal{E}) = -\text{Tr} \, \rho \log \rho + \sum_x p_x \text{Tr} \, \rho_x \log \rho_x,$$

$$= \sum_x p_x \left(\text{Tr} \, \rho_x \log \rho_x - \text{Tr} \, \rho_x \log \rho\right),$$

$$= \sum_x p_x S(\rho_x || \rho). \tag{118}$$

From (118) and the Uhlmann monotonicity (33) of the relative entropy (see chapter "Hilbert Space Methods for Quantum Mechanics", Sect. 3.2), it follows that *a quantum operation can never increase the Holevo χ quantity*: If $\mathcal{E} = \{p_x, \rho_x\}$ and $\mathcal{E}' = \{p_x, \Phi(\rho_x)\}$ then

$$\chi(\mathcal{E}') \leq \chi(\mathcal{E}). \tag{119}$$

The monotonicity of χ under quantum operations indicates that χ quantifies the amount of information encoded in a quantum system. This is because decoherence, described by a quantum operation Φ, can only retain or reduce χ. It can never increase it. (This is consistent with the fact that noise can never increase information.) In contrast, the von Neumann entropy $S(\rho)$ is not monotonic under quantum operations. We saw, for example, that (i) a *depolarizing channel* transforms a pure state into a mixed state, thereby increasing the von Neumann entropy; whereas (ii) the *amplitude damping channel* takes a mixed state to the pure state (since the excited atom decays to its ground state), thereby reducing the von Neumann entropy.

Note: In the case (ii), the decrease of $S(\rho)$ should *not* be looked upon as an information gain. This is because *every* mixed state decays to the ground state under repeated actions of the amplitude damping channel, and hence we lose the ability to distinguish between different possible preparations of the mixed state.

7.4 Capacities of a Noisy Quantum Channel

In the last section we obtained an upper bound to the maximum amount of classical information that could be sent to Bob via a *noiseless channel* by Alice, if she encoded each classical message (symbol) x (emitted by a source with probability p_x) into a quantum state ρ_x. This upper bound was given by the Holevo χ quantity

$$\chi(p_x, \rho_x) = S\left(\sum_x p_x \rho_x\right) - \sum_x p_x S(\rho_x). \tag{120}$$

Now let us consider what would happen if the channel between Alice and Bob was *noisy*. In particular, let us consider the channel to be discrete, memoryless and noisy, and let us denote it by the CPT map Φ. In this case, if Alice encodes the classical message x into the quantum state ρ_x, as before, then Bob receives the state $\Phi(\rho_x)$. Hence, the maximum amount of classical information that Bob can receive in this case is bounded above by the corresponding χ quantity:

$$\chi(\{p_x, \Phi(\rho_x)\}) = S\left(\sum_x p_x \Phi(\rho_x)\right) - \sum_x p_x S(\Phi(\rho_x)). \tag{121}$$

This observation leads to the following question:
(**Q**) What is the *classical capacity* of the discrete, memoryless, noisy channel Φ ?

In other words, how many bits of classical information can we reliably transmit per use of the noisy channel Φ? The notion of reliability that we shall use is the following: Assume that the sender, Alice, encodes each of her messages in a large "block," e.g., in a quantum state of n qubits, which she sends to the receiver through multiple (n) uses of the channel Φ. Then the transmission is said to be reliable if the probability of error in decoding the state that the receiver, Bob, gets, goes to zero as the block size (n) is made larger and larger.

Before answering this question, I would like to point out that unlike a classical channel, a quantum one has various different capacities The capacity of a classical communications channel is the maximum achievable rate (in bits per channel use) at which classical information can be transmitted through the channel, with arbitrarily low probability of error. Shannon's noisy channel coding theorem gives an explicit expression for the capacity. Hence, the capacity of a classical communications channel is unique and is given by a single numerical quantity. A quantum channel, in contrast, has various distinct capacities. This is because there is a lot of flexibility in the use of a quantum channel. The particular definition of the capacity which is applicable depends on the following:

- whether the information transmitted is classical or quantum;
- whether the sender (Alice) is allowed to use inputs *entangled* over various uses of the channel (e.g., she could have an entangled state of two qubits and send the two qubits on two successive uses of the channel) or whether she is only allowed to use product inputs.
- whether the receiver (Bob) is allowed to make collective measurements over various outputs of the channel or whether he is only allowed to measure the output of each channel use separately;
- whether the sender and the receiver have an additional resource like shared entanglement, e.g., Alice and Bob could, to start with, each have a qubit of an EPR pair.

In this chapter, we shall only consider the transmission of *classical information* through a noisy quantum channel Φ. Let us first briefly return to the example that we had considered (see Sect. 7.2). In that example, Alice encoded each equiprobable classical message $x \in \{1, 2, 3\}$ into pure states $\rho_x = |\psi_x\rangle\langle\psi_x|$, where the $|\psi_x\rangle$ were given by (90). We found that $\chi = S(\rho) = 1$ (since $\rho = \mathbb{1}/2$), whereas the mutual information (under the optimal POVM) was given by $H(X : Y) = \log 3 - 1 \simeq 0.585$ bits/qubit. Hence the Holevo bound $I_{\text{acc}} \leq \chi$ was not saturated. In *Problem 7.1* you will prove that instead of encoding each classical message into the polarization state of a single photon (i.e., into a single qubit) *if*

- Alice encodes each classical message into the polarization states of two photons (i.e., two qubits)

$$x \mapsto \widehat{\rho}_x := |\Psi_x\rangle\langle\Psi_x| \,, \tag{122}$$

where

$$|\Psi_x\rangle = |\psi_x\rangle \otimes |\psi_x\rangle \,, \tag{123}$$

with $|\psi_x\rangle$ given by (90) and sends these two qubits to Bob through two successive uses of the noiseless channel, *and if*
- Bob does a *collective measurement* on the two qubits that he receives (instead of measuring them individually),

then the mutual information is increased to approximately 0.685 bits/qubit. We hence see that encoding a classical message into *multiple qubits* and using *collective measurements* on these qubits results in a better transmission of classical information through the noiseless channel.

The same principle would hold in the case of a memoryless, noisy channel Φ. Hence, even if the channel was noisy, Alice would encode each of her messages into multiple (say n) qubits and send them to Bob through multiple (n) uses of the channel. Bob would then do a collective measurement on the qubits that he received. However, Alice has two choices. She could either encode her messages into product states or encode them into entangled states. To explain these choices consider $n = 2$.

Alice could encode her message x into a product state (as in the example of the noiseless channel previously considered)

$$\widehat{\rho}_x = |\Psi_x\rangle\langle\Psi_x| = |\psi_x\rangle\langle\psi_x| \otimes |\psi_x\rangle\langle\psi_x| , \tag{124}$$

in which case Bob would receive the two qubit state

$$\Phi^{\otimes 2}(\widehat{\rho}_x) = \Phi(|\psi_x\rangle\langle\psi_x|) \otimes \Phi(|\psi_x\rangle\langle\psi_x|) . \tag{125}$$

However, Alice could, alternatively, encode each message x into an entangled state of two qubits (say an EPR pair):

$$x \mapsto \widehat{\rho}_x = |\Psi_x\rangle\langle\Psi_x| , \tag{126}$$

where

$$|\Psi_x\rangle = \frac{1}{\sqrt{2}}|00\rangle + |11\rangle , \tag{127}$$

and send the two qubits of the EPR pair to Bob through two uses of the channel. In this case Bob would receive the state $\Phi^{\otimes 2}(\widehat{\rho}_x)$, which is not expressible as a product of two single qubit states. Hence the two different encodings by Alice lead to important differences in the mechanism of information transmission. In fact, whether using entangled states as inputs to a noisy quantum channel increases its classical capacity, is an important open question in quantum information theory.

7.5 Classical Capacity of a Quantum Channel

Let us consider the transmission of *classical information* through a quantum channel. We shall see later that any arbitrary quantum channel, Φ, can be used to transmit classical information, provided the channel is not simply a constant (i.e., provided $\Phi(\rho) \neq$ constant, for all ρ).

Consider the following scenario: Suppose Alice has a finite set $\mathcal{M}$ of classical messages, which she wants to send to Bob, through a discrete, memoryless, quantum channel Φ. She is allowed to use the channel many times. She encodes a message $M \in \mathcal{M}$ in a quantum state, say, a state $\rho_M^{(n)}$ of n qubits. She then sends this state to Bob through n uses of the channel Φ.

Bob receives the output $\sigma_M^{(n)} = \Phi^{\otimes n}\left(\rho_M^{(n)}\right)$ of the channel and does a suitable measurement on it, in order to infer what the original message was. Let his measurement be described by a POVM, E_M being the POVM element corresponding to the message M. The probability of inferring the message correctly is given by

$$\text{Tr}\left(\sigma_M^{(n)} E_M\right). \tag{128}$$

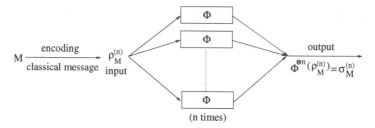

Fig. 4 Transmission of classical information through a product channel

The probability of error corresponding to the message M is thus

$$\left(1 - \mathrm{Tr}\left(\sigma_M^{(n)} E_M\right)\right),\tag{129}$$

whence the average probability of error is given by

$$p_{av}^{(n)} = \frac{\sum_{M \in \mathcal{M}}\left(1 - \mathrm{Tr}\left(\sigma_M^{(n)} E_M\right)\right)}{|\mathcal{M}|}.\tag{130}$$

The *rate of information transmission* is the number of bits of classical message that is transmitted by each qubit. It is given by

$$R := \lim_{n \to \infty} \frac{\log|\mathcal{M}|}{n}.\tag{131}$$

The transmission of classical information through the channel Φ under this scheme is said to be *reliable* if $p_{av}^{(n)} \to 0$ as $n \to \infty$. As in the classical case, a rate R (defined by (131)) is achievable if there exists a suitable encoding $M \mapsto \rho_M^{(n)}$ and a decoding given by a POVM $\{E_M\}$, for which the transmission is reliable. The capacity of the channel is defined as the maximum achievable rate, the maximum being taken over all possible encodings of the channel. We want to compute the maximum achievable rate at which Alice can send information to Bob under the above protocol. In other words, we want to find the *capacity* of the noisy quantum channel Φ, for transmission of classical information. Unfortunately, this is not known yet.

What is known, however, is the classical capacity of Φ in the special case in which Alice is allowed to encode her messages using only *product states*, i.e., states of the form $\rho_M^{(n)} = \rho_1 \otimes \rho_2 \otimes \ldots \otimes \rho_n$, where $\rho_1, \rho_2, \ldots$, are input to the channel on separate uses. Bob is allowed to decode the output of the channel by performing collective measurements over multiple uses of the channel. The capacity of the channel in this case is referred to as the *product state capacity* and is usually denoted as $C^{(1)}(\Phi)$.[8] The product state capacity is given by the celebrated Holevo–Schumacher–Westmoreland (HSW) theorem [12, 6]:

Theorem 5 (HSW) *The product state capacity of a quantum channel Φ is given by*

$$C^{(1)}(\Phi) = \chi^*(\Phi), \tag{132}$$

where $\chi^*(\Phi)$ is the Holevo capacity of the channel and is defined as follows:

$$\chi^*(\Phi) := \max_{\{p_i, \rho_i\}} \chi(\{p_i, \Phi(\rho_i)\}), \tag{133}$$

where

$$\chi(\{p_i, \Phi(\rho_i)\}) := S\left(\Phi\left(\sum_j p_j \rho_j\right)\right) - \sum_j p_j S\left(\Phi(\rho_j)\right). \tag{134}$$

The maximum in (133) is taken over all ensembles $\{p_i, \rho_i\}$ of possible input states ρ_i of the channel, with $p_i \geq 0$, $\sum_i p_i = 1$.

Information transmission through a noisy quantum channel is improved when the outputs of the channel are more distinguishable.

The Holevo *chi*-quantity $\chi(\{p_i, \Phi(\rho_i)\})$ is a measure of the distinguishability of the ensemble of output states $\Phi(\rho_i)$. It might seem natural that the distinguishability of the output states would be maximized by maximizing the distinguishability of the input states, i.e., by using a set of mutually orthogonal input states. However, this intuition has been proved to be false (see, e.g., [2, 4]), and hence, the Holevo capacity can indeed be achieved on non-orthogonal input states.[9]

The HSW theorem tells us that Alice can reliably transmit classical information to Bob, through a quantum channel Φ, using product input states, at any rate below $C^{(1)}(\Phi)$.

An interesting application of the HSW theorem is the following lemma.

Lemma 1 *Any arbitrary quantum channel Φ can be used to transmit classical information, provided the channel is not simply a constant.*

Proof This can be seen as follows. If Φ is not a constant, then there exist pure states $|\psi\rangle$ and $|\phi\rangle$ such that

$$\Phi\left(|\psi\rangle\langle\psi|\right) \neq \Phi\left(|\phi\rangle\langle\phi|\right). \tag{135}$$

---

[8] The superscript (1) is used to denote that Alice is required to use the many available copies of the channel *one at a time*, encoding her messages into product states.

[9] The maximum in (134) is potentially over an unbounded set. However, it can be shown that one can restrict the maximization to pure state ensembles containing at most d^2 elements, where d is the dimension of the Hilbert space at the input to the channel (see, e.g., [13]).

Consider the ensemble

$$\left\{ p_1 = p_2 = 1/2 \,,\ \rho_1 = |\psi\rangle\langle\psi| \,,\ \rho_2 = |\phi\rangle\langle\phi| \right\}. \tag{136}$$

From (135) and the concavity of the von Neumann entropy (see chapter "Hilbert Space Methods for Quantum Mechanics", Sect. 3.1) it follows that $\chi(\{p_i, \Phi(\rho_i)\}) > 0$. Hence, $\chi^*(\Phi) > 0$, which in turn implies that the quantum channel Φ *can* transmit classical information if the latter is encoded into quantum states which are then sent through the channel. □

The product state (or Holevo) capacity of a discrete memoryless quantum channel can be generalized to give the classical capacity of the channel, i.e., its capacity for transmission of classical information in the absence of the restriction of product state inputs.

The generalization is achieved by considering inputs which are product states over uses of blocks of n channels, but which may be entangled across different uses within the same block. The classical capacity $C_{\text{classical}}$ is obtained in the limit $n \to \infty$:

$$C_{\text{classical}} = \lim_{n \to \infty} \frac{1}{n} \chi^* \left(\Phi^{\otimes n} \right), \tag{137}$$

with

$$\chi^* \left(\Phi^{\otimes n} \right) := \max_{\{p_i, \rho_i\} : \rho_j \in B(\mathcal{H}^{\otimes n})} S \left(\Phi^{\otimes n} \left(\sum_j p_j \rho_j \right) \right) - \sum_j p_j S \left(\Phi^{\otimes n}(\rho_j) \right) \tag{138}$$

being the Holevo capacity of the block $\Phi^{\otimes n}$ of n channels. In (138) $\rho = \sum_j p_j \rho_j \in B(\mathcal{H}^{\otimes n})$. The HSW Theorem naturally leads to the following question:

(Q) *Can one increase the classical capacity of a quantum channel by using entangled input states ?*

This question is related to an important conjecture, namely, the *additivity of the Holevo capacity*, which is as follows: For two given quantum channels, Φ_1 in $\mathcal{H}_1$ and Φ_2 in $\mathcal{H}_2$:

$$\chi^*(\Phi_1 \otimes \Phi_2) = \chi^*(\Phi_1) + \chi^*(\Phi_2). \tag{139}$$

If $\Phi_1 = \Phi_2 = \Phi$, where Φ is a memoryless quantum channel, the product channel $\Phi_1 \otimes \Phi_2 \equiv \Phi^{\otimes 2}$ denotes two successive uses of the channel. Let us see how the conjecture (139) is related to the question **(Q)** above.

The Holevo capacity is superadditive, i.e., $\chi^*(\Phi_1 \otimes \Phi_2) \geq \chi^*(\Phi_1) + \chi^*(\Phi_2)$. This follows from the superadditivity of the Holevo χ quantity, which can be proved by expressing χ in terms of a relative entropy (as in (118)) and using the fact that the relative entropy is convex in each of its arguments (a consequence of (32)). See *Problem 7.3*. If the Holevo capacity is additive then $\chi^*(\Phi^{\otimes n}) = n\chi^*(\Phi)$, which implies that $C_{\text{classical}}(\Phi) = \chi^*(\Phi)$, the Holevo capacity of the channel Φ. The latter is a fixed quantity characteristic of the channel. Hence, *if* the Holevo capacity is additive then the capacity of the quantum channel to transmit classical information *cannot* be increased by using entangled inputs.

The validity of the additivity conjecture (139) had been proved for some classes of quantum channels. However, whether the additivity holds "globally," i.e., for all quantum channels, was an open problem of quantum information theory[10] until very recently. Peter Shor had showed [15] that the additivity conjecture for the Holevo capacity is equivalent to the additivity conjecture for the *minimum output entropy* of the channel. Given a channel Φ, its minimum output entropy $H_{\min}(\Phi)$ is defined as

$$H_{\min}(\Phi) = \min_{|\psi\rangle} H\left(\Phi(|\psi\rangle\langle\psi|)\right). \tag{140}$$

The additivity conjecture for the *minimum output entropy* is that for all channels Φ_1 and Φ_2, the following identity holds:

$$H_{\min}(\Phi_1 \otimes \Phi_2) = H_{\min}(\Phi_1) + H_{\min}(\Phi_2). \tag{141}$$

In fact, Shor proved that these two additivity conjectures are equivalent, in the sense that if one of them holds for all channels then the other also holds for all channels. In September 2008, Matthew Hastings proved a counterexample to the additivity of the minimum output entropy [3], and hence by Shor's equivalence, it follows that the Holevo capacity is *not* additive for all quantum channels.

Exercise 3

1. *Consider the example studied in the Sect. 7.2. We found that* $\chi = S(\rho) = 1$ *(where* $\rho = (1/3) \sum_x \rho_x \equiv \mathbb{1}/2$*), whereas the mutual information, under the optimal POVM done by Bob (discussed in the Sect. 7.2) was given by* $H(X : Y) = \log 3 - 1 \simeq 0.585$ *bits/qubit. Hence the Holevo bound* $I_{\text{acc}} \leq \chi$ *was not saturated.*

   Now let Alice encode each classical message into polarization states of two photons (i.e., two qubits)

$$x \mapsto \widehat{\rho}_x := |\Psi_x\rangle\langle\Psi_x|, \tag{142}$$

   where

$$|\Psi_x\rangle = |\psi_x\rangle \otimes |\psi_x\rangle, \tag{143}$$

   with $|\psi_x\rangle$ *given by (90). She sends these two qubits to Bob through two successive uses of the noiseless channel.*

---

[10] Recall that, in contrast, the capacity of a classical communications channel is always additive (see Exercise 2).

(a) *Compute the eigenvalues of the density matrix*

$$\rho := \frac{1}{3}\sum_{x=1}^{3}\widehat{\rho}_x \equiv \frac{1}{3}\sum_{x=1}^{3}|\Psi_x\rangle\langle\Psi_x|. \tag{144}$$

Evaluate its von Neumann entropy and show that the accessible information per qubit is less than 0.75.

Bob does a collective measurement characterized by the operators E_x, $x = 1, 2, 3$, where

$$E_x = G^{-1/2}|\Psi_x\rangle\langle\Psi_x|G^{-1/2}, \tag{145}$$

with $G := \sum_{x=1}^{3}|\Psi_x\rangle\langle\Psi_x|$. (Note: Such a measurement is called a square root measurement.)

(b) *Evaluate the operators E_x, $x = 1, 2, 3$ and show that they form a POVM. Show that the probability that the outcome of Bob's measurement is y, given that Alice's message was x is given by the following:*

$$p(y|x) = 0.9714 \quad for\ x = y,$$
$$p(y|x) = 0.0143 \quad for\ x \neq y. \tag{146}$$

2. *Use the HSW theorem to find the product state capacity of the depolarizing channel, Φ, defined by*

$$\Phi(\rho) = p\rho + (1 - p)\frac{\mathbb{1}}{2}. \tag{147}$$

3. *Prove that*

$$\chi(\Phi_1 \otimes \Phi_2) \geq \chi(\Phi_1) + \chi(\Phi_2), \tag{148}$$

where Φ_1 and Φ_2 are quantum channels (linear CPT maps) and

$$\chi(\Phi) := S\left(\sum_i p_i\Phi(\rho_i)\right) - \sum_i p_i S(\Phi(\rho_i)), \tag{149}$$

is Holevo χ quantity corresponding to the ensemble $\{p_i, \rho_i\}$ of input states to the channel Φ.

4. *What is the maximum number of classical bits of information that Alice can send to Bob by transmitting n unentangled photons to him via a memoryless, noisy, quantum channel?*
Hint: use the Holevo bound.

References

1. Cover, T.M., Thomas, J.A.: Elements of Information Theory. Wiley, New York (1991)
2. Fuchs, C.A.: Phys. Rev. Lett. **79**, 1162 (1997)
3. Hastings, M.B.: Nat. Phys. **5**, 255 (2009)
4. Hayashi, M., Imai, H., Matsumoto, K., et al.: Quantum Inf. Comput. **5**, 13 (2005)
5. Holevo, A.S.: Prob. Inf. Transm. **9**, 177 (1973)
6. Holevo, A.S.: IEEE Trans. Inf. Theory **44**, 269 (1998)
7. Lieb, E.H., Ruskai, M.B.: J. Math. Phys. **14**, 1938 (1973)
8. Nielson, M.A., Chuang, I.L.: Quantum Computation and Quantum Information. Cambridge University Press, Cambridge (2000)
9. Ohya, M., Petz, D.: Quantum Entropy and Its Use. Springer, Heidelberg (1993)
10. Preskill, J.: Lecture Notes on Quantum Computation online at http://www.theory.caltech.edu/people/preskill/ph229/
11. Schumacher, B.: Phys. Rev. A **51**, 2738 (1995)
12. Schumacher, B., Westmoreland, M.D.: Phys. Rev. A **56**, 131 (1997)
13. Schumacher, B., Westmoreland, M.D.: Phys. Rev. A **63**, 022308 (2001)
14. Shannon, C.E.: Bell Syst. Tech. J. **27**, 379, 623 (1948)
15. Shor, P.W.: Commun. Math. Phys. **246**, 543 (2004)
16. Stinespring, W.F.: Proc. Am. Math. Soc. **6**, 211 (1955)

Photonic Realization of Quantum Information Protocols

M. Genovese

1 Introduction

Quantum optics is a discipline whose main purpose is the study of electromagnetic fields and of their interactions with atoms and matter when quantum features are relevant: a part of quantum field theory concerning the electromagnetic field in interaction with fermions (quantum electrodynamics) or with atoms and matter (where the interaction is usually given by effective Hamiltonians, see chapter "Field-Theoretical Methods", Sect. 4.2).

Quantum optics is now a huge field whose description is largely beyond the purpose of these lectures. Some basic facts are contained in chapter "Field-Theoretical Methods" of this book, while for more details the reader specialistic books like [86]. Here we will only consider some very specific parts of this field, which have direct interest for applications to quantum information.

Indeed, as we will discuss, most of the realizations of quantum communication protocols have been performed by using states of the electromagnetic field. Also, interesting progresses toward quantum computation protocols (see chapter "Quantum Algorithms") have been obtained with photon-based schemes. This chapter offers an overview of these results; in particular, we will focus upon sources of single and entangled photons and experiments realized with them: this will require a description of parametric downconversion that is the most used physical phenomenon in all these realizations.

1.1 The Parametric Downconversion

The parametric downconversion (PDC), or parametric fluorescence, is a quantum effect without classical counterparts (see Fig. 1). It consists of a spontaneous decay, inside a nonlinear medium with non-zero second-order susceptibility $\chi^{(2)}$, of one photon from a pump beam (usually generated by a laser) into a couple of photons

M. Genovese (✉)
I.N.RI.M., Torino 10135, Italy, genovese@inrim.it

Genovese, M.: *Photonic Realization of Quantum Information Protocols*. Lect. Notes Phys. **808**, 215–251 (2010)
DOI 10.1007/978-3-642-11914-9_7

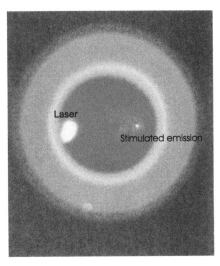

Fig. 1 Picture of type I PDC emission produced in a LiIO$_3$ crystal pumped by a UV laser beam (at 351 nm). The *bright spot* is an injected laser at 789 nm and the *small spot* is the stimulated emission at 633 nm pointing out the correlated direction

(sometimes dubbed bi-photon) conventionally, and arbitrarily, called signal and idler (for an extensive description of this phenomenon see [86]). The media useful for these applications are crystals with a significant second-order dielectric susceptibility, i.e., crystals where in the polarization vector expansion

$$P_i = \chi^1_{ij} E_j + \chi^2_{ijk} E_j E_k + \cdots , \tag{1}$$

the term χ^2_{ijk} is non-negligible and therefore one has significant contributions from the interaction of three electromagnetic fields in the Hamiltonian density

$$\frac{1}{2\mu_0} \mathbf{B}^2 + \frac{1}{2} \mathbf{E} \cdot \mathbf{D} , \quad \mathbf{D} = \varepsilon_0 \mathbf{E} + \mathbf{P} . \tag{2}$$

Among them one can mention Beta-barium borate (BBO), lithium iodate, potassium titanyl phosphate (KTP), and others.

This process obeys (phase-matching laws) energy conservation

$$\omega_0 = \omega_i + \omega_s \tag{3}$$

and (exactly, in the case of an infinite length crystal) momentum conservation

$$k_0 = k_i + k_s , \tag{4}$$

where $\omega_0, \omega_i, \omega_s$ are the frequencies and k_0, k_i, k_s are the wave vectors of pump, idler, and signal photon, respectively. Furthermore, the two photons are produced at the same time (within few tens of femtoseconds, as measured by means of an interference technique). The emission ranges from the wavelength of the pump field up to the infrared region (being limited by the absorption of the crystal).

The probability of a spontaneous decay into a pair of correlated photons is usually very low, of the order of 10^{-9} or lower (higher values can be obtained in nano-structured materials, as periodically poled crystals). Thus, with typical pump power of the order of some milliwatts, the PDC emission lies at the levels of photon counting regime. Since the photons are produced in pairs and because of the energy and momentum conservation restrictions, the detection of one photon in a certain direction and with a given energy indicates the existence of the pair correlated one, with definite energy and travelling in a well-defined direction.

To fulfill phase-matching conditions, differently polarized waves must be used, leading to two different kinds of PDC emission.[1]

In type I PDC, both photons are produced with the same polarization, ordinary in negative crystals ($n_o > n_e$) and extraordinary in positive crystals ($n_o < n_e$), orthogonal to that of the pump photon,

$$|k_0| = \frac{2\pi}{c\sqrt{\cos^2\theta/n_o^2 + \sin^2\theta/n_e^2}} , \qquad (5)$$

extraordinary in negative crystals and ordinary in positive ones.

Photons of equal wavelength are emitted on concentric cones centered on the pump laser direction (see Fig. 2, left), whose diameter depends on the angle between the pump beam and the optical axis of the crystal, the phase-matching angle θ. When projected into a plane, conjugated photons are on the same diameter and

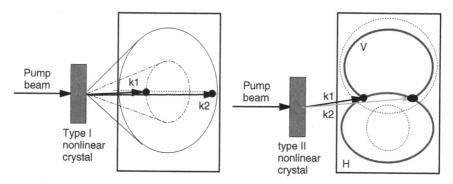

Fig. 2 *Left:* type I PDC. The two circumferences (*continuous* and *dashed*) correspond to two different wavelengths. The *spots* indicate the directions of emission of two entangled photons. *Right:* type II PDC emission. Two circumferences, which are emitted degenerate photons (*continuous line*) and correlated photons of different wavelengths (*dashed lines*), are shown. H and V denote horizontal and vertical polarization, respectively. The two spots denote the directions of entangled photons

---

[1] We limit the following discussion to uniaxial crystals.

opposite with respect to the center of the two concentric circles corresponding to their wavelengths. The regime when the phase-matching angle is such that the correlated photon pairs at half wavelength of the pump beam are emitted in the same direction is called collinear degenerate regime.

In a type II PDC, one photon has equal polarization to that of the pump photon, while the other has an orthogonal polarization. For a suitable phase-matching angle (see later), they are emitted on intersecting circumferences (see Fig. 2, right).

From a theoretical point of view, briefly, the process of PDC in a crystal with active region of volume V can be described[2] by the Hamiltonian (where the sum is only over those modes that are allowed by energy and momentum conservation):

$$
H = \frac{1}{L^3} \sum_{k,s} \sum_{k',s'} \int d\omega_0 \, E_l(\omega_0) \, \chi_{i,j,l}^{(2)}(\omega_0, \omega, \omega')
$$

$$
\times \, \varepsilon_{k,s}^* \, \varepsilon_{k',s'}^* \int_V d^3r \left(e^{-i(k_0-k-k')\cdot r} \, a(k,s) \, a(k',s') + \text{h.c.} \right), \qquad (6)
$$

where h.c. stays for *hermitean conjugate*, while k_0, k, k' are the quadri-momenta of the pump, idler, and signal photon, respectively, r is the quadri-vector space–time, $E_l(\omega_0)$ is the vector amplitude of the pump (strong enough to be treated as a classical field, but not necessarily perfectly mono-chromatic), while $a(k,s)$ and $a(k',s')$ are the annihilation operators for the produced photons (with polarization s, s'). When the pump intensity is sufficiently small, one operates in the so-called low-gain regime. In this case one is allowed to keep only the first order in the series expansion of the evolution operator acting on the vacuum (higher order terms would provide a multi-photon component), obtaining a state in the following form (l_m denotes the three sides of the nonlinear dielectric medium supposed to be a parallelepiped):

$$
|\Psi(t)\rangle = \exp\left(-\frac{i}{\hbar} \int H(t')\, dt' \right) |0\rangle
$$

$$
\simeq |0\rangle - \frac{i}{\hbar L^3} \sum_{k,s} \sum_{k',s'} \int d\omega_0 \, E_l(\omega_0) \chi_{i,j,l}^{(2)}(\omega_0, \omega, \omega') \, (\varepsilon_{k,s}^*)_i (\varepsilon_{k',s'}^*)_j \qquad (7)
$$

$$
\times \prod_{m=1}^{3} \frac{2\sin\left((k_0-k-k')_m l_m/2\right)}{(k_0-k-k')_m \, l_m} \, \delta(\omega + \omega' - \omega_0) \, |k,s\rangle \, |k',s'\rangle \,,
$$

where the term $\text{sinc}(x) = \sin(x)/x$ follows from integration over the crystal volume and keeps explicitly into account the non-exact phase-matching due to the finite dimension of the nonlinear medium.

Thus correlated wavelengths are slightly spread around the perfect phase-matching condition (3). Perfect conservation of energy, $\delta(\omega + \omega' - \omega_0)$, derives

---

[2] The form of this "effective" Hamiltonian can be deduced from (rather involved) microscopic calculations, e.g., see [116].

from assuming a plane wave as pump beam (with infinite temporal extension). The sum over k, k' does not allow factorization into a product of signal and idler states: the state described by (7) is therefore entangled (see chapter "Bipartite Quantum Entanglement"). As we shall describe in detail, it can be used in various quantum information tasks.

1.2 Heralded Photon Sources

A first interesting application of PDC is that, due to the property of strong temporal and spatial correlation of emission of the photon pairs, one can realize a "heralded single photon source," where the observation of one photon of the pair "certifies" the presence of the correlated one at a specific time, frequency and in a determined direction. From an experimental point of view, one needs to identify two correlated photon directions by spatial (pinholes) and spectral (interference filters) selection and use the observation of one photon as a trigger signal, certifying the presence of the correlated one.

These experimental achievements represent a first fundamental step toward applications in quantum information; indeed, they allow the generation of single qubits to be used in quantum information protocols. Incidentally, these schemes also provide absolute sources of light that can be used for quantum metrology, e.g., for calibrating detectors [22, 20] (indeed, the quest for accurately calibrated detectors is an important issue in many quantum information protocols).

Anyway, this technique has some limitations. First of all, the PDC emission is not deterministic and therefore one obtains one photon in a certain temporal window only with a finite probability. Furthermore, one also has a non-zero probability that two or more photons might be emitted within a same temporal window. In principle, this problem could be solved by using photo-detectors able to discriminate among more photons. Nevertheless, at the moment, photo-detectors well suited for this purpose are not available. Indeed, one would need a congruous linearity in the internal current amplification process: each single electron produced by the different photons in the primary step of the detection process (either ionization or promotion to a conduction band) should experience the same average gain and this gain should have sufficiently low spread. The fulfillment of both these requisites is necessary for the charge integral of the output current pulse to be proportional to the number of detected photons. Few examples exist of photo-detectors that can operate as photon counters and each one has some drawback. Among these, photomultiplier tubes (PMTs) [142] and hybrid photo-detectors [59] present low quantum efficiency, since the detection starts with the emission of an electron from the photocathode. Solid-state detectors with internal gain, in which the nature of the primary detection process ensures higher efficiency, are still under development. Highly efficient thermal photon counters have also been realized as prototypes, though their operating conditions are still extreme (cryogenic conditions) to be suited for common use [68, 98, 35]. Also, methods based on statistical sampling for reconstructing the density matrix of an optical state [91, 146, 106, 141] (as quantum tomography) are not very useful for this purpose.

For this reason PDC "heralded photon sources" do not represent a perfect solution for realizing a single photon source and some alternatives are under development as quantum dots [109, 83] and color centers in diamonds [10, 140, 72, 58, 118]. Further methods based on two-photon emission from semiconductor structure [108] or single photon emission by controlled molecular fluorescence [23] and from nanotubes [61] are still at a very seminal level, far from applicative possibilities.

Nevertheless, these systems present various drawbacks as well: low collection efficiency and relatively large multi-photon component for nano-diamonds, difficulty in producing quantum dots with defined characteristics, and so on. Thus, almost all experiments in quantum information requiring single photon sources and performed up to now have been built with PDC sources.

2 Photon Entanglement

As discussed in chapters "Quantum Cryptography" and "Quantum Algorithms", entanglement represents a fundamental physical resource for quantum information protocols. For pure states, a state of two or more particles is entangled when it cannot be factorized into a tensor product of single-particle states, namely, as stated in the original Schrödinger definition [114], it describes a compound systems whose subsystems are not probabilistically independent (see chapter "Bipartite Quantum Entanglement").

In this section we will present some of the sources of entangled states of photons, which have been realized in the last years. Their use is widespread and they will recur in various experimental realizations of quantum information protocols, some of them being the focus of the next paragraphs.

Historically, before being applied in quantum information, such sources had been used for studies concerning the foundations of quantum mechanics; in particular, to test local realism through Bell inequalities (see chapter "Quantum Probability and Quantum Information Theory", Sect. 2). Inequalities that, albeit born within these basic field, later found widespread applications in quantum information.

Therefore, before discussing in more detail entangled photon sources, we present an introduction to Bell inequalities that complements the more abstract one presented in chapter "Quantum Probability and Quantum Information Theory" (for a review paper, see [47]).

2.1 The Bell Inequalities

Bell inequalities were introduced for testing possible local realistic alternatives to standard quantum mechanics (see chapter "Quantum Probability and Quantum Information Theory", Sects. 2.4 and 3.6).

This problem was raised in 1935 by Einstein, Podolsky, and Rosen [37]. With the purpose of discussing if quantum mechanics is a complete theory, they introduced the concept of *element of reality* according to the following definition: If, without disturbing in any way a system, one can predict without any uncertainty the value of a physical quantity, then there is an element of physical reality corresponding to this quantity. They formulated also the reasonable hypothesis that, because of special relativity, any non-local action should be forbidden, with the consequence that a measurement performed on a subsystem cannot influence a measurement on another subsystem when they are space-like separated. Their conclusions were that either one of their hypothesis was wrong or quantum mechanics was not a complete theory, in the sense that not every element of physical reality had a counterpart in the theory.

In the following, we elaborate on this topic following Bohm's formulation. Let us consider a singlet state of two spin 1/2 particles

$$|\psi_0\rangle = \frac{|\uparrow\rangle|\downarrow\rangle - |\downarrow\rangle|\uparrow\rangle}{\sqrt{2}},$$

(8)

where $|\uparrow\rangle$ and $|\downarrow\rangle$ represent a single particle of spin up and down, respectively. This state is manifestly entangled. The total spin of the pair in this state is zero; however, before any spin measurement is performed, the spin components of each one of the two particles are undefined. If we let the two particles get spatially separated and then measure the z-component of the spin of the first particle, we also know the z-component of the spin of particle 2 (being opposite to the one of particle 1) without disturbing in anyway this second particle. Thus the z-component of the spin of the second particle is an element of reality according to the previous definition.

But, since the singlet state is invariant under rotations, we could refer to any other axis (as x or y, etc.): thus we can argue that any other spin component of particle 2 is an element of reality. However, spin components on different axes are incompatible variables in quantum mechanics, to which one cannot simultaneously assign definite values. Thus, Eistein, Podolsky, and Rosen argued that quantum mechanics is not a complete theory, since it is not able to predict all elements of reality.

This statement was the starting point of the development of the so-called *local hidden variable* (LHV) theories, i.e., of theories based on the idea that it may exist a deterministic (and local) theory describing nature so that quantum mechanics is only a statistical version of a more fundamental structure. The situation is somehow alike to statistical thermodynamics: this theory describes probabilistically systems consisting of many particles, each of them behaving in a perfectly deterministic way according to the classical equations of motion.

In little more detail, in a hidden variable theory every particle has a perfect assigned value for each observable, determined by a hidden variable x. A statistical ensemble of particles has a certain distribution $\rho(x)$ of the hidden variable and thus the expectation value of an observable A is given by the following:

$$\langle A \rangle = \int dx \, \rho(x) \, A(x).$$

(9)

Of course, considering the huge success of quantum mechanics in predicting many different experimental data, the average $\langle A \rangle$ given by (9) must reproduce the quantum mechanical predictions.

The next fundamental step in discussing possible LHV extensions of quantum mechanics was the 1964 discovery of Bell [8] that any *realistic LHV theory* must satisfy certain inequalities, which can be violated by quantum mechanics, thus allowing an experimental test of the validity of quantum mechanics against LHV theories. Demonstration of Bell inequalities is quite a simple algebraic exercise. We consider here the form proposed by Clauser and Horne (many different forms of Bell inequalities exist, all of them substantially equivalent). Let us consider a source emitting a pair of entangled particles; the first particle goes to detector 1 and the second to detector 2. Let us suppose that before being detected by detector i, a certain property θ_i is measured on the impinging particle. For example, if the particles are entangled in spin, then θ_i is the angle defining the direction (with respect to the z-axis) along which we are going to measure the spin; for polarization-entangled pairs of photons it represents the setting of a polarizer and so on. The Clauser–Horne sum reads as follows:

$$CH = P(\theta_1, \theta_2) - P(\theta_1, \theta_2') + P(\theta_1', \theta_2) + P(\theta_1', \theta_2') - P(\theta_1') - P(\theta_2) , \quad (10)$$

where

1. $P(\theta_1, \theta_2)$ represents the joint probability of observing a particle in 1 with the selection θ_1 and, in coincidence, a particle in 2 with the selection θ_2 (apices denote other angles' choices);
2. $P(\theta_i)$ denotes the probability of observing a single particle at i with selection θ_i.

If these probabilities derive from a local hidden variable theory, calling the hidden variable x and denoting by $\rho(x)$ the probability distribution for the hidden variable, we have

$$P(\theta_i) = \int dx \, \rho(x) \, P(\theta_i, x) , \quad (11)$$

and

$$P(\theta_i, \theta_j) = \int dx \, \rho(x) \, P(\theta_i, \theta_j, x) . \quad (12)$$

If the theory is local, then the outcomes of the measurement in 1 cannot depend on the choice of θ_2 and vice versa. Therefore, we have the following:

$$P(\theta_1, \theta_2, x) = P(\theta_1, x) \cdot P(\theta_2, x) . \quad (13)$$

In order to demonstrate the Clauser–Horne inequality, we construct a simple algebraic relation for four variables: x, x', which lie between 0 and X $(X \leq 1)$ and y, y' which lie between 0 and Y $(Y \leq 1)$:

$$xy - xy' + x'y + x'y' - x'Y - yX \leq 0 . \quad (14)$$

In fact, for $x < x'$, one can rewrite the right-hand part of (14) as

$$x(y - y') + (x' - X)y + (y' - Y)x' \leq (x' - X)y + x(y - Y), \quad (15)$$

which is negative. On the other hand for $x \geq x'$, one rewrites the right-hand part of (14) as

$$(x - X)y + (y - Y)x' + (x' - x)y', \quad (16)$$

which is also negative. By substituting $P(\theta_1, x) = x$, $P(\theta_1', x) = x'$, $P(\theta_2, x) = y$, $P(\theta_2', x) = y'$, and $X = 1$, $Y = 1$, we finally obtain

$$P(\theta_1, \theta_2) - P(\theta_1, \theta_2') + P(\theta_1', \theta_2) + P(\theta_1', \theta_2') - P(\theta_1') - P(\theta_2) \leq 0. \quad (17)$$

This relation must hold in any LHV theory; on the other hand, if the probabilities are evaluated according to the rules of quantum mechanics, for a suitable choice of the parameters θ_i, the inequality is violated.

Let us notice that the proof does not require that the LHV theory be deterministic, namely that the outcomes of a measurement be fixed by the hidden variables, but only the less restrictive request that the single measurement probability be determined by them.

The obvious relevance of this result is that, at least in principle, one can now exclude any local realistic hidden variable theory just observing a violation of this (or another equivalent) inequality. On the other hand, of course, testing Bell inequalities does not allow to exclude non-local hidden variable theories (where non-locality must, however, be such as not to allow faster-than-light communication of information in order to be compatible with special relativity) [47].

Unluckily, experimental verifications of the above inequality are not an easy task. Experimentally, one measures the number $N(\theta_1, \theta_2)$ of coincidences, while in (17) $P(\theta_1, \theta_2) = N(\theta_1, \theta_2)/N$ appears, where N is the total number of pairs emitted by the source, which is not really measurable because a relevant fraction of the pairs is usually lost. Anyway, when considering the ratio

$$R = \frac{N(\theta_1, \theta_2) - N(\theta_1, \theta_2') + N(\theta_1', \theta_2) + N(\theta_1', \theta_2')}{N(\theta_1') + N(\theta_2)}, \quad (18)$$

N cancels between numerator and denominator and for a hidden variable theory, it is always $R \leq 1$, while in quantum mechanics R can reach the value 1.207.

Before concluding this paragraph, it is worth mentioning that various other inequalities have been derived in the course of the years. Here we only mention the one obtained by Clauser, Horne, Shimony, and Holt [32] which is one of the most often used in experiments:

$$S = \left| C(a, b) - C(a, c) \right| + C(b', b) + C(b', c) \leq 2, \quad (19)$$

where $C(a, b)$ is the expected value for joint measurements whose outcomes are distributed according to the probabilities in (12).

2.2 First Sources of Entangled States of Photons

Since the end of the 1960s, many interesting experiments have been devoted to testing Bell inequalities, leading to a substantial agreement with quantum mechanics, hence disfavoring LHV theories. However, up to now, no experiment has yet been able to exclude such theories in an absolute manner. In fact, so far, due to the low total detection efficiency, one has always been forced to introduce, at least, one further additional hypothesis, stating that the observed sample of particle pairs is a faithful sub-sample of the initial set of pairs. This problem is known as *detection or efficiency loophole* and remains the main limitation to conclusive tests of local realism. Indeed if the hidden variables determine not only the result of the measurement but also if the particle is detected or not then LHV models in agreement with present experiments can be built [47].

Many different systems have been considered in the literature (as entangled pairs of ions,[3] $K\overline{K}$, $\Lambda\overline{\Lambda}$, etc.) for realizing tests of Bell inequalities, but up to now most experiments have been realized with entangled photons since all these other systems present some drawbacks [47].

The past search for an efficient source of entangled photons for Bell inequality tests can now be looked at as the development of sources that would later be used in quantum information protocols, whereby the violation of Bell inequalities represents their figure of merit.

Historically, the first experiments were performed, in the 1970s and 1980s, by using polarization-entangled photon pairs produced either in cascade atom decays (visible light) or in positronium decays (gamma rays).[4]

This series of experiments culminated in 1982 with the celebrated Orsay's experiment [3]. Here space-like separation between the two detections was obtained by the use of rapid acousto-optic switches operating at 50 MHz, which were selecting different paths for the incident photons in a way that no communication of the selected basis for the polarization analysis was possible between the two different parts of the apparatus. Thus the two photo-detections after polarization selection were really two non-causally connected events. Bell inequalities were violated (modulo the detection loophole) more than 5 standard deviations. Nevertheless, collection efficiency was very low (with coincidences ranging between 0 and $40\,s^{-1}$ against a typical rate of production of pairs of $5 \times 10^7\,s^{-1}$). This low value, even smaller than the previous ones, was mainly due to the necessity of reducing the divergence of the

---

[3] A recent experiment [60] based on the use of Be ions has reached very high efficiencies (around 98%), but in this case the two subsystems (the two ions) are not really separated systems during the measurement and the test cannot be considered a real implementation of a loophole-free test of Bell inequalities, even if it represents a relevant progress in this sense.

[4] In this case detection efficiency was very high, but the measurement of polarization was indirect.

beams in order to get a good switching. Thus, detection loophole was very far from being eliminated.

In the 1990s, a large improvement on this problem was made possible by bright sources of entangled states of photons based on PDC. These sources are also the ones which are nowadays applied in quantum information protocols and will be discussed in the next two paragraphs, after a short introduction to the quantum description of a beam splitter (BS).

2.3 The Quantum Beam Splitter

A beam splitter is a semitransparent mirror that allows combination of two optical beams: a necessary element for building interferometers that will be described in the next paragraph.

Let us denote with 0 and 1 the inputs ports and 2 and 3 the outputs (see Fig. 3). The quantum description of a BS uses input operators a_i $(i = 0, 1)$ and output ones a_i $(i = 2, 3)$ that are actually creation and annihilation operators of Bosonic modes (see chapter "Field-Theoretical Methods", Sect. 2.1). The tricky point is that even if the input beam only concerns port 1, one cannot neglect port 0 (vacuum) if one wants to preserve the usual commutators. Thus,

$$a_2 = ta_1 + ra_0, \ a_3 = ta_0 + ra_1, \tag{20}$$

where r and t are the reflectivity and transmissivity, respectively (that we suppose to be the same on the two sides). The requests that the canonical commutation relations (CCR) hold

$$\left[a_2, a_2^\dagger\right] = 1, \ \left[a_2, a_3^\dagger\right] = 0 \tag{21}$$

lead to the relations

$$|t|^2 + |r|^2 = 1, \quad tr^* + rt^* = 0. \tag{22}$$

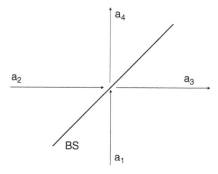

Fig. 3 The "quantum" beam splitter

In the following we will also often mention the polarizing beam splitter (PBS): this is a system where, in a certain basis (for example, horizontal and vertical polarization), one polarization always exits one port and the other polarization the other port. Again the same rules for vacuum inputs apply (see Exercise 6).

2.4 Phase–Momentum Entanglement

The type I PDC bi-photon state described by (8) presents a phase and momentum entanglement that can be directly exploited by using two separated interferometers according to the scheme proposed in [43] and realized as in [105, 15, 75, 128]. Franson's scheme [43] consists in placing two Mach–Zehnder interferometers, whose effect can be easily calculated (see exercises) from (20), each on the path of one of the two entangled photons (see Fig. 4).

If the long arm of the interferometers for the idler and signal photons has a tunable phase ϕ_i and ϕ_s with respect to the short one, the final state is

$$\Psi_{fr} = \frac{1}{2}\left(|s_1\rangle |s_2\rangle + |l_1\rangle |l_2\rangle \, e^{i(\phi_1+\phi_2)} + e^{i\phi_1} |l_1\rangle |s_2\rangle + e^{i\phi_2} |s_1\rangle |l_2\rangle\right), \quad (23)$$

where the subscripts 1, 2 refer to the photon entering the first and the second interferometers, while s, l denote short and long path, respectively. After the interferometers the two photons are addressed to detectors; if both have followed the short or the long path, they arrive in coincidence, otherwise they are lost for the coincidence window.

The detected photon pairs generate a coincidence rate

$$R_c \propto \eta_i\eta_s\langle\Psi_{fr}|a_i^\dagger a_s^\dagger a_i a_s|\Psi_{fr}\rangle = \frac{1}{4}\,\eta_i\eta_s\left(1 + \cos(\phi_1 + \phi_2)\right), \quad (24)$$

where $\eta_i\eta_s$ are the detector efficiencies on idler and signal paths, respectively. The striking fact about this equation is that it can be modulated with 100% certainty

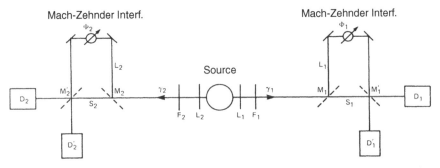

Fig. 4 Franson scheme for generating entangled states of photons. A source emits two energy–time entangled photons that after having crossed a Mach–Zehnder interferometer are detected by D_1 or D_1' and D_2 or D_2', respectively

using either of the widely separated phase plates (i.e., optical elements introducing a phase shift in the propagating electromagnetic wave). This "non-local" effect was the one suggested in [43] for testing Bell inequalities, where the parameters to be set are the phases $\phi_{i,s}$.

Of course these entangled states are also perfectly suited for quantum information applications. In particular phase entanglement is very well suited for quantum communication since it can be propagated in fiber without *birefringence effects* that affect polarization and thus entanglement.

The specific applications to quantum communication will be discussed later; instead, we now consider in some more detail the experimental realization of this kind of sources. For example, in the setup of [15], one of the first realizations of this kind of schemes, a BBO crystal was pumped by an argon ion laser beam in collinear regime producing photons pairs at 916 nm. A beam splitter separated the pair in two orthogonal directions, addressed to the two interferometers. Detectors were silicon avalanche photodiodes (cooled at $-25°$) with a measured efficiency of 16%, a relatively small value that nowadays is ameliorated by a factor 3–4. A 7 standard deviation violation of Clauser–Horne inequality (17) was observed.

Various more experiments [80, 30, 105, 62, 121] were realized with this kind of sources. In particular, some experiments are worth mentioning where a long distance entanglement transmission was performed. Among them one can quote the one by Tapster et al. [123] where a pair of entangled photons at 820 nm and 1.3 μm was produced in a lithium iodate crystal pumped by an argon laser at 501.7 nm. The shorter wavelength photon was immediately addressed to a single-mode fiber interferometer, while the other passed through a 4.3 km single-mode communications fiber before reaching the interferometer (1.3 μm is a well-suited wavelength for propagation in communication fibers). A 86.9% visibility (largely exceeding the 71% classical limit) was obtained.

A separation of more than 10 km was later obtained in [126, 127, 122, 119]. This experiment is particularly interesting for Bell inequality tests since (together with another one based on polarization entanglement realized at the same time [135]) it definitively closes the locality loophole (i.e., the request of having space-like separation between measurements) with an observed violation of the CHSH inequalities (19) violation of $S = 2.92 \pm 0.18$. Indeed, it allowed to eliminate also the small doubts pertaining to the non-random choice of the path of the Orsay experiment (doubts due to the fact that the polarization measurement choice was driven by a periodic signal).

Also another kind of entanglement (*time-bin*), used for applications to quantum communication, was realized with interferometers and PDC in pulsed regime. The scheme is based on placing on the pump beam a Mach–Zehnder interferometer (whose path length difference is large compared to the pump pulse length) before the nonlinear system where a polarization-entangled pair is generated. The pump photon can follow the short or the long path originating the superposition [16]

$$|\Psi_p\rangle = \frac{1}{\sqrt{2}}\left(|s\rangle + e^{i\phi}|l\rangle\right),\tag{25}$$

where $|s\rangle$ and $|l\rangle$ denote the photon that has followed the short and the long path, respectively, and ϕ denotes the phase difference between the two paths. After the PDC process one achieves the entangled state

$$\frac{1}{\sqrt{2}} \left(|s\rangle \, |s\rangle + e^{i\phi} \, |l\rangle \, |l\rangle \right) . \qquad (26)$$

In [16] a high visibility, 84%, was measured, again clearly exceeding the upper limit, 71%, for separable states. Later [124] the robustness of this entanglement to decoherence was shown for a 11 km fiber propagation.

In summary, phase-entanglement sources have been realized with high visibility and they represent a very important element for realizing long-distance quantum communication in fiber as we will discuss later.

2.5 Polarization Entanglement

In the previous paragraph, we have described sources of phase-entangled photons with important applications in fiber transmission. On the other hand, when dealing with quantum computation protocols or with communication in open space, polarization entanglement represents an importance resource.

Recently, several bright sources of polarization-entangled states of two photons have been produced: by type II PDC [45, 77] or by superimposing two type I PDC emissions, in this case using two thin adjacent crystals [137] or two crystals with an optical condenser between them [18] or by inserting them in an interferometer [66, 65, 41]. All of them can be used for generating all four Bell states, $\boldsymbol{\Phi}_\pm$ and $\boldsymbol{\Psi}_\pm$.

The schemes with type II PDC are based on the fact that, as already mentioned, in this case PDC-correlated photons are produced with orthogonal polarizations. In the collinear regime the two degenerate photons are emitted in two tangent cones. By selecting the intersection point of them, the two orthogonally polarized correlated photons exit in the same direction and can be separated by a beam splitter, generating an entangled state when one post-selects (i.e., only keeps into account) events where photons have left the beam splitter on different paths (taking therefore only a 50% of original pairs).

This scheme is rather simple and has allowed, for example, to measure a 10 standard deviation violation of Clauser–Horne inequality (17). Nevertheless, with the purpose of building very bright sources it is convenient to work in the non-collinear regime [76].

Indeed, in type II PDC, when the angle θ between pump and crystal optical axis is decreased, the two cones separate from each other entirely. On the other hand, if θ is increased, they intersect: therefore along two correlated intersections a, b one superimposes the probability amplitudes of generating a H (V) or V (H) photon in direction a (b). However, this bi-photon state is not yet entangled, since, due to birefringence in the nonlinear crystal, ordinary and extraordinary photons propagate with different velocities and different directions inside it. Therefore, longitudinal and transverse walk-offs (i.e., the optical beam displacement due to birefringence) must be compensated for restoring indistinguishability between the two polarizations and really generating an entangled state. This is usually achieved by inserting some birefringent medium (as quartz) along the optical path of photons.

The first realization of the non-collinear type II PDC scheme appeared in [45, 77], where a pump beam at 351 nm (150 mW) pumped a 3 mm long BBO crystal. The transverse walk-off was estimated to be negligible compared to coherent pump beam width. On the other hand, the longitudinal walk-off (385 fs) was larger than the coherence time, determined by interference filters and irises, and was compensated by an additional BBO crystal. All four Bell states were generated. A very significant violation, 102 standard deviations, of CHSH inequality was achieved, $S = -2.6489 \pm 0.0064$, showing the efficiency of this source.

More recently some very bright sources have been obtained [74, 71, 34, 4] reaching up to $77\,\mathrm{s}^{-1}$ coincidence counts for a milliwatt pump power [72] and a traditional crystal and even up to a measured coincidence flux of $300\,\mathrm{s}^{-1}$ for milliwatt of the pump [71] by using a periodically poled KTP crystal (i.e., a crystal where the susceptibility was periodically modulated giving a constructive interference in the emission).

Concerning transmission of polarization entanglement in open air, by using a type II source producing 20,000 entangled pairs per second (with a violet diode laser at 405 nm and 18 mW power as a pump) it was possible to transmit entanglement for more than 600 m with a clear violation of CHSH inequality, $S = 2.41 \pm 0.10$ [4]. More recently this result was extended up to 13 km [99], with a CHSH violation $S = 2.45 \pm 0.09$ and then to 144 km [129, 130].

As already mentioned, in alternative to the use of type II PDC, one can superimpose the emissions (with orthogonal polarizations) of two type I PDC crystals whose optical axes are rotated by 90° [57]: if the optical distance between the two crystals is shorter than the coherence length of pump laser, one generates a (non-maximally) entangled state:

$$|\psi_{\mathrm{NME}}\rangle = \frac{|H\rangle|H\rangle + f|V\rangle|V\rangle}{\sqrt{1 + |f|^2}} . \tag{27}$$

The explicit value of the parameter f can be tuned according to the specific choices in the setup. Incidentally, for both the type II and the present scheme a certain care must be addressed to the phase between the two components: moving in angle or frequency, change it and eventually allows to switch between different Bell states [19, 21].

A first realization of this scheme is based on superimposing the emission of two thin adjacent type I crystals. A little more in detail, in [137], an argon laser beam at 351 nm pumped two adjacent BBO crystals 0.59 mm long with optical axes oriented orthogonally. A rotatable half-wave plate on the pump beam before the crystals allowed to tune the laser beam polarization and therefore the parameter f of the generated state (27). A large violation of Bell inequalities (for maximally entangled states), $S = 2.7007 \pm 0.0029$, was observed. Furthermore, the source was rather bright, giving a $21{,}000\,\mathrm{s}^{-1}$ coincidence rate for 150 mW pump power (an order of magnitude larger than previous type II sources).

Alternatively, the two emissions can be superimposed by using an optical condenser [17]. In principle, this scheme allows a very precise superposition of the whole parametric fluorescence even with long crystals allowing higher intensities. A little more in detail, in [17] (see Fig. 5) a 351 nm argon laser beam pumped two crystals of LiIO₃, 250 mm apart (a distance smaller than the coherence length of the pumping laser). The PDC emission from the first crystal was focused into the second one by an optical condenser (two plano-convex lenses). A hole, drilled into the center of the lenses, allowed transmission of the pump radiation without absorption. A small quartz plate ($5 \times 5 \times 5\,\mathrm{mm}$) in front of the first lens of the condenser compensated the displacement of the pumping beam at the exit of the first crystal deriving from birefringence. Finally, a half-wavelength plate immediately after the condenser rotated the polarization of the pump beam that excited in the second crystal a spontaneous emission which was cross-polarized with respect to the first one.

The achieved coincidence rate was analogous to the one of the previous scheme. A test of Bell inequalities by using non-maximally entangled states (which, incidentally, allow a reduction on the quantum efficiency limit for a detection loophole-free test with respect to maximally entangled ones) led to a clear violation of (18), $R = 1.082 \pm 0.006$.

The setups discussed up to now were realized in the continuous wave (cw) regime. Nevertheless, for timing reasons, the pulsed regime is preferable for many quantum information protocols. Thus, recently, many studies were devoted to produce sources in this regime. When the pump pulses are very short (typically hundreds of femtoseconds), amplitudes for photon pairs produced at different depth inside the crystal become distinguishable, reducing two-photon interference

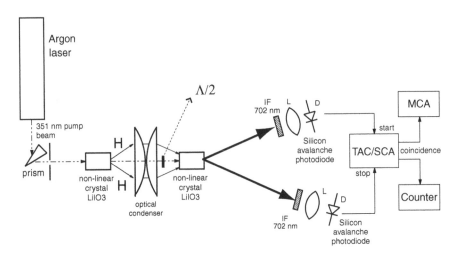

Fig. 5 Sketch of a bright source of polarization-entangled photons realized by superimposing two type I PDC emissions. The detection apparatus is also shown

visibility [64]. This problem required to use either thin ($\approx 100\,\mu$m) nonlinear crystals [116] or narrow band spectral filters (for increasing coherence length) in front of detectors [55, 54, 33]. However, these solutions significantly reduce the available flux of entangled photon pairs.

With the purpose of overcoming these limits, bright sources in pulsed regime were obtained in interferometric schemes (as by pumping with a femtosecond mode-locked laser two type I BBO crystals inserted in a Mach–Zehnder interferometer [67] or in a scheme with a polarising beam splitter [103]). Another bright source in femtosecond pulsed regime was also obtained by addressing back the PDC emission and the pump beam, both with the polarization rotated $\pi/2$ by two passages through a $\lambda/4$ wave plate, to the same type I crystal by means of a spherical mirror [5, 31]. A 213 σ violation of Bell inequality was observed.

Before concluding this list of sources of bipartite photon-entangled states, we mention that a scheme based on an interferometer in the continuous wave regime has been realized as well [41]. The setup consisted in a Mach–Zehnder interferometer where the laser beam, split by a first beam splitter, pumped two identical type II crystals inserted each in a different arm (A,B) originating in the state

$$\frac{|H_A(\omega_s)\rangle|V_A(\omega_i)\rangle + |H_B(\omega_s)\rangle|V_B(\omega_i)\rangle}{\sqrt{2}}. \tag{28}$$

After rotating the polarization with a half-wave plate on one of the arms, the two emissions were recombined on a polarizing beam splitter producing the following entangled state:

$$\frac{|H_1(\omega_s)\rangle|V_2(\omega_i)\rangle + |V_1(\omega_s)\rangle|H_2(\omega_i)\rangle}{\sqrt{2}}, \tag{29}$$

where 1, 2 refer to the two PBS output ports. This source, whose practical implementation was based on a single crystal with counter-propagating pump beams, was able to produce a flux of $12,000\,\text{s}^{-1}$ entangled photon pairs, for milliwatt of pump beam, and 100 σ violation of the CHSH inequality.

Finally, it is worth mentioning that an extension of these sources also allowed to generate entangled states of more than two photons. Let us sketch the proposal in [96, 14] to generate a GHZ *polarization-entangled state* (see chapter "Bipartite Quantum Entanglement", Example 4)

$$\Psi_{\text{GHZ}} = \frac{1}{\sqrt{2}}\left(|H\rangle|H\rangle|H\rangle + |V\rangle|V\rangle|V\rangle\right). \tag{30}$$

The scheme consisted in transforming two pairs of polarization-entangled photons produced simultaneously in a type II crystal pumped by a high-intensity UV 200 fs pulse into three entangled photons by using post-selection [145]. In more detail, in some rare event two entangled pairs $(|H\rangle|V\rangle - |V\rangle|H\rangle)/\sqrt{2}$ were produced by the same pulse. The selection of the desired state was then obtained by inspecting a posteriori the four-fold coincidence recording obtained by the apparatus in Fig. 6.

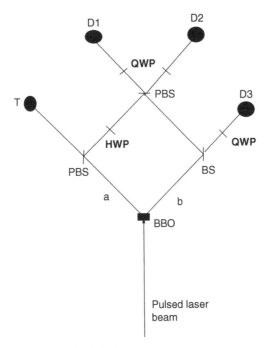

Fig. 6 Setup for generating a GHZ pair [96]

The photon registered at detector T is always horizontally (H) polarized and thus its partner in b must be vertically (V) polarized. The photon reflected at the polarizing beam splitter in arm a is always V, being turned into equal superposition of V and H by a $\lambda/2$ wave plate, and its partner in arm b must be H. Therefore if all four detectors click at the same time, the two photons at detectors D_1 and D_2 must have been either VV or HH. Thus the photon at D_3 was H or V, respectively. The indistinguishability of both cases was obtained by using narrowband filters (4 nm) to increase coherence time to about 500 fs. Extensions of this scheme were then realized for entangling from 4 [147, 136], 5 [148] up to 6 photons [84].

3 Optical Realizations of Quantum Information Protocols

In line of principle, every two-level system can be used for implementing qubits (see chapter "Quantum Probability and Quantum Information Theory"). A short list (of systems that have already had experimental implementations) includes the following:

- ions (where the CNOT gate was implemented [82, 112]);
- nuclear magnetic resonance (factorization of $N = 15$ by means of Shor's algorithm [132]) (see chapter "Quantum Algorithms");

- solid-state devices as quantum dots or superconductor devices (realized proofs of principle of logical gates);
- cavity QED, where a controlled phase gate is obtained by coupling Rydberg atoms with photons in a cavity [104] (see chapter "Physical Realizations of Quantum Information").

As previously discussed, among the various proposals photons represent one of the most interesting system for implementing a qubit realization, since

- one has relatively efficient single-photon sources at disposal;
- efficient bright sources of (few) entangled photons do exist;
- photons can travel large distances keeping their coherence properties;
- single-photon gates are easily implemented by beam splitters, phase shifters, etc.

On the other hand, the weak (medium-mediated) interaction between photons poses serious drawbacks to the implementation of two-photon quantum gates (that also poses limits on distinguishability of Bell states as we will discuss in Sect. 5.6). This strongly limits the possibilities of building a universal set of quantum optical gates and thus the chances of realizing an optical implementation of quantum computation (see chapter "Quantum Algorithms"). Furthermore, various quantum communication protocols, as dense coding and quantum teleportation (see chapter "Quantum Probability and Quantum Information Theory"), rely on the so-called Bell measurement, i.e., the discrimination between the four maximally entangled states of two qubits (see chapter "Quantum Probability and Quantum Information Theory", Sect. 6.4 and chapter "Bipartite Quantum Entanglement", Sect. 2). A complete deterministic Bell state discrimination is possible only if the two photons effectively interact; otherwise some of the Bell states cannot be discriminated. As we will see, many quantum information protocols realized up to now are affected by this problem.

4 Optical Quantum Computation Protocols

4.1 Kerr Controlled Phase Gate

Generally speaking, any quantum information processor performs a unitary transformation, which can always be decomposed in terms of single- and two-qubit operations (see chapter "Quantum Algorithms", Sect. 2).

This means that if one wants to implement quantum information protocols using photons, controlled conditional two-photon quantum dynamics is a necessary tool. A simple, yet fundamental, example of conditional quantum dynamics is provided by the so-called quantum phase gate, which operates on a single-photon state, a phase shift whose value is conditioned upon the state of the other photon. This gate together with single qubit gates, that also are easily realizable with beam splitters and phase shifters, represents a universal set of quantum logical gates.

In the case of photonic qubits, an implementation of the quantum phase gate would be in principle trivial, because it is a direct consequence of the cross-phase modulation taking place in nonlinear Kerr media (i.e., on the reciprocal effect on the phase of the two electromagnetic fields). Kerr effect is indeed a phenomenon where the presence of an electromagnetic field modifies the refraction index of a medium and, therefore, affects the propagation of a second electromagnetic field.

If the interaction is sufficiently strong the effect happens at single-photon level, being described by the interaction hamiltonian

$$H_{\mathrm{Kerr}} = g\,\hbar\,n_1\,n_2\,, \tag{31}$$

where n_1, n_2 are the number operators for the two fields. Thus field 2 evolves accordingly to (for simplicity we neglect free-field evolution)

$$\frac{\mathrm{d}a_2(t)}{\mathrm{d}t} = \frac{\mathrm{i}}{\hbar}\left[H, a_2\right] = \mathrm{i}g\left[n_1 n_2, a_2\right] = -\mathrm{i}\,gn_1 a_2, \quad a_2(t) = \mathrm{e}^{-\mathrm{i}gn_1 t}a_2(0). \tag{32}$$

Namely, it acquires a phase that depends on the number of photons of field 1. Unfortunately, the practical implementation of such a dynamics is problematic, because one needs very large nonlinearities together with negligible photon absorption, which are usually incompatible in standard Kerr devices. However, recent achievements in ultra-slow light propagation in a cold gas of atoms opened the way for the realization of significant conditional phase shifts also between two traveling single-photon pulses. In fact, an extremely slow group velocity (and even its "freezing") is obtained as a consequence of *electromagnetically induced transparency* (EIT), a quantum optical phenomenon where due to interference introduced by driving two levels of a three-level atomic system with a driving electromagnetic field the medium becomes effectively transparent to a probe field corresponding to the transition between one of these levels and the third one. For example, a scheme for realizing a controlled phase shift with rubidium atoms was proposed in [94]. Very recently, a step in this direction has been realized by observing EIT at single-photon level [38]. Nevertheless, the road toward the realization of a two-qubit gate by this method looks still hard and long. Incidentally, we would also like to mention that EIT has also been studied for realizing quantum memories [100, 36, 28, 38, 2].

4.2 Linear Optics Probabilistic CNOT

An alternative would be the realization of a two-qubit gate based on linear optics. In general no deterministic universal two-qubit gate can be realized by linear optics; however, a probabilistic one (i.e., working only with a non-unity probability) can be made with linear optical elements only [70].

This possibility seems rather interesting and in the following we present as an example the setup implemented in [101]. For further, recent progresses see [93, 42, 149, 46, 92] as well. In general fidelities above 80% have been reached.

A linear optics CNOT can be achieved by combining the sub-elements that we are now going to discuss. The first of them is the quantum parity check, whose function is to transfer the value of the input qubit to the output one, provided that its value is the same as that of the second input qubit. If the two inputs are different (odd parity) the device produces no output. A photon is absorbed (measured) in any event. This operation is realized by mixing two photons in a *polarizing beam splitter* (PBS) oriented in H, V basis and accepting the output of mode 2 only when one and only one photon on the other output of PBS is measured by a polarization-sensitive detector (in the 45° basis).

In more detail, let us input an unknown state

$$\Psi_{\text{in}} = \alpha|H\rangle + \beta|V\rangle \tag{33}$$

on one side and a single-photon state

$$\frac{|H\rangle + |V\rangle}{\sqrt{2}} \tag{34}$$

on the other side, after PBS we obtain (2 and c denote the two PBS outputs the following):

$$\frac{|45\rangle_c \left(\alpha|H\rangle + \beta|V\rangle\right)_2 + |-45\rangle_c \left(-\alpha|H\rangle + \beta|V\rangle\right)_2}{2} \tag{35}$$

$$+\frac{\left(\alpha|H\rangle_2|V\rangle_2 + \beta|H\rangle_c|V\rangle_c\right)}{\sqrt{2}}, \tag{36}$$

where the second term leads to unsuccessful cases when the detector receives zero or two photons.

On the other hand, by accepting the output when one and only one photon is received by the detector D_c^{45} and no photon by D_c^{-45} we obtain the correct output, with probability 0.25. This probability of success can be increased to 0.5 by accepting also the case with one and only one photon observed by D_c^{-45} and no photon by D_c^{45}, but actively imparting an additional phase shift to the output. Also, it is worth noticing that no knowledge on the output polarization is acquired.

Then, a quantum encoder can be implemented by inputting to one port of the parity check a two-photon entangled state of the form

$$|\Phi_+\rangle = \frac{|H\rangle|H\rangle + |V\rangle|V\rangle}{\sqrt{(2)}}. \tag{37}$$

In this case, a successful detection post-selects, with 0.5 probability, the state

$$\alpha|H\rangle_2|H\rangle_b + \beta|V\rangle_2|V\rangle_b. \tag{38}$$

A second gate to be constructed for building a (non-destructive) CNOT is the so-called *destructive* CNOT, whose goal is to flip the polarization state of target photon if and only if the control one is vertically polarized. This device is realized by mixing a target input photon Ψ_{in} with a second input photon on a PBS oriented in 45° basis. If the control photon is vertical one has (d and 3 denote the two outputs)

$$\frac{\alpha\left(|45\rangle_d\,|45\rangle_3 \,-\, |-45\rangle_d\,|-45\rangle_3\right) + \beta\left(|45\rangle_d\,|45\rangle_3 \,+\, |-45\rangle_d\,|-45\rangle_3\right)}{2}$$

$$+ \frac{\alpha\left(|45\rangle_3\,|-45\rangle_3 \,-\, |45\rangle_d\,|-45\rangle_d\right) + \beta\left(|45\rangle_3\,|-45\rangle_3 \,+\, |45\rangle_d\,|-45\rangle_d\right)}{2}\,,$$

(39)

where the second term originates unsuccessful cases. Rewriting the amplitudes (in the first term) in H basis, one has

$$\frac{|H\rangle_d\left(\alpha\,|V\rangle_3 \,+\, \beta\,|H\rangle_3\right) + |V\rangle_d\left(\alpha|H\rangle_3 \,+\, \beta\,|V\rangle_3\right)}{2}\,.$$

(40)

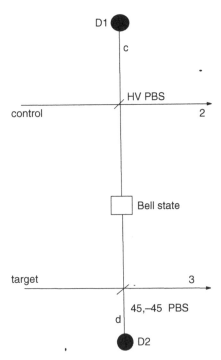

Fig. 7 Probabilistic CNOT. Detectors $D1$ and $D2$ are polarization sensitive (i.e., a PBS plus two photo-detectors)

Further, by accepting only the outcomes in which detector D_d^H receives one and only one photon and D_d^V no photons, we have the correct output $\alpha|V\rangle_3 + \beta|H\rangle_3$ with probability 0.25. Again, success probability can be increased to 0.5 by accepting also outputs where D_d^V receives one and only one photon and D_d^H no photons and performing a unitary operation on the output. Finally, one can consider the case where the control-qubit was horizontally polarized for checking that in this case the output is equal to target input.

The previous logical operations can then be combined for realizing the CNOT gate. The quantum encoder (see Fig. 7) copies the value of control-qubit in one output (2 in the figure) and in the input of the destructive CNOT. Thus the output from CNOT (3 in the figure) contains the desired result of the logical operation, while the value of control qubit is preserved in the other output mode. A simple calculation shows that this device works effectively as a CNOT with 0.25 success probability.

Recently, integrated quantum optics circuits including two-qubit gate have also been realized [120]. In conclusion of this paragraph we would also like to mention recent developments toward an experimental realization of linear optics cluster computation [44, 131].

5 Quantum Communication

As we have discussed in the former section qubit realization based on photons can offer a scheme for realizing quantum computation protocols, a possibility that is now investigated among many other proposals. On the other hand, when considering the transmission of quantum information, photons are the obvious carriers. In fact, photons can travel long distances both in air and in fiber and they do it at the speed of light.

As already described in chapter "Quantum Probability and Quantum Information Theory", quantum communication allows either protocols that have no classical equivalent as teleportation and dense coding or protocols allowing absolutely secure transmission of information in chapter "Quantum Cryptography". In the following we will describe some experimental realization of these quantum communication protocols. However, as in chapter "Quantum Probability and Quantum Information Theory", Sect. 6.1, we begin by discussing *quantum impossibilities*, that is, what quantum communication is not able to do, in particular *faster-than-light transmission* and *cloning of unknown quantum states*.

5.1 No Faster-than-Light Transmission by Using EPR Correlations

The non-locality of quantum entangled states does not allow any transmission of information faster than light and thus does not conflict with special relativity, albeit many opposite claims due to many different, also rather influent, authors.

Before discussing the general theorem, we consider some specific examples that provide a clear insight into why quantum non-locality does not allow super-luminal communication and give somehow a first indication as to what can or cannot be done in quantum communication.

Let an entangled pair be shared between two observers, Alice and Bob. Alice performs on her particle a spin test along the z direction: what she obtains is a perfectly casual sequence of 1 and -1, each outcome with probability 1/2. If also Bob performs the same test (in principle they could be space-like separated, whence, in a certain reference frame, Alice's measurement comes before Bob's, while in another reference frame the opposite occurs), quantum mechanics predicts that Bob's results are perfectly correlated with Alice's ones: every time Alice observes a 1 Bob obtains -1 and vice versa. Anyway, if there is no classical communication between them, the only knowledge on Bob's side is that contained in a random sequence of 1 and -1. Therefore, he cannot in any way know if Alice has performed a measurement or not, his results forming in both cases a random sequence. Thus, Alice and Bob cannot communicate any piece of information in this way.

However, the situation could change if Bob would be able to "clone" each particle he receives. For example, let us suppose that he posseses an apparatus creating 4N photons that are exact copies of an arbitrary input photon. If Alice chooses between performing a test along the z axis or along a basis at $45°$ with respect to the z axis and she obtains either 1 or -1 as outcomes, then Bob can use his cloning apparatus for detecting which result she has obtained. In fact, he can clone his particle and send N copies to four different apparatuses measuring the spin along z, $-z$, or the two conjugated directions of the second basis, respectively. When Alice has got outcome 1 (-1) in the z basis, Bob gets, in the same basis, N (0) particles giving outcome -1 and 0 (N) with outcome 1. Instead, in the other basis, he would end up with $N/2$ particles giving outcomes 1 and -1. Exactly the same result would be obtained, *mutatis mutandis*, had Alice chosen to perform the test in the other basis. Namely, in this case, if Alice obtains outcomes 1 (-1), Bob would observe, using the second basis, N (0) photons giving outcome -1 and 0 (N) with outcome 1, whereas in the z basis he would observe $N/2$ particles providing both the outcomes 1 and -1. Therefore, by simply inspecting which detector detects zero events, Bob would be able to know Alice's outcome. Of course, such a knowledge could be used to suitably engineer signal transmission. Nevertheless, quantum mechanics laws do not allow such a scheme, since it has been demonstrated that cloning an arbitrary unknown quantum state is impossible [138].

The demonstration of this theorem is rather simple. Let us suppose that we dispose of a cloning machine, that from an unknown quantum state $|\Psi_1\rangle$ and a second "target" state $|\Psi_0\rangle$ produces a copy of the first. The action of the cloning machine can be described by a unitary operator U through

$$|\Psi_1\rangle \otimes |\Psi_0\rangle \mapsto U\Big(|\Psi_1\rangle \otimes |\Psi_0\rangle\Big) = |\Psi_1\rangle \otimes |\Psi_1\rangle . \tag{41}$$

If we assume to apply the same cloning machine to an arbitrary second state $|\Psi_2\rangle$, the result would be

$$|\Psi_2\rangle \otimes |\Psi_0\rangle \mapsto U\Big(|\Psi_2\rangle \otimes |\Psi_0\rangle\Big) = |\Psi_2\rangle \otimes |\Psi_2\rangle. \tag{42}$$

Furthermore, taking the inner product of the equalities in the two former equations we obtain the following:

$$\langle \Psi_1 | \Psi_2 \rangle = (\langle \Psi_1 | \Psi_2 \rangle)^2 \; ; \tag{43}$$

however, this relation requires that either $|\Psi_1\rangle = |\Psi_2\rangle$ or $|\Psi_1\rangle$ is orthogonal to $|\Psi_2\rangle$: the cloning machine can clone only orthogonal states and therefore a universal cloning machine is impossible. Incidentally, this theorem is also at the basis of the security of QKD (see chapter "Quantum Probability and Quantum Information Theory", Sect. 6.1 and chapter "Quantum Cryptography", Sect. 3.1).

The former argument gives a clear hint of why quantum non-locality cannot be used for transmitting faster-than-light communications. In effect, a general theorem, demonstrated at the beginning of the 1980s [50], shows the absolute impossibility of using quantum non-locality for faster-than-light transmission. Here is a sketch of the proof.

Suppose to perform a projective measurement (see chapter "Quantum Probability and Quantum Information Theory", Sect. 6.4) on the subsystem S_1 of a composed system $S_1 + S_2$ described by a statistical operator ρ_{12}. This corresponds to projecting onto a specific eigenstate $|s\rangle$, an operation described by a projection operator $P_s^1 = |s\rangle\langle s|$. When Alice performs a non-selective measurement on S_1 (i.e., where one keeps all the outcomes), the statistic operator ρ_{12} transforms according to the following:

$$\rho_{12} \rightarrow \rho'_{12} = \sum_s P_s^1 \rho_{12} P_s^1 \; . \tag{44}$$

All the information on the subsystem S_2 is contained in the reduced statistical operator ρ_2, obtained by taking the partial trace on the Hilbert space $\mathcal{H}_1$ (corresponding to S_1):

$$\rho_2 = \mathrm{Tr}_1(\rho'_{12}) = \mathrm{Tr}_1\Big(\sum_k P_k^1 \rho_{12} P_k^1\Big) = \sum_k \mathrm{Tr}_1\Big(P_k^1 \rho_{12} P_k^1\Big), \tag{45}$$

where we have used the fact that trace is a linear operation. By using the properties of the trace (see chapter "Hilbert Space Methods for Quantum Mechanics", Sect. 1.5), it follows that

$$\rho_2 = \sum_k \mathrm{Tr}_1\Big(P_k^1 \rho_{12} P_k^1\Big) = \sum_k \mathrm{Tr}_1\Big(P_k^1 \rho_{12}\Big) = \mathrm{Tr}_1\Big(\sum_k P_k^1 \rho_{12}\Big) = \mathrm{Tr}_1 \rho_{12} \; . \tag{46}$$

This is exactly the same reduced density operator that one would have obtained without any measurement on the subsystem S_1: it is not possible to distinguish whether a measurement on system 1 has been made or not, by performing measurements on the system 2 only.

More recently, this theorem has been extended to more general kinds of measurement [49] and to the case of approximate cloning [24] as well, showing with absolute generality the impossibility of faster-than-light communication by exploiting quantum mechanical non-locality.

5.2 Dense Coding

Let us now describe what quantum mechanics allows to do beyond classical mechanics, namely let us consider some quantum communication protocols without classical equivalent. We begin with the simplest one: the dense coding.

As already seen in chapter "Quantum Probability and Quantum Information Theory", Sect. 7.2, dense coding is a quantum communication protocol addressed to transmit 2 bits of information by manipulating just one single qubit. The idea is to use the entanglement of a Bell state, for example, of

$$|\Psi_+\rangle = \frac{|0\rangle |1\rangle + |1\rangle |0\rangle}{\sqrt{2}} . \tag{47}$$

One of the particles is kept by Alice, the other is sent to Bob who performs one of the following actions:

1. an identity operation;
2. a state flip: $|0\rangle \rightarrow |1\rangle$, $|1\rangle \rightarrow |0\rangle$;
3. a state-dependent phase shift (i.e., a phase shift differing by π for the two qubits);
4. the previous two operations together.

This results in generating the whole set of Bell states (see chapter "Bipartite Quantum Entanglement", step 1 of Exercise 3)

$$|\Phi_{+,-}\rangle = \frac{|0\rangle |0\rangle \pm |1\rangle |1\rangle}{\sqrt{2}} , \quad |\Psi_{+,-}\rangle = \frac{|0\rangle |1\rangle \pm |1\rangle |0\rangle}{\sqrt{2}} . \tag{48}$$

Thus if Bob sends back its particle to Alice, she can obtain two bits by measuring the Bell state of the pair.

The first experiment demonstrating this scheme was realized in [89] by using polarization entangled photons produced by type II PDC. The operations 1–4 were easily realized by using wave plates (one $\lambda/2$ and one $\lambda/4$). As discussed before, the difficult part is the detection of Bell states since no linear optical system allows a perfect identification of all the four states. The scheme used by [89] was based on combining the two photons on a BS followed by a PBS (in H, V basis) on every output port with a detector on every branch (in the following denoted by D_H and D_V

for the first PBS and D'_H and D'_V for the second, where H, V denote the polarization of the photons impinging on a given detector).

Since only Ψ_- has an antisymmetric spatial wave function, only this state is registered by coincidence detection of different outputs of the beam splitter (i.e., coincidences between D_H and D'_V or between D'_H and D_V). In the other cases the photons exit from the same output of the BS. Ψ_+ is then easily recognized since it gives rise to different polarizations of the photons and thus they reach two different detectors as well (D_H and D_V or their equivalent on the other branch). A high reliability (larger than 90%) for identification of these two states was reached when both photons were observed. On the other hand, the Φ states cannot be distinguished.

This represents an emblematic example of Bell state measurement with linear optics: in the simplest scheme one can recognize only 2 states over 4. Thus, the experiment gives a proof of principle of the scheme, but cannot represent a complete realization of it.

A complete Bell-state discrimination could only be obtained either with linear optics assisted by hyper-entanglement (i.e., entanglement in more degrees of freedom as polarization and momentum) [134] or with non-unity probability, by exploiting a non-deterministic scheme as the one described in the previous section.

5.3 Teleportation

As described in chapter "Quantum Probability and Quantum Information Theory", Sect. 7.1, teleportation is a protocol where an unknown quantum state Ψ is measured in a laboratory (Alice) together with a member of a bipartite entangled state; then, by applying a unitary operation on the other member of the entangled state according to the result of this measurement (communicated by a classical channel), Ψ is reconstructed in a remote laboratory (Bob). This protocol has been successfully reproduced with photons and (more recently) with atoms. In this paragraph we describe one of the first two realizations [14] and briefly mention the other [11].

The setup proposed in [13] is a modification of the one described in the previous paragraph: two polarization-entangled pairs of photons (Ψ_-) are produced by a pump crossing go and back (i.e., coming back to the crystal after a reflection) a type II crystal. One of the photons of the first pair is detected realizing a heralded single photon source, whose polarization is fixed with a suited choice of retardation wave plates. This is the state to be teleported. A Bell state measurement of this heralded photon (photon 1) and one member of the second pair (photon 2) is then performed with the method described in the former paragraph. Of course this requires a precise setting of the arrival time of the two photons at the measurement apparatus, which is obtained by acting on the mirrors addressing the two photons to the BS where they are combined. Finally, the second member of the entangled pair (photon 3) was submitted to a polarization measurement (by means of a PBS and two *avalanche photo-detectors (APD)*).

Because of the limitations discussed in the previous paragraph and for simplicity of realization of the setup, the measurement was restricted to Ψ_- projection only. Furthermore, Bob did not perform any unitary operation, but only correlation among photons 1 and 3 was measured. The results of this measurement clearly showed the correlation between photons 1 and 3, certifying that if the unitary operation would have been performed then photon 1 would have been teleported successfully. In 45° basis one observed a strong dip, $(63 \pm 2)\%$ visibility, for opposite polarizations of photons 1 and 3 when the right delay between them was set.

It is also worth to mention that to rule out any classical explanation for these results [13], a four-fold coincidence measurement for the case of teleportation of the $+90°$ polarization states, i.e., for a non-orthogonal state to the $+45°$, has been performed obtaining a $(70 \pm 3)\%$ visibility without background subtraction. These results correspond to a fidelity above 80% for the teleported state. Also, at the same time of [13], another teleportation experiment was performed by exploiting only two photons [11]: the EPR pair is realized by path entanglement and the polarization state of one of the photons of the EPR pair is used as the state to be teleported. Drawbacks similar to those of the former experiment were also present.

More recent experiments involved a complete Bell-state measurement [68, 111] or the realization of the final unitary transformation [51]. More precisely, complete teleportation has been obtained in a scheme analog to the first one described in this paragraph either by using a probabilistic CNOT gate [111] or by using *sum frequency generation (SFG)*.

In the first case, one applies to the Bell states the operator $U = (\mathbf{H} \otimes 1)U_{\text{CNOT}}$, where $\mathbf{H}$ denotes the *Hadamard gate* and U_{CNOT} the unitary operator implementing the CNOT gate; whence the easily measurable states

$$U\Phi^+ = |HH\rangle \,, \ U\Phi^- = |VH\rangle \,, \ U\Psi^+ = |HV\rangle \,, \ U\Psi^- = |VV\rangle \qquad (49)$$

follow. In the second case, either two $H(V)$ photons are converted into a single H (V) photon, type I SFG, or a $H-V$ photon pair is converted into a single photon by type II SFG. The four Bell states are then distinguished by polarization measurements on the up-converted photons.

Finally, in [51] the final unitary transformation at Bob's side was obtained by using an electro-optic Pockels cell driven by Alice's classical transmission after having retarded Bob's photon in fiber.

In conclusion of this section, it should also be mentioned that, recently, a quantum state encoded in light was teleported to an atomic ensemble containing 10^{12} caesium atoms [117].

5.4 Quantum Swapping

Another interesting quantum communication protocol is quantum swapping, where entanglement between two photons, both part of a different entangled pair, is established by performing a joint Bell measurement on the other two photons of the two pairs. The experimental scheme for swapping has been again obtained by a modification of the setup described in the two previous paragraphs [63].

The simple modification is that now the first photon of the first EPR pair (0) is not used as trigger for heralding the other, but it is measured together with photon 3 showing how they became entangled by a Bell measurement on photons 1–2. In [63] the entanglement between photons 0 and 3 was then checked by measuring the CHSH inequality (19). The measured value $S = 2.421 \pm 0.091$, largely violating the classical limit, shows the success of the protocol.

5.5 Quantum Cryptography

Among the recent applications of quantum mechanics to technology the possibility of transmitting absolutely confidential messages has received much attention, due to the possibility of creating a key for encoding and decoding secret messages by transmitting single quanta between two parties (see chapter "Quantum Cryptography"). The underlying principle of *quantum key distribution* (QKD) is that nature prohibits to gain information on the state of a quantum system without disturbing it. Since the original proposal of quantum cryptography [138], many different protocols for such a transmission have been proposed [53].

In the BB84 scheme [9] single photons are transmitted from Alice to Bob, preparing them at random in four partly orthogonal polarization states (at 0° and 90°, −45° and 45°, for example). Bob selects the basis for measuring the photon's polarization at random as well. Then Alice and Bob communicate on a classical channel the basis they have used (but not the results of course): when they have used the same basis Bob knows what polarization was selected by Alice and can build a key, for example, by attributing a 0 when the photon is found with vertical polarization and 1 when horizontally polarized. If a spy (Eve) tries to intercept the message, she inevitably introduces errors, which Alice and Bob can detect by comparing a sub-sample of the generated key.

In the Ekert protocol [39] entangled pairs are used. Both Alice and Bob receive one particle of the entangled pair. Then they perform a measurement choosing among at least three different directions. Again, Alice and Bob communicate on a classical channel the basis they have used: if measurements were performed along parallel axes, they are used for generating the secret key. The other measurements can be used for a test of Bell inequalities. If Eve tries to eavesdrop the message, she inevitably affects the entanglement between the two particles leading to a reduction of the violation of the Bell inequalities, which allows Alice and Bob to recognize the presence of the spy.

Quantum key distribution is now at the stage of prototypes demonstrating the feasibility of long distance communications to the point that QKD is now ready for commercial applications, whose success will depend on the real commercial request of absolutely secure communications. Various possible protocols are available, based both on entangled and on non-entangled states. Furthermore, one can use either single photon states or multi-photon (as coherent) ones. The maximum distance that can be reached with a certain source and a certain protocol is the one where the quantum bit error (Qber) introduced by the channel reaches the limit for having a secure distilled rate above zero between Alice and Bob. In protocols that do not require entanglement, as BB84, in practice one can use either PDC-heralded photons or an attenuated laser beam (that well approximates a weak coherent state (see chapter "Field-Theoretical Methods", Sect. 2.3). The difference between the two sources is in the two or more photons component, which is larger for coherent states. Since having a higher photon number component means that Eve can eventually eavesdrop these photons in excess without affecting the transmission between Alice and Bob, this reflects in a longest distance for a secure communication for the heralded photon sources. For example [53], if one compares what is expected for BB84 QKD in fiber at 1,550 nm the limit decreases from about 120 km for a single-photon source to about 80 km for a coherent state with on average 0.1 photons.

Nevertheless, more recently, protocols in continuous variables were proposed and some of them implemented that can in principle allow long-distance QKD with intense coherent or squeezed states (see chapter "Field-Theoretical Methods", Sect. 2.4). For example, in [56] a scheme was presented based on modulating in phase and amplitude coherent states at Alice's side and subsequently performing a homodyne detection at Bob's side. More precisely, Alice sends a coherent state (of a few hundreds of photons) modulated in phase and amplitude $|x + i\,p\rangle$, and Bob randomly chooses to measure either quadrature $x = a + a^\dagger$ or $p = i\,(a^\dagger - a)$. The set of correlated variables, to be used for building the key, is established by informing Alice on a public channel of which quadrature was measured for each run. An eventual eavesdropping would introduce some noise that allows Alice and Bob to identify it. Experimental implementation yielded a net key transmission of about 1.7 Mbits for a loss-free line.

A different situation is met when one wants to use protocols based on entangled states. In this case one must refer to either phase or polarization-entangled photon sources described previously. Here we want also to mention that recently it has been shown how quantum communication based on codification in d-dimensional ($d > 2$) Hilbert spaces (*qudits*) presents a higher security than the traditional qubit schemes [7, 6, 12, 27, 25, 48]. Realization of qudits has been obtained, for example, by using multi-arm interferometers (an extension of time-bin scheme) [125], codification in bi-photon pairs (*HH, HV, VV qutrits*, or *ququats* if using two different wavelengths such that HV and VH are distinguishable) [88, 26, 29], in orbital angular momentum states (generated by holograms) [133], etc. Nevertheless, at the moment experimental transmissions of qudits are not yet competitive with traditional qubit ones. For what concerns experimental QKD based on single photon qubit transmission, that at the moment is the one allowing the longest communications, the use of encryption in polarization is particularly well suited for communication in open space, since air does not substantially present birefringence. For example, by using weak coherent states transmissions up to 23.4 km were reached in open air in daylight [73] with a 2 kbit per second rate. Very recently, [129, 130], violation of CHSH inequality was even observed up to a 144 km transmission between La Palma and Tenerife, where the receiver was a 1 m Ritchey–Chrétien telescope (the link optical efficiency was on at best -25 dB). Furthermore, a BB84 QKD (28 bit/s) was established by transmitting attenuated coherent states [113]; the use of a decoy state protocol was fundamental for reaching the security level, which derives by the request that the (random) error rate introduced by the background is smaller than the threshold for the used protocol.

In this kind of experiments the background due to diffuse sunlight is strongly attenuated by using both spectral selection (by interference filter) and temporal selection (by opening a detection window when the transmission is expected). An important point to be emphasized is that reaching such a distance in open air means that a transmission toward a satellite is now technically possible, since losses would be limited to the lower part of atmosphere [1]. Researches in this sense are now carried on. On the other hand, the control of birefringence in fiber is by far more difficult [53].

However, a QKD with polarization-entangled states produced by PDC (type II crystal source) was realized between a bank and the Vienna city hall by using a 1.45 km optical fiber [103], and a QKD network, including also a part exploiting transmission of polarized photons in fiber, has been built among several research institutes in Boston [40]. Furthermore, recent studies [19, 17] show interesting effects at level of restoration of entanglement in fiber when the temporal spread of correlation function is considered, results that could find application to polarization-based QKD in fiber.

Nevertheless, even if these achievements show that polarization encryption can be used for QKD in fiber as well, phase encryption looks more promising for a long distance communication in this case, due to smaller and more controllable fiber effects on the transmission. Among the most interesting results with these schemes we can quote the one of [87], where time-bin entanglement was distributed in fiber over 50 km. The use of an active phase stabilization with a frequency-stabilized laser and feedback loop allowed a long-term stability. A 15 standard violation of CHSH inequality and a Qber of 11.5% demonstrate the validity of the scheme for QKD purposes. Recently QKD even over 100 km in fiber was obtained by using interferometric schemes [89, 68], with a recent record distance of 184.6 km [107] by using cryogenic detectors (TES) with high quantum efficiency (65%).

Finally, we would like to hint at recent results concerning simple and rapid QKD protocols. Here the quest is for schemes as simple as possible in order to be easily implementable in a commercial version, even if relaxing security requests with respect to an eavesdropping realistically implementable with available technologies. For example, in [144] a very simple scheme is proposed which is able to transmit at more than 10 kHz transmission rate (raw detection rate) at a long distance in fiber (various tens of kilometers). The security analysis was limited to intercept-resend attacks. In this scheme Alice's source consists of an intensity-modulated cw laser, which prepares either a pulse with mean number α or a vacuum state. The logical bits are

$$|0\rangle = |\alpha\rangle|0\rangle \,, \quad |1\rangle = |0\rangle|\alpha\rangle \,, \tag{50}$$

where the order of the states denotes two subsequent pulses. Coherence (i.e., the presence of an eventual eavesdropper) is checked by the transmission of *decoy* pulses $|\alpha\rangle|\alpha\rangle$, whose coherence is checked by Bob through a Michelson interferometer, whose states are randomly addressed by a beam splitter (due to laser coherence there is a well-defined phase between any two non-empty pulses).

At the end of the exchange Bob tells which bits he has detected in the data line and which were observed by a detector of the interferometer that cannot fire when a decoy state was transmitted (due to interference). Then Alice tells Bob which bits he must remove from his raw key, since they correspond to decoy states and estimates the break of coherence (and thus the eventual presence of an eavesdropper) by using the data from the interferometer.

5.6 Other Quantum Protocols with Photons

Before concluding, we mention that various other interesting protocols have been realized by using the sources of entangled photons that we have described in Sect. 1.2. To review all of them is largely beyond the purposes of these lectures. In the following, we simply hint at a few of them.

A first example is entanglement distillation (see chapter "Bipartite Quantum Entanglement", Sect. 3.1); this may be obtained by acting through glass slabs with polarization-dependent reflectivity on polarization-entangled states produced by the two type I crystal scheme [79] and entanglement purification (i.e., extraction of maximally entangled pure states from a mixture) [97] (see chapter "Bipartite Quantum Entanglement", Sect. 3.1).

A second interesting achievement is approximate quantum cloning. As previously discussed, cloning an arbitrary unknown state is not possible in quantum mechanics. Nevertheless, an approximate quantum cloning can be realized with a fidelity up to 83%. PDC can be used as "amplifier" for cloning a seed state. In extreme synthesis, the scheme is based on applying the type II PDC Hamiltonian $k(a_H^\dagger b_V^\dagger - b_H^\dagger a_V^\dagger)$ ($a^\dagger, b^\dagger$ corresponding to two different spatial modes) to a seed (the state to be cloned), e.g. (because of rotational invariance of the Hamiltonian which is sufficient to consider one particular initial polarization)

$$-ikt\left(a_H^\dagger b_V^\dagger - b_H^\dagger a_V^\dagger\right)|1_V, 0\rangle = -ikt\left(\sqrt{2}|2, 0\rangle_a\, |0, 1\rangle_b - |1, 1\rangle_a\, |1, 0\rangle_b\right). \quad (51)$$

The two photons in mode a are the clones. Inspection of the output state (51) shows that with probability 2/3 both photons are vertically polarized, i.e., they are perfect clones, while they have opposite polarization with probabilities 1/3; in such a case the probability of picking a vertical photon is 1/2. Thus overall fidelity is $F = 2/3 \cdot 1 + 1/3 \cdot 1/2 = 5/6$, i.e., the optimal one. Experimentally, fidelities (see chapter "Hilbert Space Methods for Quantum Mechanics", Sect. 3.3) up to 81% have been obtained [81, 110].

As a further example we would like to hint at a first realization of decoherence-free subspaces. In [143, 95] it was shown how some particular state can be unaffected by the Hamiltonian describing decoherence, leading to a passive stabilization of quantum information. In [80] an example of this stabilization was shown by considering propagation of polarization-entangled photons in the state Ψ_- in a birefringent medium (the Bell states were produced by using the two adjacent type I PDC crystals' scheme). The effect of the medium was to transform $|H\rangle$ in $e^{i\theta_0}|H\rangle$ and $|V\rangle$ in $e^{i\theta_1}|V\rangle$, thus

$$|\Psi_-\rangle = \frac{|H\rangle\,|V\rangle - |V\rangle\,|H\rangle}{\sqrt{2}} \tag{52}$$

only acquired an irrelevant global phase $e^{i\theta_0+\theta_1}$. Since Ψ_- is invariant for basis transformation, this property remains valid in any basis (on the other hand, this is not the case of Ψ_+). The decoherence-free propagation of Ψ_- was then tested by checking the fidelity for the final state, which was measured to be $F = 0.97\pm0.04$ to be compared with $F = 0.51\pm0.03$, $F = 0.35\pm0.02$, and $F = 0.54\pm0.02$ for Φ_+, Φ_-, and Ψ_+, respectively. Finally, we would like to mention that correlations in PDC find applications also in disciplines related to quantum information as quantum imaging [85] and quantum metrology [52, 20]. Quantum imagining is devoted to overcome classical imaging limits by exploiting the properties of quantum states in analogy to the attempt of quantum information to overcome limits of classical one. Among the applications of quantum imaging actually under investigation we can mention the detection of weak amplitude objects beyond the standard quantum limit, entangled two-photon microscopy, image amplification by PDC, measurements of small displacements, quantum optical lithography, and teleportation of optical images. On the other hand, quantum metrology is addressed to exploit properties of quantum systems for improving calibration of devices and realization of standards of unities. Altogether one can envisage that the future of our society will be strongly influenced by the development of quantum technologies.

Exercise 1

1. *Plot the curve of PDC emission in function of angle and wavelength from a BBO crystal pumped by a 351 nm laser beam with an angle 31° with the crystal optical axis (type I).*
   Hint: find the wavelength dependence, Sellmeier equations, of the refraction index, e.g., on the Internet, and use it in the momentum conservation equation for deriving the emission angle of a given wavelength. Consider refraction at the exit surface of the crystal.
2. *Repeat Exercise 1 for the PDC emission from a BBO crystal pumped by a 351 nm laser beam with an angle 49° with the crystal optical axis (type II).*
3. *Calculate the rate of downconversion ($\langle\Psi|E^-E^+|\Psi\rangle$, where E is the electromagnetic field operator).*

4. *Evaluate which is the output of a beam splitter (BS) where two identical single photons enter simultaneously two different ports.*
   Result: the photons always exit the same port
5. *Evaluate the output of a Michelson interferometer by using the quantum description of a BS.*
   Hint: a_2 and a_3 are reflected back to the BS with a certain phase, Θ_2 and Θ_3, respectively. The final outputs are therefore described by $a_5 = ta_2e^{i\Theta_2} + ra_3e^{i\Theta_3}$ and a_6. Express a_5 and a_6 in functions of a_0 and a_1 and evaluate $\langle n_5 \rangle$ and $\langle n_6 \rangle$ for various inputs. As a check, show that $\langle n_5 \rangle + \langle n_6 \rangle = \langle n_1 \rangle$ when port 0 is entered by vacuum.
6. *Evaluate the output of a Mach–Zehnder interferometer by using the quantum description of a beam splitter.*
   Hint: proceed as in the former exercise by writing output operators in terms of input ones.
7. *Describe the quantum PBS.*
8. *Homodyne detection:*
   (a) show that by combining on a BS the optical state to be measured with a local oscillator (an intense coherent state with fixed phase) one can measure the quadrature variables x, p.
   Hint: evaluate the average photon number of the two outputs of the BS when the intensity of the LO field is much higher than the one of the state to be measured. Show that by changing the phase of the LO the difference between these two values gives the average value of both the quadrature observables.
   (b) by considering the heralded single-photon sources described in the text suggest a scheme for having the local oscillator locked in phase to the single photons.
9. *Suppose Alice transmits 10^5 photons per second, that the transmission of the channel is T, that quantum efficiency of Bob detector is $\eta = 0.5$, and that the background counts per second are $B = 100$. If the error threshold for the used protocol is $E = 12\%$, what is the minimal transmission T for having a secure QKD?*
10. *Evaluate the probabilities given in the paragraph following Eq. (51)*
11. *Show that the state Ψ_- is invariant for basis transformations and discuss why this property allows to define it as a decoherence-free state for certain Hamiltonians.*

References

1. Antonietti, N., et al.: quant-ph 0609049 and references therein
2. Appel, J., et al.: Phys. Rev. Lett. **100**, 093602 (2008) and references therein
3. Aspect, A., et al.: Phys. Rev. Lett. **49**, 1804 (1982)
4. Aspelmeyer, M., et al.: Science **301**, 621 (2003)
5. Barbieri, M., et al.: quant-ph 0303018
6. Bechmann-Pasquinucci, H., Peres, A.: Phys. Rev. Lett. **85**, 3313 (2000)
7. Bechmann-Pasquinucci, H., Tittel, W.: Phys. Rev. A **61**, 062308 (2000)

8. Bell, J.S.: Physics **1**, 195 (1965)
9. Bennet, C.H., Brassard, G.: Proceedings of the IEEE International Conference on Computers, Systems, and Signal Processing, Bangalore, India, p. 175. IEEE, New York (1984)
10. Beveratos, A., et al.: Eur. Phys. J. D **18**, 191 (2002)
11. Boschi, D., et al.: Phys. Rev. Lett. **80**, 1121 (1998)
12. Bourennane, M., et al.: Phys. Rev. A **63**, 062303 (2001)
13. Bouwmeester, D., et al.: Nature **390**, 575 (1997)
14. Bouwmeester, D., et al.: Phys. Rev. D **82**, 1345 (1999)
15. Brendel, J., et al.: Europhys. Lett. **20**, 275 (1992)
16. Brendel, J., et al.: Phys. Rev. Lett. **82**, 2594 (1999)
17. Brida, G.: Phys. Rev. A **75**, 015801 (2007)
18. Brida, G., et al.: Phys. Lett. A **299**, 121 (2002)
19. Brida, G., Chekhova, M., Genovese, M., et al.: Phys. Rev. Lett. **96**, 143601 (2006)
20. Brida, G., Genovese, M., Gramegna, M.: Laser Phys. Lett. **3**, 115 (2006) and references therein
21. Brida, G., Genovese, M., Krivitsky, L.A., et al.: Phys. Rev. A **76**, 053807 (2007)
22. Brida, G., Genovese, M., Novero, C.: J. Mod. Opt. **47**, 2099 (2000)
23. Brunel, C., et al.: Phys. Rev. Lett. **83**, 2722 (1999)
24. Bruss, D., et al.: Phys. Rev. A **62**, 062302 (2000)
25. Bruss, D., Macchiavello, C.: Phys. Rev. Lett. **88**, 127901 (2002)
26. Burlakov, A.V., et al.: Phys. Rev. A **60**, R4209–R4212 (1999)
27. Cerf, N.J., et al.: Phys. Rev. Lett. **88**, 127902 (2002)
28. Chanelière, T., et al.: Nature **438**, 833 (2005)
29. Chekhova, M.V., et al.: Phys. Rev. A **70**, 053801 (2004)
30. Chiao, R.Y., Kwiat, P.G., Steinberg, A.M.: Quant. Semicl. Opt. **7**, 259 (1995)
31. Cinelli, C., et al.: quant-ph 0406148
32. Clauser, J., et al.: Phys. Rev. Lett. **23**, 880 (1969)
33. Di Giuseppe, G., et al.: Phys. Rev. A **56**, R21 (1997)
34. Di Giuseppe, G., et al.: Phys. Rev. A **66**, 013801 (2002)
35. Di Giuseppe, G., Sergienko, A.V., Saleh, B.E.A., et al.: Quantum information and computation. In: Donkor, E., Pirich, A.R., Brandt, H.E. (eds.) Proceedings of the SPIE, vol. 5105, p. 39. SPIE, Bellingham (2003)
36. Duan, L.M., et al.: Nature **414**, 413 (2001)
37. Einstein, A., Podolsky, B., Rosen, N.: Phys. Rev. **47**, 77 (1935)
38. Eisaman, M.D., et al.: Nature **438**, 837 (2005)
39. Ekert, A.K.: Phys. Rev. Lett. **67**, 661 (1991)
40. Elliot, C., et al.: quant-ph 0503058
41. Fiorentino, M.: Phys. Rev. A **69**, 041801(R) (2004)
42. Fiorentino, M., Wong, N.C.: Phys. Rev. Lett. **93**, 070502 (2004)
43. Franson, J.P.: Phys. Rev. Lett. **62**, 2205 (1989)
44. Gao, W., et al.: arXiv:0905.2103
45. Garuccio, A.: In: Greenberger, D. (ed.) Fundamental Problems in Quantum Theory. New York Academy of Sciences, New York (1995)
46. Gasparoni, S., et al.: quant-ph 0404107
47. Genovese, M.: Phys. Rep. **413**, 319 (2005)
48. Genovese, M., Novero, C.: Eur. J. Phys. D **21**, 109 (2002)
49. Ghirardi, G.C., et al.: Europhys. Lett. **6**, 95 (1988)
50. Ghirardi, G.C., Rimini, A., Weber, T.: Lett. Nuovo Cim. **27**, 293 (1980)
51. Giacomini, S., et al.: Phys. Rev. A **66**, 030302 (2002)
52. Giovannetti, V., et al.: Science **306**, 1330 (2004)
53. Gisin, N., et al.: Rev. Mod. Phys. **74**, 145 (2002)
54. Grice, W.P., et al.: Phys. Rev. A **57**, R2289 (1998)
55. Grice, W.P., Walmsley, I.A.: Phys. Rev. A **56**, 1627 (1997)

56. Grosshans, F., et al.: Nature **421**, 238 (2003)
57. Hardy, L.: Phys. Lett. A **161**, 326 (1992)
58. Hayat, A., et al.: Nat. Phot. **2**, 238 (2008)
59. Hergert, E.: Single Photon Detector Workshop. NIST, Gaithersburg (2003)
60. Hiskett, P.A., et al.: New. J. Phys. **8** 193 (2006)
61. Hogele, A., et al.: Phys. Rev. Lett. **100**, 217401 (2008)
62. Horne, M.A., Shimony, A., Zeilinger, A.: Phys. Rev. Lett. **62**, 2209 (1989)
63. Jennewein, T., et al.: Phys. Rev. Lett. **88**, 017903 (2001)
64. Keller, T.E., Rubin, M.H.: Phys. Rev. A **56**, 1534 (1997)
65. Kim, Y.: Phys. Rev. A **63**, 062301 (2001)
66. Kim, Y., et al.: Phys. Rev. A **63**, 060301(R) (2001)
67. Kim, Y., et al.: Phys. Rev. Lett. **86**, 1370 (2001)
68. Kim, J., Takeuchi, S., Yamamoto, Y., et al.: Appl. Phys. Lett. **74**, 902 (1999)
69. Kimura, T.: Jpn. J. Appl. Phys. **43**, L1217 (2004)
70. Knill, E., et al.: Nature **409**, 46 (2001)
71. Kuklewicz, C.E., et al.: Phys. Rev. A **69**, 013807 (2004)
72. Kurtsiefer, C., et al.: Phys. Rev. Lett. **85**, 290 (2000)
73. Kurtsiefer, C., et al.: Nature **419**, 450 (2002)
74. Kurtsiefer, C., Oberparlaiter, M., Weinfurter, H.: Phys. Rev. A **64**, 023802 (2001)
75. Kwiat, P.G., et al.: Phys. Rev. A **41**, 2910 (1990)
76. Kwiat, P.G., et al.: Phys. Rev. A **49**, 3209 (1994)
77. Kwiat, P.G., et al.: Phys. Rev. Lett. **75**, 4337 (1995)
78. Kwiat, P.G., et al.: Science **290**, 498 (2000)
79. Kwiat, P., et al.: Nature **409**, 1014 (2001)
80. Kwiat, P.G., Steinberg, A.M., Chiao, R.Y.: Phys. Rev. A **47**, R2472 (1993)
81. Lamas-Linares, A., et al.: Science **296**, 712 (2002)
82. Leibfried, D., et al.: Nature **422**, 412 (2003)
83. Lodal, P., et al.: Nature **430**, 654 (2004)
84. Lu, C.-Y., et al.: Nat. Phys. **3**, 91 (2007)
85. Lugiato, L.A., et al.: J. Opt. B **4**, S176 (2002)
86. Mandel, L., Wolf, E.: Optical Coherence and Quantum Optics. Cambridge University Press, Cambridge (1985)
87. Marcikic, I., et al.: quant-ph/0404124
88. Maslennikov, G.A., et al.: J. Opt. B **5**, S530 (2003)
89. Mattle, K., et al.: Phys. Rev. Lett. **76**, 4656 (1996)
90. Mo, X., et al.: quant-ph/0412023
91. Munroe, M., Boggavarapu, D., Anderson, M.E., et al.: Phys. Rev. A **52**, R924 (1995)
92. O'Brien, J.: Science **318**, 1567 (2007)
93. O'Brien, J.L., et al.: Nature **426**, 264 (2003)
94. Ottaviani, C., et al.: Phys. Rev. Lett. **90**, 197902 (2003)
95. Palma, G., Suominen, K., Ekert, A.: Proc. R. Soc. Lond. A **452**, 567 (1996)
96. Pan, J.W., et al.: Nature **403**, 515 (2000)
97. Pan, J., et al.: Nature **423**, 417 (2003)
98. Peacock, A., Verhoeve, P., Rando, N., et al.: Nature **381**, 135 (1996)
99. Peng, C.-Z., et al.: quant-ph0412218
100. Phillips, D.F., et al.: Phys. Rev. Lett. **86**, 783 (2001)
101. Pittman, T.B., et al.: Phys. Rev. Lett. **88**, 257902 (2002)
102. Poh, H., et al.: arXiv 0905.3849
103. Poppe, A., et al.: Opt. Express **12**, 3865 (2004)
104. Raimond, M., et al.: Rev. Mod. Phys. **73**, 565 (2001)
105. Rarity, J.G., Tapster, P.R.: Phys. Rev. Lett. **64**, 2495 (1990)
106. Raymer, M., Beck, M.: In: Paris, M.G.A., Řeháček, J. (eds.) Quantum States Estimation. Lect. Notes Phys. **649**. Springer, Heidelberg (2004)

107. Rowe, M.A., et al.: Nature **409**, 791 (2001)
108. Sanaka, K., et al.: Indistinguishable photons from independent semiconductor single-photon devices. arXiv:0903.1849
109. Santori, C., et al.: Nature **419**, 594 (2002)
110. Scarani, V.: Rev. Mod. Phys. **77**, 1225 (2005)
111. Schmid, C.: Communication Presented at "Recent advances in Foundations of Quantum Mechanics and Quantum Information. Ad memoriam of Carlo Novero", Turin (May 2006).
112. Schmidt-Kaler, F., et al.: Nature **422**, 408 (2003)
113. Schmitt-Manderbach, T., et al.: Phys. Rev. Lett. **98**, 010504 (2007)
114. Schrödinger, E.: Proc. Camb. Philol. Soc. **31**, 555 (1935)
115. Sergienko, A.V., et al.: Phys. Rev. A **60**, R2622 (1999)
116. Shen, Y.R.: Nonlinear Optics. Wiley, New York (1984)
117. Sherson, J., et al.: Nature **443**, 557 (2006)
118. Simpson, D.A., et al.: arXiv:0903.3807
119. Stefanov, A., et al.: Phys. Rev. A **63**, 022111 (2001)
120. Stephens, A.M., et al.: Phys. Rev. A **78**, 032318 (2008)
121. Strekalov, D.V., et al.: Phys. Rev. A **54**, R1 (1996)
122. Stucki, D., et al.: Appl. Phys. Lett. **87**, 194108 (2005)
123. Tapster, P.R., Rarity, J.G., Owens, P.C.M.: Phys. Rev. Lett. **73**, 1923 (1994)
124. Thew, R.T., et al.: Phys. Rev. A **66**, 062304 (2002)
125. Thew, R.T., et al.: Phys. Rev. Lett. **93**, 010503 (2004)
126. Tittel, W., et al.: Phys. Rev. Lett. **81**, 3563 (1998)
127. Tittel, W., et al.: Phys. Rev. A **59**, 4150 (1999)
128. Tittel, W., Brendel, J., Zbinden, H., et al.: Phys. Rev. Lett. **81**, 3563 (1998)
129. Ursin, R., et al.: quant-ph 0607182
130. Ursin, R., et al.: Nat. Phys. **3**, 481 (2007)
131. Vallone, G., et al.: Phys. Rev. A **78**, 042335 (2008)
132. Vandersypen, L.M.K., Chuang, I.L.: Rev. Mod. Phys. **76**, 1037 (2004)
133. Vaziri, A., et al.: Phys. Rev. Lett. **89**, 240401 (2002)
134. Walborn, S.P., et al.: Phys. Rev. A **68**, 042313 (2003)
135. Weihs, G., et al.: Phys. Rev. Lett. **81**, 5039 (1998)
136. Weinfurther, H., Zukowski, M.: Phys. Rev. A **64**, 010102 (2001)
137. White, A.G., et al.: Phys. Rev. Lett. **83**, 3103 (1999)
138. Wiesner, S.: Sigact News **15**, 78 (1983)
139. Wootters, W.K., Zurek, W.H.: Nature **299**, 802 (1982)
140. Wu, E., et al.: Opt. Express **14**, 1296 (2006)
141. Zambra, G., et al.: Phys. Rev. Lett. **95**, 063602 (2005)
142. Zambra, G., Bondani, M., Spinelli, A.S., et al.: Rev. Sci. Instrum. **75**, 2762 (2004)
143. Zanardi, P., Rasetti, M.: Phys. Rev. Lett. **79**, 3306 (1997)
144. Zbinden, H., et al.: J. Phys. A **34**, 7103 (2001)
145. Zeilinger, A., et al.: Phys. Rev. Lett. **78**, 3031 (1997)
146. Zhang, Y., Kasai, K., Watanabe, M.: Opt. Lett. **27**, 1244 (2002)
147. Zhao, Z., et al.: Phys. Rev. Lett. **91**, 180401 (2003)
148. Zhao, Z., et al.: quant-ph0402096
149. Zhao, Z., et al.: quant-ph 0404129

Physical Realizations of Quantum Information

F. de Melo and A. Buchleitner

1 Introduction

When we want to consider physical realizations of quantum information models and protocols, we have to identify specific (experimental) settings, which allow to implement the fundamental building blocks of quantum information processing. Thus, first we have to identify these building blocks and the requirements they imply for an experimental realization.

The key ingredient in all quantum information processing is the *coherent* superposition and evolution of quantum eigenstates $|\psi_j\rangle$ of some system Hamiltonian H. For reasons of simplicity and because of the analogy with classical binary coding, one, in general, thinks of

- superpositions of the eigenstates $|0\rangle$ and $|1\rangle$ of two-level systems and
- superpositions of many-particle states

$$|j_1\rangle \otimes \cdots \otimes |j_k\rangle \otimes \cdots \otimes |j_N\rangle = |j_1, \ldots, j_k, \ldots, j_N\rangle, \quad \text{with } j_k \in \{0, 1\}, \quad (1)$$

of N-particle systems, composed of single particles with two discrete levels $|0\rangle$ and $|1\rangle$.

The levels $|0\rangle$ and $|1\rangle$, in our present context also known as the "computational basis states," are the discrete eigenstates of some single-particle[1] Hamiltonian H and can be energy eigenstates of, e.g., an atom, as shown in Fig. 1, or position eigenstates in a double-well potential, as in Fig. 2, or anything such as photon

---

F. de Melo (✉)

Physikalisches Institut der Albert-Ludwigs-Universität, Freiburg, Germany,
fernando.demelo@physik.uni-freiburg.de

A. Buchleitner (✉)

Physikalisches Institut der Albert-Ludwigs-Universität, Freiburg, Germany,
abu@uni-freiburg.de

[1] Or rather, of some single physical entity. It is indeed possible to encode the two-level system in more than one particle, forming a "logical qubit" [53].

de Melo, F., Buchleitner, A.: *Physical Realizations of Quantum Information*. Lect. Notes Phys. **808**, 253–276 (2010)
DOI 10.1007/978-3-642-11914-9_8 © Springer-Verlag Berlin Heidelberg 2010

Fig. 1 The two ("computational") energy eigenstates of an atom – in short a "two-level atom." In the laboratory, experimentalists make sure to isolate these two states from the, in general, many-energy eigenstates of an atom, ion, or molecule, by suitable realization of the interaction Hamiltonian which generates the desired unitary time evolution

Fig. 2 An alternative realization of a "qubit" or two-level system, favored by solid state physicists – a double-well potential [29]. The two levels are defined by the *left* and *right* eigenstates $|L\rangle$ and $|R\rangle$, which are obtained as linear superpositions of the energy eigenstates of the double-well potential

polarization states dealt with in chapter "Photonic Realization of Quantum Information Protocols".

In quantum optics and atomic physics, we tend to think of energy eigenstates, while solid states prefer the double-well picture (which incarnates what is known as the *spin-Boson Hamiltonian* [29]). Note that, a priori, there is no obvious reason (except familiarity and simplicity, in some sense) why one should not use discrete multi-state systems (i.e., one-particle systems with a Hilbert space dimension larger than 2) or even systems with continuous spectra.

A "quantum information processor" will consist of a set of two-level systems – *qubits* – which

- can be prepared in an arbitrary (collective) initial state

$$|\psi_0\rangle = \sum_{j_1,\dots,j_N=0}^{1} c_{j_1,\dots,j_N} |j_1,\dots,j_N\rangle; \qquad (2)$$

- can be coupled to each other (e.g., through nearest-neighbor coupling between adjacent sites);
- undergo a coherent/unitary time evolution (what, practically, will be tantamount to the execution of some algorithm); and
- can be read out in their final state.

In other words, the fundamental requirements for successful quantum information processing are

- quantum state preparation,
- coherent control,
- quantum state readout/tomography,

and all this for *composite, multipartite* quantum systems (in quantum information jargon: "quantum registers") [16].

Since quantum information processing develops its full potential – as compared to classical information processing – only in the limit of *large* quantum registers ($N \gg 1$), these requirements have to be met in this very limit, i.e., any experimental technique which succeeds on the single-particle level has to fulfill the condition of *scalability* (i.e., $1 \to N$), in order to have any relevance under the perspective of quantum information processing.

This means that we are targeting at coherent control of many-particle quantum dynamics, on a mesoscopic or rather macroscopic scale![2]

Note that this implies – notwithstanding the focus on two-level constituents of a quantum register – high-dimensional Hilbert spaces (of dimension 2^N), high spectral densities, and, a priori, a high sensitivity with respect to perturbations.

Given the requirements of initialization, control, readout, and scalability, one needs to choose a suitable physical entity to represent a single qubit and registers thereof. One has to guarantee single- or (coupled) multi-qubit coherence on the timescales which are required for performing a certain task, which need to be shorter than the timescales on which detrimental effects such as the coupling to uncontrolled degrees of freedom manifest themselves.[3]

The arguably most advanced control on the microscopic properties of matter is realized in quantum optics, which deals with the interaction of the quantum constituents of matter with quantized electromagnetic field modes. The fact that quantum information processing actually has an experimental perspective is a consequence of the tantalizing experimental progress in quantum optics during the last

---

[2] "Mesoscopic" systems [40] are usually understood as composite quantum systems which mediate between the microscopic and the macroscopic worlds, being sufficiently *small* for quantum effects to emerge. As a matter of fact, one of the origins of this research area is the continuing miniaturization of integrated circuitry used in traditional, "classical" computers, what implies that the quantum granularity of matter – manifest, e.g., through conductance fluctuations [17] across mesoscopic conductors, and usually assumed to be smoothed on the macroscopic scale, due to decoherence effects – has to be taken into account. Note that the line of thought is somewhat exactly opposite in our present context: Here, the question is not how much we have to reduce the size of a device to witness quantum effects, but rather whether there exists a fundamental size limit beyond which quantum effects cannot be observed anymore [2].

[3] The coupling to uncontrolled degrees of freedom implies decoherence, noise, and decay, the characteristic features of "open system dynamics." For example, strong resonant driving of an ionic two-level system may induce residual, non-resonant coupling to a third atomic level. If the latter remains unobserved in the specific experimental setting, what is equivalent of tracing over this part of the ionic Hilbert space, decoherence is induced on the two-dimensional subspace. Such a mechanism is believed to limit the maximally achieved gate fidelity, e.g., in the ion trap experiments in Innsbruck [44]. Possible strategies to reduce such detrimental effects are error correction and decoherence-free subspaces [50, 30].

decades [35, 25, 22, 45, 9]. Mesoscopic and/or solid state devices start only now to progress in the same direction of deterministic quantum control [56, 51, 32]. Remember, however, that only a solid state/semiconductor approach seems to bear the actual potential of scalability [47].[4] Also note that coherent control may play an important role in photochemistry [18] and biological processes [5, 8], thus the recent blossoming of "molecular quantum computing"; see [53] for an example.

2 A Single Qubit in Interaction with the Radiation Field

In this chapter, we start out with the dynamics and the control of a single qubit, as the elementary constituent of a quantum register. The control can be exerted by a classical or a quantized electromagnetic field, what will naturally lead us to "entanglement" (see chapter "Bipartite Quantum Entanglement") between the atomic and the field degrees of freedom. This will then be used as the fundamental tool for entangling atoms or for executing simple quantum logic operations in chapter "Quantum Probability and Quantum Information Theory". As much as single qubit control is concerned, we closely follow the presentation in [31, 33, 13, 26], which the interested reader should consult for more (technical) details.

2.1 A Two-Level System Subject to a Classical Field – The Semiclassical Model

Consider a two-level system (e.g., an atom – we shall see later what a "two-level atom" is) under periodic forcing by an oscillating electric field $F \cos(\omega t)$, with the linearly polarized field parallel to the $\hat{x}$-direction, $F || \hat{x}$.

The energy eigenstates of the unperturbed atom are determined through

$$H_0 |j\rangle = E_j |j\rangle , \quad j = 0, 1 , \tag{3}$$

and the total Hamiltonian reads

$$H = H_0 + V = H_0 + \underbrace{e\, x\, F\, \cos(\omega t)}_{V(t)} . \tag{4}$$

With the ansatz

$$|\psi\rangle = c_0 e^{-i E_0 t/\hbar} |0\rangle + c_1 e^{-i E_1 t/\hbar} |1\rangle , \tag{5}$$

and the assumption

---

[4] That is why proposals of "silicon quantum computing" [28] earned a lot of attention.

$$\langle j|V|j \rangle = 0, \tag{6}$$

for parity reasons,[5] we obtain the equations of motion for the amplitudes of both levels:

$$i\hbar \dot{c}_0 = c_1 \langle \phi_0|V|\phi_1 \rangle, \tag{7}$$

$$i\hbar \dot{c}_1 = c_0 \langle \phi_1|V|\phi_0 \rangle, \tag{8}$$

where $|\phi_j\rangle = e^{-i E_j t/\hbar}|j\rangle$. More explicitly, with (4),

$$
\begin{aligned}
\langle \phi_0|V|\phi_1 \rangle &= e^{-i(E_1-E_0)t/\hbar} \langle 0|V|1 \rangle \\
&= e^{-i\omega_0 t} e\, F \langle 0|x|1 \rangle \cos(\omega t) \\
&= e^{i\omega_0 t} \mathcal{V} \hbar \cos(\omega t),
\end{aligned}
\tag{9}
$$

ergo[6]

$$i\dot{c}_0 = c_1 e^{-i\omega_0 t} \mathcal{V} \cos(\omega t), \tag{10}$$

$$i\dot{c}_1 = c_0 e^{+i\omega_0 t} \mathcal{V}^* \cos(\omega t), \tag{11}$$

where we introduced the short-hand notation

$$\mathcal{V} = e\, F \langle 0|x|1 \rangle. \tag{12}$$

Given (10) and (11), also the evolution equation for the statistical operator of the two-level system ($\sigma = |\psi\rangle\langle\psi|$) can be derived: With

$$\sigma_{00} = |c_0|^2 = c_0 c_0^*, \quad \sigma_{11} = |c_1|^2 = c_1 c_1^* - \text{"populations,"} \tag{13}$$

$$\sigma_{01} = c_0 c_1^*, \quad \sigma_{10} = c_0^* c_1 = \sigma_{01}^* - \text{"coherences,"} \tag{14}$$

one finds[7]

$$\dot{\sigma}_{11} = -\dot{\sigma}_{00} = -i \cos(\omega t) \left[\sigma_{01} \mathcal{V}^* e^{i\omega_0 t} - \sigma_{10} \mathcal{V} e^{-i\omega_0 t} \right], \tag{15}$$

$$\dot{\sigma}_{01} = \dot{\sigma}_{10}^* = i \mathcal{V} \cos(\omega t) e^{-i\omega_0 t} [\sigma_{00} - \sigma_{11}]. \tag{16}$$

Rewriting (15) and (16) with $\cos(\omega t) = (e^{i\omega t} + e^{-i\omega t})/2$ leads to the appearance of terms $\sim e^{i(\omega_0-\omega)t}$ and $\sim e^{i(\omega_0+\omega)t}$ (we shall encounter them again later). Assuming

---

[5] Exercise: Why/when is this a good assumption?

[6] Exercise: Convince yourselves that the norm $|c_0|^2 + |c_1|^2 = 1$ is conserved for all times by (10) and (11).

[7] Verify this, as an exercise!

that $\omega_0 \simeq \omega$ (i.e., "near-resonant driving," as the jargon says), we can neglect the rapidly (with frequency $\omega_0 + \omega$) oscillating terms ("rotating wave approximation," in short "RWA") and obtain

$$\dot{\sigma}_{11} = -\frac{i}{2}\left[\sigma_{01}\mathcal{V}^* e^{i(\omega_0-\omega)t} - \sigma_{10}\mathcal{V}e^{-i(\omega_0-\omega)t}\right], \tag{17}$$

$$\dot{\sigma}_{01} = \frac{i}{2}\mathcal{V}e^{i(\omega-\omega_0)t}[\sigma_{00} - \sigma_{11}]. \tag{18}$$

The above equations can be solved with the ansatz [31]

$$\sigma_{jj} = \sigma_{jj}^{(0)}\exp(\lambda t), \quad j = 0, 1, \tag{19}$$

$$\sigma_{01} = \sigma_{01}^{(0)}\exp\left(-i(\omega_0 - \omega)t\right)\exp(\lambda t), \tag{20}$$

$$\sigma_{10} = \sigma_{10}^{(0)}\exp\left(i(\omega_0 - \omega)t\right)\exp(\lambda t). \tag{21}$$

After substitution into (17) and (18), this leads to a secular equation with roots

$$\lambda_1 = 0, \quad \lambda_2 = i\Omega, \text{ and } \lambda_3 = -i\Omega, \tag{22}$$

where $\Omega = \sqrt{(\omega_0 - \omega)^2 + |\mathcal{V}|^2}$. These are then used to express the general solution

$$\sigma_{ij} = \sigma_{ij}^{(1)} + \sigma_{ij}^{(2)}\exp(i\Omega t) + \sigma_{ij}^{(3)}\exp(-i\Omega t), \tag{23}$$

as a linear superposition of the associated eigensolutions, with coefficients $\sigma_{ij}^{(k)}$, $k = 1, 2, 3$ (to be determined by the specific initial conditions).[8] This implies another condition of validity for the RWA: It must also be consistent with the system dynamics, i.e. $(\omega_0 + \omega)^{-1}$ must be much smaller than the typical timescale of system's evolution: By virtue of (23), this requires $(\omega_0 + \omega)^{-1} \ll \Omega$.

For the specific initial condition[9] $\sigma_{11} = 0$ and $\sigma_{01} = 0$ it follows[10]

$$\sigma_{11}(t) = \frac{|\mathcal{V}|^2}{\Omega^2}\sin^2\left(\frac{1}{2}\Omega t\right), \tag{24}$$

---

[8] Note that, for the terms given by (20) and (21), $\sigma_{ij}^{(k)}$ also incorporates the "trivial" time dependence $\exp\left(\mp i(\omega_0 - \omega)t\right)$.

[9] This uniquely defines the initial state – why?

[10] Exercise: Derive this result with the above ansatz!

$$\sigma_{01}(t) = e^{-i(\omega_0 - \omega)t} \frac{\mathcal{V}}{\Omega^2} \sin\left(\frac{1}{2}\Omega t\right) \tag{25}$$

$$\times \left[-(\omega_0 - \omega)\sin\left(\frac{1}{2}\Omega t\right) + i\Omega\cos\left(\frac{1}{2}\Omega t\right)\right],$$

and, at exact resonance, $\omega_0 = \omega$:

$$\sigma_{11}(t) = \sin^2\left(\frac{1}{2}|\mathcal{V}|t\right), \tag{26}$$

$$\sigma_{01}(t) = i\frac{\mathcal{V}}{|\mathcal{V}|}\sin\left(\frac{1}{2}|\mathcal{V}|t\right)\cos\left(\frac{1}{2}|\mathcal{V}|t\right). \tag{27}$$

Equations (17) and (18) are also known as "optical Bloch equations" (here in their simplest form). Their solution (27) reflects the "Rabi oscillation" of the population between levels $|0\rangle$ and $|1\rangle$. Obviously, the amplitude and frequency of these oscillations depend on the detuning $\Delta = \omega_0 - \omega$. The presence of $|\mathcal{V}|^2$ in the expression for Ω in (27) expresses a shift of the coupled atom–field system's eigenfrequencies (or eigenenergies) with respect to the unperturbed frequencies ω and ω_0. This shift is known as the "dynamical" or "AC[11] Stark shift" [31, 33, 39],[12] which is also the origin of light-induced potentials used to trap neutral atoms at the focus of laser beams [45]. Remember that a potential energy gradient implies a force. Hence, when an atom is exposed to a laser field with spatial variation of its intensity, it will experience an AC shift-induced gradient or shift of its electronic eigenenergies, tantamount of a potential gradient, leading to a force [13].

2.2 Dynamics on the Bloch Sphere

The optical Bloch equations (17) and (18) have a geometrical interpretation, which becomes apparent upon substituting

$$\tilde{\sigma}_{00} = \sigma_{00}, \quad \tilde{\sigma}_{11} = \sigma_{11}, \tag{28}$$

$$\tilde{\sigma}_{01} = \sigma_{01}e^{i(\omega_0 - \omega)t}, \quad \tilde{\sigma}_{10} = \sigma_{10}e^{-i(\omega_0 - \omega)t} = \tilde{\sigma}_{01}^*. \tag{29}$$

---

[11] "AC" for "alternating current," since induced by an oscillating electromagnetic field.

[12] Exercise: Have a look at [12], and there at the sections on avoided level crossings. Interpret the AC Stark shift in terms of the quantities which characterize an avoided level crossing. In [12], the avoided crossing results from solving the stationary eigenvalue equation for the two-level Hamiltonian $H = H_0 + W$, with $H_0 = E_0|0\rangle\langle 0| + E_1|1\rangle\langle 1|$ and $W = W_{00}|0\rangle\langle 0| + W_{11}|1\rangle\langle 1| + W_{10}|1\rangle\langle 0| + W_{01}|0\rangle\langle 1|$, when the eigenvalues $E_\pm$ of H are plotted as a function of the detuning $\delta = (E_1 - E_0)/2$. For the identification of that treatment with our present problem, identify δ with our present definition of the detuning Δ, as the frequency mis-match between the driving field frequency and the driven atomic transition.

This transforms (17) and (18) into

$$\dot{\tilde{\sigma}}_{11} = -\frac{i}{2}\left[\tilde{\sigma}_{01}\mathcal{V}^* - \tilde{\sigma}_{10}\mathcal{V}\right], \quad \dot{\tilde{\sigma}}_{00} = -\dot{\tilde{\sigma}}_{11}, \tag{30}$$

$$\dot{\tilde{\sigma}}_{01} = \frac{i}{2}\mathcal{V}\left[\tilde{\sigma}_{00} - \tilde{\sigma}_{11}\right] + i\,\Delta\,\tilde{\sigma}_{01}, \quad \dot{\tilde{\sigma}}_{10} = \dot{\tilde{\sigma}}_{01}^*. \tag{31}$$

Now we introduce the new variables

$$
\begin{aligned}
u &= \tilde{\sigma}_{10} + \tilde{\sigma}_{01} && -\text{real part of coherence,}\\
v &= i\,(\tilde{\sigma}_{10} - \tilde{\sigma}_{01}) && -\text{imaginary part of coherence,}\\
w &= \tilde{\sigma}_{11} - \tilde{\sigma}_{00} && -\text{population exchange,}
\end{aligned}
\tag{32}
$$

the time derivatives of which turn out to be, with the help of (31) and the assumption that $\mathcal{V}$ be real,[13]

$$\dot{u} = -\Delta\,v, \tag{33}$$

$$\dot{v} = -\mathcal{V}w + \Delta\,u, \tag{34}$$

$$\dot{w} = \mathcal{V}v. \tag{35}$$

The "Bloch vector"

$$\mathbf{S} = u\hat{x} + v\hat{y} + w\hat{z}, \tag{36}$$

which has unit modulus, because of

$$|\mathbf{S}|^2 = |u|^2 + |v|^2 + |w|^2 = |c_0|^2 + |c_1|^2 = 1, \tag{37}$$

$$\frac{d|\mathbf{S}|^2}{dt} = 0. \tag{38}$$

evolves on the surface of the unit sphere, according to the evolution equation[14]

$$\dot{\mathbf{S}} = \mathbf{\Omega} \times \mathbf{S}, \text{ with } \mathbf{\Omega} = \mathcal{V}\hat{x} + \Delta\,\hat{z} = (\mathcal{V}, 0, \Delta), \tag{39}$$

as illustrated in Fig. 3. The center of this sphere corresponds to the maximally mixed state, since $\mathbf{S} = 0$ if and only if $w = u = v = 0$, i.e., $\tilde{\sigma}_{00} = \tilde{\sigma}_{11}$ and $\tilde{\sigma}_{01} = \tilde{\sigma}_{10} = 0$, what implies $\tilde{\sigma} = 1/2\,(|0\rangle\langle0| + |1\rangle\langle1|)$.

---

[13] Exercise: Derive this result.
[14] Mind the analogy of (39) with $d\mathbf{L}/dt = \mathbf{\Omega} \times \mathbf{L}$, $\mathbf{L}$ the angular momentum, $\mathbf{\Omega}$ the precession frequency in the classical dynamics of rigid bodies (e.g., motion of a top) [34].

Fig. 3 The Bloch sphere, spanned by the unit vectors $\hat{x}$, $\hat{y}$, and $\hat{z}$. For a pure state $\sigma = |\psi\rangle\langle\psi|$, the tip of the Bloch vector S moves on the unit sphere. If $|S| < 1$, the state described by S is mixed

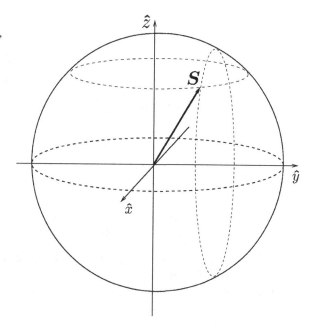

2.3 Some Special Cases and Implications

Let us briefly discuss some special cases and some implications of the above and the jargon which comes with it:

- $w = 1$, i.e., $\tilde{\sigma}_{11} = 1$, $\tilde{\sigma}_{00} = 0$: "complete inversion" (of the population of the field-free atomic eigenstates).
- $w = -1$, i.e., $\tilde{\sigma}_{11} = 0$, $\tilde{\sigma}_{00} = 1$: "atom in the ground state."
- $\Delta = 0$: This implies $Q = V\hat{x}$, thus S rotates in the plane defined by $\hat{x}$.
- If V is time dependent through a time-dependent amplitude $F(t)$ of the driving field (see (4)), then the Bloch vector rotates by an angle

$$\theta(t) = \int_0^t V(t')\,dt',\tag{40}$$

which is also called "the area under the pulse" (remember that, according to (12), $V(t)$ is equivalent to $F(t)$, up to a constant). This allows to initialize the qubit in arbitrary states on the Bloch sphere. Accordingly, one speaks of "$\pi/2$-pulses," for $\theta(t) = \pi/2$, which are nowadays routinely used to prepare states of the type $|\psi_0\rangle = (|0\rangle + |1\rangle)/\sqrt{2}$, and of "$\pi$ pulses," $\theta(t) = \pi$, to prepare $|\psi_0\rangle = |1\rangle$, from the ground state $|0\rangle$; see Fig. 4.

A "2π-pulse" leaves the atomic state unchanged. (For more details, see [33], Chap. 15.)

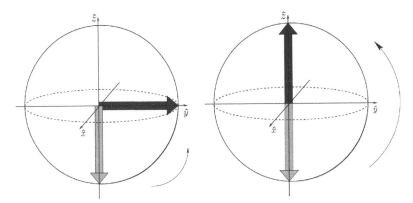

Fig. 4 Evolution of the Bloch vector on the Bloch sphere, under a $\pi/2$-pulse (*left*) and under a π-pulse (*right*). Initial and final states are represented by the *pale* and *black arrows*, respectively

- If we want to initialize several qubits, we could do this *locally* with such type of transformation. Technically, this requires the "single-qubit addressability" in any experimental setup and has important consequences, e.g., for the experimental geometry [36, 37].

2.4 The Jaynes–Cummings Model

We now revisit the atom–field interaction, but this time also the field is quantized (as described in chapter "Field-Theoretical Methods", Sect. 3.1). More precisely, we will consider the interaction of a two-level atom (whatever this may be ...) with *a single mode* of the quantized electromagnetic field, or, rather, the energy exchange between the associated degrees of freedom (one for the atom, one for the field mode– on a more abstract level, the latter is simply a quantized, one-dimensional harmonic oscillator).

From an atomic physicist's point of view, one typically is interested in the evolution of the electronic state of the atom under the external field – from a quantum optician's point of view one is interested in the excitation dynamics of the field mode. Both perspectives are perfectly legitimate and have equal right, but give rise to a zoology of different physical phenomena, by simply changing the various parameters which determine the atom–field dynamics.

Note that two-level atoms do not really exist in nature, but they can be "engineered" with high precision (what is crucial for quantum information purposes) in the lab,[15] in particular when the field mode's frequency is near-resonant with the selected atomic transition [25] (Fig. 5).

The state vector of the atom at time t has the form

$$|\psi_{\text{atom}}\rangle = c_u(t)|u\rangle + c_d(t)|d\rangle \tag{41}$$

---

[15] See [46, 23] for a detailed account of the involved approximations.

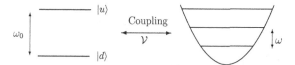

Fig. 5 The abstract skeleton of the Jaynes–Cummings picture: A two-level atom with upper and lower energy eigenstates $|u\rangle$ and $|d\rangle$, respectively, is coupled to a single mode of the quantized electromagnetic field – which, formally, is nothing else but a quantized harmonic oscillator (see chapter "Field-Theoretical Methods", Sect. 2). In the laboratory, such single field modes can be defined and isolated against the electromagnetic vacuum with stunning precision in micromaser experiments [35, 25]. Preparation of the harmonic oscillator in an energy eigenstate then corresponds to the experimental preparation [6, 7, 54] of a Fock state of the radiation field (with well-defined occupation number), and it is a highly nontrivial task, essentially because Fock states are extremely non-classical states of the field [31, 33]. An important additional assumption for the Jaynes–Cummings approximation, (62), is the near-resonant coupling between atom and field, i.e., the atomic transition frequency ω_0 should be close to the photon frequency ω, to justify the neglect of anti-rotating terms in the rotating wave approximation (see the text for details)

(where "u" stands for "up" and "d" for down) and the state of the field, in the Fock basis (see chapter "Field-Theoretical Methods", Sect. 2.4),

$$|\psi_{\text{field}}\rangle = \sum_n c_n(t)|n\rangle . \tag{42}$$

The collective atom–field state is the tensor product of both,

$$|\psi_{\text{atom+field}}\rangle = \sum_n \left[c_{n,u}(t)|u\rangle \otimes |n\rangle + c_{n,d}(t)|d\rangle \otimes |n\rangle \right] , \tag{43}$$

and, in general, not separable (this means, see chapter "Bipartite Quantum Entanglement", Sect. 2, "entangled").

We write the total atom–field Hamiltonian as

$$H = H_0^{\text{atom}} + H_0^{\text{field}} + V , \tag{44}$$

with

$$H_0^{\text{atom}} = \hbar \left(E_u |u\rangle\langle u| + E_d |d\rangle\langle d| \right) , \tag{45}$$

$$H_0^{\text{field}} = \hbar\omega \left(a^\dagger a + \frac{1}{2} \right) , \tag{46}$$

$$V = \hbar g \left(a + a^\dagger \right) (\sigma_- + \sigma_+) , \tag{47}$$

where we use the atomic lowering and raising operators σ_- and σ_+, defined as suitable linear combinations of the Pauli matrices σ_x, σ_y, and σ_z,[16]

---

[16] Exercise: Express the Bloch vector S, (36), in terms of the Pauli matrices.

$$\sigma_- = |d\rangle\langle u| = \begin{pmatrix} 0 & 0 \\ 1 & 0 \end{pmatrix} = \frac{1}{2}\left(\sigma_x - i\,\sigma_y\right), \tag{48}$$

$$\sigma_+ = |u\rangle\langle d| = \begin{pmatrix} 0 & 1 \\ 0 & 0 \end{pmatrix} = \frac{1}{2}\left(\sigma_x + i\sigma_y\right), \tag{49}$$

$$\sigma_x = \begin{pmatrix} 0 & 1 \\ 1 & 0 \end{pmatrix}, \quad \sigma_y = \begin{pmatrix} 0 & -i \\ i & 0 \end{pmatrix}, \quad \sigma_z = \begin{pmatrix} 1 & 0 \\ 0 & -1 \end{pmatrix}, \tag{50}$$

as well as the atom–field coupling constant g given by

$$g = -\frac{\langle d|x|u\rangle}{\hbar} F \sin(kz). \tag{51}$$

This latter expression is given in units of the electronic charge e (thus the negative sign), assumes the polarization of the quantized field mode along the $\hat{x}$-axis, while the field's wave vector k points along the $\hat{z}$-direction. The expression for g is completely analogous to (9), with the only amendment of an explicit spatial dependence, which we will, however, immediately omit again. If we assume that the wavelength of the field ($\lambda = 2\pi/k$) is large with respect to the typical dimension of the atom (of the order of Angstroms), this is justified. This approximation is known as the "dipole approximation" [31].

The terms $(a + a^{\dagger})$ and $(\sigma_- + \sigma_+)$ in (47) represent the time-dependent part of the driving field and the induced transitions $|u\rangle \to |d\rangle$ and $|d\rangle \to |u\rangle$, respectively. Thus, (47) represents a coupling term $V = -e\,F(R, t)\cdot r$ (with R the position of the atomic nucleus) which can be derived from the Schrödinger equation for an electron under the influence of a vector potential A.

Note that (47) contains the following operator products:

- $a\sigma_-$ – This mediates the transition $u \to d$ and the absorption of a photon, corresponding to a total energy loss $\sim 2\hbar\omega$;
- $a\sigma_+$ – This expresses the transition $d \to u$ and the absorption of a photon ("stimulated absorption"), the total energy is conserved;
- $a^{\dagger}\sigma_-$ – This enforces the transition $u \to d$, together with the emission of a photon ("stimulated emission"), again without net cost of energy;
- $a^{\dagger}\sigma_+$ – which, together with an energy gain of $\sim 2\hbar\omega$, describes the emission of a photon and the transition $d \to u$.

Transformation of (47) to the interaction picture[17] yields the new interaction term (with (45) and (46))

---

[17] Here is a short reminder: Given the Schrödinger equation

$$i\hbar|\dot{\psi}\rangle = H|\psi\rangle, \text{ with } H = H_0 + V, \tag{52}$$

the "trivial" time evolution (which we suppose to be known) generated by H_0 (which we assume to be autonomous, i.e., time-independent) can be transformed away by defining

$$V_I = \hbar g \exp\left(i\omega\left(a^\dagger a + \frac{1}{2}\right)t\right)(a+a^\dagger)\exp\left(-i\omega\left(a^\dagger a + \frac{1}{2}\right)t\right)$$
$$\exp\left(i\begin{pmatrix} E_u & 0 \\ 0 & E_d \end{pmatrix}t\right)\begin{pmatrix} 0 & 1 \\ 1 & 0 \end{pmatrix}\exp\left(-i\begin{pmatrix} E_u & 0 \\ 0 & E_d \end{pmatrix}t\right). \tag{56}$$

With the identities[18]

$$\exp\left[i\omega\left(a^\dagger a + \frac{1}{2}\right)t\right](a+a^\dagger)\exp\left[-i\omega\left(a^\dagger a + \frac{1}{2}\right)t\right] \tag{57}$$
$$= a\exp(-i\omega t) + a^\dagger \exp(i\omega t)$$

and

$$\exp\left[i\begin{pmatrix} E_u & 0 \\ 0 & E_d \end{pmatrix}t\right]\begin{pmatrix} 0 & 1 \\ 1 & 0 \end{pmatrix}\exp\left[-i\begin{pmatrix} E_u & 0 \\ 0 & E_d \end{pmatrix}t\right] \tag{58}$$
$$= \sigma_+ \exp\left[i(E_u-E_d)t\right] + \sigma_- \exp\left[-i(E_u-E_d)t\right]$$

the interaction term turns into

$$V_I = \hbar g\{a\sigma_-\exp\left[-i(\omega+\omega_0)t\right] + a\sigma_+\exp\left[-i(\omega-\omega_0)t\right]$$
$$+ a^\dagger\sigma_-\exp\left[i(\omega-\omega_0)t\right] + a^\dagger\sigma_+\exp\left[i(\omega+\omega_0)t\right]\}. \tag{59}$$

In (59) the energy conserving terms $a\sigma_+$ and $a^\dagger\sigma_-$ carry a *slow time dependence*, provided $\omega \simeq \omega_0$,

$$\sim \exp[\pm i(\omega-\omega_0)t], \tag{60}$$

and the energy non-conserving terms carry a rapid time dependence

$$\sim \exp[\pm i(\omega+\omega_0)t]. \tag{61}$$

Upon integration of the Schrödinger equation in the interaction picture, the rapidly oscillating terms acquire a denominator $1/(\omega+\omega_0)$, as opposed to the slow terms

$$|\tilde{\psi}(t)\rangle = T^{-1}|\psi(t)\rangle, \text{ where } T = e^{-iH_0 t/\hbar}; \tag{53}$$
$$\tilde{V}(t) = T^{-1}VT, \quad \tilde{H}_0 = H_0. \tag{54}$$

The time evolution of $|\tilde{\psi}\rangle$ is now given by

$$i\hbar|\dot{\tilde{\psi}}\rangle = \tilde{V}|\tilde{\psi}\rangle \tag{55}$$

and describes the time evolution induced by the perturbation V, relative to the unperturbed, trivial dynamics induced by H_0.

[18] Which the keen reader should prove, as an exercise.

which get a "resonance denominator" $1/(\omega - \omega_0)$. The latter are therefore dominant in the time evolution, for near-resonant driving and $1/(\omega + \omega_0) \ll g\sqrt{n+1}$, what once again motivates the rotating wave approximation RWA,[19] i.e., to drop the rapidly oscillating terms.[20] We thus arrive at the final form of the *Jaynes–Cummings interaction*:

$$V_I = \hbar g \left\{ a\,\sigma_+ \exp\left(-i\,(\omega - \omega_0)\,t\right) + a^\dagger\,\sigma \exp\left(i\,(\omega - \omega_0)\,t\right) \right\}. \qquad (62)$$

2.5 Solutions for the Field Initially in a Fock State

Let us now investigate a bit closer the dynamics induced by V_I, when the quantized field is initially prepared in a Fock state $|n\rangle$ – this will reveal as one of the corner stones of a large number of fundamental implementations of quantum information processing (see chapter "Field-Theoretical Methods").

The interaction V_I mediates only transitions from $|d, n+1\rangle = |d\rangle \otimes |n+1\rangle$ to $|u, n\rangle = |u\rangle \otimes |n\rangle$ and back, and the general state vector thus has the form

$$|\psi(t)\rangle = c_{u,n}(t)|u, n\rangle + c_{d,n+1}|d, n+1\rangle. \qquad (63)$$

Solution of the Schrödinger equation (in the interaction picture) then leads to the following expressions for the time-dependent amplitudes:

- For $c_{u,n}(0) = 0$, $c_{d,n+1}(0) = 1$, at resonance $\Delta = \omega - \omega_0 = 0$:

$$c_{u,n}(t) = -i \sin\left(g\sqrt{n+1}\,t\right), \qquad (64)$$

$$c_{d,n+1}(t) = \cos\left(g\sqrt{n+1}\,t\right); \qquad (65)$$

- for $c_{u,n}(0) = 1$, $c_{d,n+1}(0) = 0$, $\Delta = 0$:

$$c_{u,n}(t) = \cos\left(g\sqrt{n+1}\,t\right), \qquad (66)$$

$$c_{d,n+1}(t) = -i \sin\left(g\sqrt{n+1}\,t\right). \qquad (67)$$

Consequently, for (66) and (67), the populations in the upper/lower state are given by

---

[19] See also our earlier discussion following (15) and (16). Once again, $g\sqrt{n+1}$ fixes the typical system timescale, as will become evident hereafter.
[20] Note that this is a special variant of the secular approximation, which is ubiquitous, e.g., also in the derivation of master equations or in nonlinear resonance analysis in classical mechanics.

$$P_u(t) = |\langle u|\psi(t)\rangle|^2 = \cos^2\left(g\sqrt{n+1}\,t\right), \tag{68}$$

$$P_d(t) = \sin^2\left(g\sqrt{n+1}\,t\right), \tag{69}$$

and one finds[21] for

- the expectation value of the field amplitude

$$\langle F\rangle = \langle\psi|\left(a^\dagger + a\right)|\psi\rangle = 0; \tag{70}$$

- the photon number

$$\langle\psi|a^\dagger a|\psi\rangle = \sin^2\left(g\sqrt{n+1}\,t\right); \tag{71}$$

- the photon number variance

$$\langle\psi|\Delta^2 n|\psi\rangle = \frac{1}{4}\sin^2\left(2g\sqrt{n+1}\,t\right). \tag{72}$$

This very nicely illustrates how energy – in the form of photons – is exchanged between the atom and the field modes (see Fig. 6): The atom, initially in the upper state and exposed to the resonant single-mode field, undergoes Rabi oscillations. After a $\pi/2$ evolution, the atom is in a balanced coherent superposition of $|u\rangle$ and $|d\rangle$, and the expectation value of the photon number is $n+1/2$, since, at that instance, the atom already released a photon in the field mode, with probability $1/2$. Consequently, the photon number variance is maximal at this very moment. At phase π, the atom underwent a π-pulse, resides in its ground state, and *deterministically* delivered a photon into the field mode. Accordingly, the photon number in the field has increased by 1, and the photon number variance vanishes, due to the deterministic character of the process. As time passes, the exchange of one energy quantum continues periodically. Note that this even happens when the field mode initially resides in its vacuum state, i.e., initially, $n = 0$ – then we speak of "vacuum Rabi oscillations" (which were observed experimentally [7]).[22]

---

[21] Convince yourselves, as an exercise.

[22] The "Floquet picture" [19, 49, 41, 58], where the field is once again treated classically, allows a completely analogous treatment as the "dressed state picture" [14] expressed in (43) and (47), with (essentially) the same tensor structure $\mathcal{H}_{atom} \otimes \mathcal{H}_{field}$ of the underlying Hilbert space. In particular, it defines the appropriate framework when dealing with more complicated atomic/molecular specimens. The essential approximation with respect to the dressed state picture is the neglect of the change of the field state upon emission/absorption of a photon by the atom. The Floquet picture naturally incorporates co- and anti-rotating terms (the latter are neglected in the Jaynes–Cummings picture). It therefore defines an excellent framework to assess the limitations of RWA, as well as of the two-level approximation. A very nice discussion of the contribution of the anti-rotating terms can be found in Sect. III.B of [49].

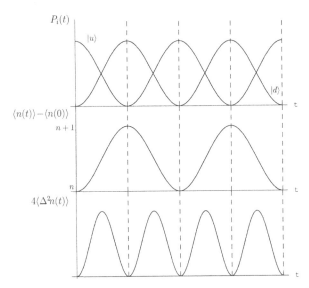

Fig. 6 The Jaynes–Cummings interaction (62) couples only neighboring photon (Fock) states of the field. Thus, if the field is initially prepared in a single Fock state, the Rabi oscillation of the atomic population $P_i(t)$, $i = u, d$ (*top*), is associated with the periodic oscillation of the field population between Fock states $|n\rangle$ and $|n + 1\rangle$. (*Middle*): The excitation energy of the atom – due to the resonance condition, $\omega_0 = \omega$, equal to a single quantum of the field excitation, i.e., a photon – is periodically exchanged between the atomic and the field degrees of freedom. Consequently, the photon number variance (*bottom*) passes through a maximum each time the atom (or the field) is in a balanced coherent superposition of $|u\rangle$ and $|d\rangle$ ($|n\rangle$ and $|n + 1\rangle$). Note that this elementary process can also be used to transfer quantum information between a quantum system with discrete degrees of freedom (the atomic qubit or a string of atomic qubits) and a system with continuous degrees of freedom (the harmonic oscillator) [57]

3 Qubit Entanglement Through the Jaynes–Cummings Interaction

The Jaynes–Cummings interaction (44) describes the dynamics of a two-level system resonantly coupled to a quantized harmonic oscillator.

3.1 Atom–Atom Entanglement in Cavity QED

This has been one of the first demonstrations of the controlled entanglement of two qubits [24]. The idea is the following: The two qubits are realized by two two-level atoms, which interact, one *after* the other, with a single-mode quantized field which is resonant with the atomic transition frequency, thus described by the Jaynes–Cummings interaction. If we arrange things such that the first atom exits the interaction region with the field in a balanced coherent superposition of the upper and the lower states, the mode will be in a balanced superposition of n and $n + 1$

photons (remember Fig. 6); hence the first atom and the field will be in a maximally entangled state (see chapter "Bipartite Quantum Entanglement"). The second atom will then in some sense recover the information from the field – which was left there by the first atom – and, at the end, the two atoms will be maximally entangled, while the mode will remain as a disentangled spectator. In this approach, the mode simply plays the role of an intermediate information storage (kind of a "data bus"). We could also say that the atoms, while not directly interacting with each other, interact *through* the field – a scenario which is quite generic, and encountered in many, quite distinct physical contexts [48, 1].

Specifically (see Fig. 7),

- the atom–field system (indices 1, 2 denote atoms 1 and 2) is initially prepared in the state

$$|\psi_0\rangle = |u_1\rangle \otimes |d_2\rangle \otimes |0\rangle . \tag{73}$$

- The velocity of atom 1 is chosen such that its interaction time t_1 with the field mode satisfies

$$\Omega t_1 = \frac{\pi}{2} , \quad \Omega = 2g\sqrt{n+1}|_{n=0} , \tag{74}$$

where Ω is the vacuum Rabi frequency. According to (66) and (67), the atom leaves the cavity – which defines and confines the field mode – with equal probability $1/2$ in the upper and the lower atomic states.
- The total atom–field state after the first and before the second atom's passage across the cavity consequently reads

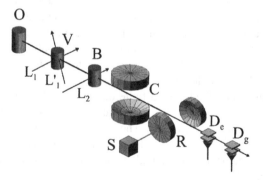

Fig. 7 (courtesy of Serge Haroche) Experimental setup for atom–atom entanglement, mediated by a quantized field mode of a high-quality ("high Q") microwave resonator [24] – marked by C in this plot. The atoms are provided by an atomic jet from an oven O, velocity selected in the area V (by lasers L_1 and L_1'), excited to the Rydberg levels 50 or 51 in zone B (by laser L_2), before they interact with the quantized field mode deconfined and confined by C. Within the low-quality analyzing cavity R, a $\pi/2$-pulse can be applied to the atoms, for diagnostic purposes. Finally, D_e and D_g are field ionization detection units, where the final state of the atoms ("e" for excited state and "g" for ground state) after exit from C can be read out

$$|\psi'\rangle = \frac{1}{\sqrt{2}}\Big(|u_1\rangle \otimes |d_2\rangle \otimes |0\rangle - i\,|d_1\rangle \otimes |d_2\rangle \otimes |1\rangle\Big). \qquad (75)$$

- If now the velocity of the second atom is chosen such that $t_2 = 2t_1$, i.e., $\Omega t_2 = \pi$, then
  - the second atom will leave the cavity in $|d_2\rangle$, provided the first atom left the field mode empty,
  - the second atom leaves the cavity in $|u_2\rangle$, with unit probability, if the first atom did deposit a photon in the field (due to (65), with $n + 1 = 1$).

- The final atom–field state, after exit of the second atom from the cavity, reads

$$|\psi\rangle = \frac{1}{\sqrt{2}}\Big(|u_1\rangle \otimes |d_2\rangle \otimes |0\rangle - |d_1\rangle \otimes |u_2\rangle \otimes |0\rangle\Big),$$
$$= \frac{1}{\sqrt{2}}\Big(|u_1\rangle \otimes |d_2\rangle - |d_1\rangle \otimes |u_2\rangle\Big) \otimes |0\rangle. \qquad (76)$$

$|\psi\rangle$ is maximally entangled in the atomic degrees of freedom and separable in atoms and field.[23]

While the above procedure is conceptually almost trivial, the experimental "mise en œuvre" is an art, since all the individual steps demand the precise experimental control of *single quantum objects*!

- Atom–atom entanglement is here realized through a two-step process, where we (a) entangle atom 1 with the field and (b) entangle atom 2 with atom 1 while disentangling the field from atom 1.
- To ensure that the atoms undergo the desired $\pi/2$ and π pulses, the individual atomic velocities have to be controlled very well: Indeed, in the experiment, the atoms, which are extracted from an atomic jet, are selected at velocities $v_1 = 337$ ms^{-1} and $v_2 = 432$ ms^{-1}, with an uncertainty of only $\Delta v \simeq \pm 0.4$ ms^{-1}!
- The above theoretical model completely neglects incoherent processes which might perturb the entanglement creation process. However, all cavities which confine a single-mode field have a finite damping coefficient, i.e., the photons which are trapped in such cavities will leak out with a certain probability, on a characteristic timescale τ_{cav} [25]. The experiment uses a "high-quality cavity," with "high Q,"[24] and $\tau_{\mathrm{cav}} = 112$ μs. Thus, a photon survives for approx. 0.1 ms in the resonator, and the entanglement must therefore be created on a much shorter timescale.
- Furthermore, atoms in excited electronic states do emit spontaneous photons – the quantum manifestation of the radiation emitted by an accelerated charge

---

[23] Exercise: Why, in the final state detection with detectors $D_{e,g}$, is it important to measure the population of the excited state *before* that of the ground state?
[24] The Q factor is proportional to τ_{cav} [43].

[4]. Once again, the experimentalists have to prepare atomic states with spontaneous emission rates τ_{spon}^{-1} as small as possible, to make sure that the information encoded in the atomic excitation is not lost spontaneously (and this means inadvertently) on the timescale of the entangling process. In the experiment that we presently focus on, this was achieved by preparing the atoms in circular Rydberg states with principal quantum numbers $n_d = 50$ (the d state) and $n_u = 51$ (the u state).[25] The efficient production of circular Rydberg states is in itself a masterpiece [27, 15, 21, 20], since such states are defined by maximal values of angular momentum and angular momentum projection, $\ell = m - 1$, respectively. Therefore, one first needs to pump maximal excitation into the angular momentum degree of freedom. Once this is done, one has to protect the fragile creature against any kind of stray field, since such Rydberg states exhibit a large polarizability, and therefore react very sensitively to any weak perturbation.

- Once the experimentalists succeed to perform the necessary single qubit $\pi/2$ and π pulses on timescales much faster than $\tau_{cav,spon}$, they have reached the "strong coupling regime." This was accomplished many years ago with cavity QED [52, 25] and ion traps [42], though only very recently with hybrid qubits [56].

- Not surprisingly, given all the above experimental complications and unavoidable error sources, the experiments could not demonstrate maximal entanglement, which would imply conditional probabilities $p_{u,d} = p_{d,u} = 1/2$ and $p_{u,u} = p_{d,d} = 0$ to observe the second atom in $|u\rangle(|d\rangle)$ provided the first atom exited in $|d\rangle(|u\rangle)$.[26] Instead, the experiment produced $p_{u,d} = 0.44$, $p_{d,u} = 0.27$, $p_{d,d} = 0.23$, and $p_{u,u} = 0.06$. This simply reminds us of the challenge ahead when we want to build a large-scale quantum computer where we have to control the coherent dynamics of a *large* number of qubits: Perfection is really hard to achieve in the macroscopic world.

3.2 Realizing a CNOT Gate with Trapped Cold Atoms

Much as in the cavity QED setting, the Jaynes–Cummings model can also be implemented with N ions loaded into a linear ion trap, where each individual ion can be addressed individually by a laser (see Fig. 8) [11]. The ions in the trap repel each other through Coulomb forces and therefore undergo collective motion in their

---

[25] Circular Rydberg states [27, 15, 21, 20] belong to the most "classical" eigenstates of single electron atoms: The electronic density is localized along an eccentricity zero classical Kepler trajectory and looks much like a swimming ring centered around the nucleus, in the plane defined by the quantization axis.

[26] Note that, in principle, the simple verification of these conditional probabilities does not prove entanglement generation. The same results would be expected for the state $\rho = (|u\rangle\langle u| \otimes |d\rangle\langle d| + |d\rangle\langle d| \otimes |u\rangle\langle u|)/2$ – which is separable. This is the reason why in a "Bell experiment" one has to perform the measurement in different bases (see chapter "Quantum Probability and Quantum Information Theory", Sect. 6.4 and references therein). In the present experiment, the coherence between $|u_1\rangle \otimes |d_2\rangle$ and $|d_1\rangle \otimes |u_2\rangle$ was therefore explicitly verified in a second, complementary measurement.

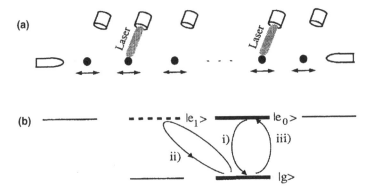

Fig. 8 (courtesy of Ignacio Cirac) Experimental scheme suggested in [11] (and realized in [44]) for the realization of a CNOT gate with trapped ions. (**a**) The experimental geometry, with the *horizontal arrows* indicating the ionic translational degree of freedom, and (**b**) the electronic level structure of an individual ion (one and the same for all ions). e and g label the ionic excited and ground states, respectively, and the suffix $q = 0, 1$ of $|e_q\rangle$ labels the (degenerate) magnetic sub-level of the excited state. Transitions between $|g\rangle$ and $|e_0\rangle/|e_1\rangle$ are driven by linearly/circularly polarized laser beams, respectively. Single ion addressability is required to perform the local operations on the individual ions

external (translational) degree of freedom (along the trap axis, which is weakly confined by a shallow harmonic potential). The dynamics of ion n ($\in \{1, \ldots, N\}$), with degenerate excited states $|e_0\rangle$ and $|e_1\rangle$ (see Fig. 8), under its external addressing lasers (which drive the transitions $g \leftrightarrow e_q, q = 0, 1$) is described by the effective (Jaynes–Cummings) Hamiltonian[27] (in the interaction picture!)

$$H_{n,q} = \frac{\Omega_{\text{eff}}}{2} \left(|e_q\rangle_n \langle g| a + |g\rangle_n \langle e_q| a^\dagger \right), \tag{77}$$

with Ω_{eff} the effective Rabi frequency. Rather than excitations of the radiation field – photons – in our previous example in Sect. 3.1, a and $a^\dagger$ do here annihilate and create, respectively, excitations of the collective center-of-mass motion in the translational degree of freedom – phonons. Hence, the energy is here exchanged between the electronic degrees of freedom of the ions and one single (quantized) mode of their collective translational dynamics, with a characteristic frequency Ω_{eff} [11, 42]. We will now convince ourselves that Hamiltonian (77) suffices to implement a CNOT gate between two ions, using the phonons as a "data bus." To see this, we have to remember that a CNOT gate performs the mapping $|\epsilon_1\rangle \otimes |\epsilon_2\rangle \rightarrow |\epsilon_1\rangle \otimes |\epsilon_1 \oplus \epsilon_2\rangle$, with $\oplus$ the addition modulo 2, and $\epsilon_1, \epsilon_2 \in \{0, 1\}$.

---

[27] This Hamiltonian implicitly assumes the "Lamb–Dicke approximation" [10, 55], which requires that the ions be spatially localized to dimensions smaller than the addressing laser wavelength, i.e., that the position uncertainty Δx of the center-of-mass wavefunction of the atom in the trap potential is much smaller than λ_L, the laser wavelength.

The time evolution generated by $H_{n,q}$ induces the following mapping under a $k\pi$-pulse (i.e., $\Omega_{\text{eff}}t = k\pi$, by virtue of (64), (65), (66), and (67)):

$$U|g\rangle_n|0\rangle \rightarrow |g\rangle_n|0\rangle \,, \tag{78}$$

$$U|g\rangle_n|1\rangle \rightarrow \cos\left(\frac{k\pi}{2}\right)|g\rangle_n|1\rangle - i\sin\left(\frac{k\pi}{2}\right)|e_q\rangle_n|0\rangle \,, \tag{79}$$

$$U|e_q\rangle_n|0\rangle \rightarrow \cos\left(\frac{k\pi}{2}\right)|e_q\rangle_n|0\rangle - i\sin\left(\frac{k\pi}{2}\right)|g\rangle_n|1\rangle \,. \tag{80}$$

With the three-step process

(i) π-pulse on the mth ion, $U_m^{k=1,q=0}$, on the 0-transition;
(ii) 2π-pulse on the nth ion, $U_n^{k=2,q=1}$, on the 1-transition;
(iii) π-pulse on the mth ion, $U_m^{k=1,q=0}$, on the 0-transition;

i.e., the complete transformation

$$U_{m,n} = U_m^{1,0} U_n^{2,1} U_m^{1,0} \,, \tag{81}$$

we can thus create the mapping [28]

$$|g\rangle_m|g\rangle_n|0\rangle \rightarrow |g\rangle_m|g\rangle_n|0\rangle \rightarrow |g\rangle_m|g\rangle_n|0\rangle \rightarrow |g\rangle_m|g\rangle_n|0\rangle \,, \tag{82}$$

$$|g\rangle_m|e_0\rangle_n|0\rangle \rightarrow |g\rangle_m|e_0\rangle_n|0\rangle \rightarrow |g\rangle_m|e_0\rangle_n|0\rangle \rightarrow |g\rangle_m|e_0\rangle_n|0\rangle \,, \tag{83}$$

$$|e_0\rangle_m|g\rangle_n|0\rangle \rightarrow -i|g\rangle_m|g\rangle_n|1\rangle \rightarrow i|g\rangle_m|g\rangle_n|1\rangle \rightarrow |e_0\rangle_m|g\rangle_n|0\rangle \,, \tag{84}$$

$$|e_0\rangle_m|e_0\rangle_n|0\rangle \rightarrow -i|g\rangle_m|e_0\rangle_n|1\rangle \rightarrow -i|g\rangle_m|e_0\rangle_n|1\rangle \rightarrow -|e_0\rangle_m|e_0\rangle_n|0\rangle \,. \tag{85}$$

By linearity, this implies

$$|g\rangle_m \left(|g\rangle_n \pm |e_0\rangle_n\right)|0\rangle \rightarrow |g\rangle_m \left(|g\rangle_n \pm |e_0\rangle_n\right)|0\rangle \,, \tag{86}$$

$$|e_0\rangle_m \left(|g\rangle_n \pm |e_0\rangle_n\right)|0\rangle \rightarrow |e_0\rangle_m \left(|g\rangle_n \mp |e_0\rangle_n\right)|0\rangle \,. \tag{87}$$

Hence, with a further local, single ion transformation (modulo normalization),

$$V^n : |g\rangle_n \rightarrow |g\rangle_n - |e_0\rangle_n \,,$$
$$|e_0\rangle_n \rightarrow |g\rangle_n + |e_0\rangle_n \,, \tag{88}$$

$$V_-^n : |g\rangle_n + |e_0\rangle_n \rightarrow |e_0\rangle_n \,,$$
$$|g\rangle_n - |e_0\rangle_n \rightarrow |g\rangle_n \,, \tag{89}$$

---

[28] Note that this evolution already implements a CPHASE (conditional phase) gate.

the operation $V_{-}^{n} U_{m,n} V^{n}$ produces the desired CNOT gate:

$$|g\rangle_m |g\rangle_n |0\rangle \rightarrow |g\rangle_m (|g\rangle_n - |e_0\rangle_n) |0\rangle \rightarrow |g\rangle_m (|g\rangle_n - |e_0\rangle_n) |0\rangle$$
$$\rightarrow |g\rangle_m |g\rangle_n |0\rangle , \tag{90}$$

$$|g\rangle_m |e_0\rangle_n |0\rangle \rightarrow |g\rangle_m (|g\rangle_n + |e_0\rangle_n) |0\rangle \rightarrow |g\rangle_m (|g\rangle_n + |e_0\rangle_n) |0\rangle$$
$$\rightarrow |g\rangle_m |e_0\rangle_n |0\rangle , \tag{91}$$

$$|e_0\rangle_m |g\rangle_n |0\rangle \rightarrow |e_0\rangle_m (|g\rangle_n - |e_0\rangle_n) |0\rangle \rightarrow |e_0\rangle_m (|g\rangle_n + |e_0\rangle_n) |0\rangle$$
$$\rightarrow |e_0\rangle_m |e_0\rangle_n |0\rangle , \tag{92}$$

$$|e_0\rangle_m |e_0\rangle_n |0\rangle \rightarrow |e_0\rangle_m (|g\rangle_n + |e_0\rangle_n) |0\rangle \rightarrow |e_0\rangle_m (|g\rangle_n - |e_0\rangle_n) |0\rangle$$
$$\rightarrow |e_0\rangle_m |g\rangle_n |0\rangle , \tag{93}$$

where the phonon mode remains unchanged after the operation, as in the cavity QED scheme for atom–atom entanglement studied in the previous chapter.

We thus succeeded to realize the CNOT gate, one of the cornerstones of quantum computation. This type of gate is of primordial importance since it creates entanglement between two qubits. Together with single qubit operations, it forms a set of universal gates [3] (see chapter "Quantum Algorithms") – that is, with such set of gates any quantum logic operation or algorithm can be implemented . . . even those yet to be discovered.

Acknowledgments Partial support by VolkswagenStiftung is gratefully acknowledged. F. de Melo also acknowledges financial support by Alexander von Humboldt Foundation.

References

1. Amthor, T., Reetz-Lamour, M., Westermann, S., et al.: Phys. Rev. Lett. **98**, 023004 (2007)
2. Arndt, M., Hornberger, K., Zeilinger, A.: Phys. World **18**, 35 (2005)
3. Barenco, A., Bennett, C.H., Cleve, R., et al.: Phys. Rev. A **52**, 3457 (1995)
4. Bethe, H.A., Salpeter, E.E.: Quantum Mechanics of One- and Two-Electron Atoms. Springer, Berlin (1957)
5. Briegel, H.J., Popescu, S.: Entanglement and intra-molecular cooling in biological systems? – a quantum thermodynamic perspective. arXiv.org:0806.4552
6. Brune, M., Haroche, S., Lefevre, V., et al.: Phys. Rev. Lett. **65**, 976 (1990)
7. Brune, M., Schmidt-Kaler, F., Maali, A., et al.: Phys. Rev. Lett. **76**, 1800 (1996)
8. Cai, J., Popescu, S., Briegel, H.J.: Dynamic entanglement in oscillating molecules. arXiv.org:0809.4906
9. Chuu, C.S., Schreck, F., Meyrath, T.P., et al.: Phys. Rev. Lett. **95**, 260403 (2005)
10. Cirac, J.I., Blatt, R., Parkins, A.S., et al.: Phys. Rev. Lett. **70**, 762 (1993)
11. Cirac, I., Zoller, P.: Phys. Rev. Lett. **74**, 4091 (1995)
12. Cohen-Tannoudji, C.C., Diu, B., F. Laloë: Mécanique quantique I. Hermann, Paris (1973)
13. Cohen-Tannoudji, C., Dupont-Roc, J., Grynberg, G.: Atom–Photon Interactions. Wiley, New York (1992)

14. Cohen-Tannoudji, C., Haroche, S.: J. Phys. **30**, 153 (1969)
15. Delande, D., Gay, J.-C.: Europhys. Lett. **5**, 303 (1988)
16. Dittrich, T., Hänggi, P., Ingold, G.-L., et al.: Quantum Transport and Dissipation. Wiley-VCH, Weinheim (1998)
17. Di Vincenzo, D.P.: The physical implementation of quantum computation. In: Braunstein, S.L., Lo, H.-K. (eds.) Scalable Quantum Computers: Paving the Way to Realization, pp. 1–14. Wiley-VCH, Weinheim (2001)
18. Engel, G.S., Calhoun, T.R., Read, E.L., et al.: Nature **446**, 782 (2007)
19. Floquet, M.G.: Ann. Sci. École Norm. Sup. **12**, 47 (1883)
20. Gallagher, T.F.: Rydberg Atoms. Cambridge University Press, Cambridge (1994)
21. Gay, J.-C., Delande, D., Bommier, A.: Phys. Rev. A **39**, 6587 (1988)
22. Greiner, M., Mandel, O., Esslinger, T., et al.: Nature **415**, 39 (2002)
23. Guerlin, C., Bernu, J., S. Deléglise, et al.: Nature **448**, 889 (2007)
24. Hagley, E., X. Maî tre, G. Nogues, et al.: Phys. Rev. Lett. **79**, 1 (1997)
25. Haroche, S.: Cavity quantum electrodynamics. In: Dalibard, J., Raimond, J.M., Zinn-Justin, J. (eds.) Fundamental Systems in Quantum Optics. Les Houches École d'été de physique théorique, vol. 53. Elsevier, Amsterdam (1992)
26. Haroche, S., Raimond, J.-M.: Exploring the Quantum: Atoms, Cavities, and Photons. Oxford University Press, Oxford (2006)
27. Hulet, R.G., Kleppner, D.: Phys. Rev. Lett. **51**, 1430 (1983)
28. Kane, B.E.: Nature **393**, 133 (1998)
29. Leggett, A.J., Chakravarty, S., Dorsey, A.T., et al.: Rev. Mod. Phys. **59**, 1 (1987)
30. Lidar, D.A., Whaley, K.B.: Decoherence-free subspaces and subsystems. In: Benatti, F., Floreanini R. (eds.) Irreversible Quantum Dynamics. Lect. Notes Phys. 622, pp. 83–120. Springer, Heidelberg (2003)
31. Loudon, R.: The Quantum Theory of Light. Clarendon Press, Oxford (1982)
32. Majer, J., Chow, J.M., Gambetta, J.M., et al.: Nature **449**, 443 (2007)
33. Mandel, L., Wolf, E.: Optical Coherence and Quantum Optics. Cambridge University Press, Cambridge (1995)
34. Marion, J.B.: Classical Dynamics of Particles and Systems. Academic, New York (1970)
35. Meschede, D., Walther, H., Müller, G.: Phys. Rev. Lett. **54**, 551 (1985)
36. Mintert, F., Wunderlich, C.: Phys. Rev. Lett. **87**, 257904 (2001)
37. Mintert, F., Wunderlich, C.: Erratum: Ion-trap quantum logic using long-wavelength radiation. Phys. Rev. Lett. **87**, 257904 (2001)
38. Mintert, F., Wunderlich, C.: Erratum: Ion-trap quantum logic using long-wavelength radiation. Phys. Rev. Lett. **91**, 29902 (2003)
39. Mollow, B.R.: Phys. Rev. **188**, 1969 (1969)
40. Richter, K.: Semiclassical Theory of Mesoscopic Quantum Systems. Springer Tracts in Modern Physics. Springer, Berlin (2000)
41. Ritus, V.I.: Zh. Eksp. Teor. Fiz. **51**, 1492 (1966)
42. Roos, C., Zeiger, T., Rohde, H., et al.: Phys. Rev. Lett. **83**, 4713 (1999)
43. Sargent III, M., Scully, M.O., Lamb, W.E., Jr.: Laser Physics. Addison-Wesley, Reading (1974)
44. Schmidt-Kaler, F., Häffner, H., Riebe, M., et al.: Nature **422**, 408 (2003)
45. Schrader, D., Dotsenko, I., Khudaverdyan, M., et al.: Phys. Rev. Lett. **93**, 150501 (2004)
46. Schuster, D.I., Houck, A.A., Schreier, J.A., et al.: Nature **445**, 515 (2007)
47. Science and Technology. Orion's Belter: The Economist, p. 77, 17 Feb 2007
48. Shatokhin, V., Müller, C.A., Buchleitner, A.: Phys. Rev. Lett. **94**, 043603 (2005)
49. Shirley, J.H.: Phys. Rev. B **138**, 979 (1965)
50. Shor, P.W.: Phys. Rev. A **52**, R2493 (1995)
51. Sillanpää, M.A., Park, J.I., Simmonds, R.W.: Nature **449**, 438 (2007)
52. Sirko, L., Buchleitner, A., Walther, H.: Opt. Commun. **78**, 403 (1990)
53. Tesch, C.M., de Vivie-Riedle, R.: Phys. Rev. Lett. **89**, 157901 (2002)

54. Varcoe, B.T.H., Brattke, S., Weidinger, M., et al.: Nature **403**, 743 (2000)
55. Vogel, W., Welsch, D.-G., Wallentowitz, S.: Quantum Optics – An Introduction. Wiley-VHC, Weinheim (2001)
56. Wallraff, A., Schuster, D.I., Blais, A., et al.: Phys. Rev. Lett. **95**, 060501 (2005)
57. Wellens, T., Buchleitner, A., Maassen, H., et al.: Phys. Rev. Lett. **85**, 3361 (2000)
58. Zeldovich, Y.B.: JETP (Soviet Physics) **24**, 1006 (1967)

Quantum Cryptography

D. Bruß and T. Meyer

1 Introduction

The Greek words "*kryptos*" ≡ "hidden" and "*logos*" ≡ "word" are the etymological sources for "*cryptology*," the science of secure communication. Within cryptology, one distinguishes *cryptography* (or "code-making") and *cryptanalysis* (or "code-breaking"). The aim of cryptography is to ensure secret or "secure" communication between a sender, traditionally called Alice, and a receiver, called Bob. The encryption and decryption of a so-called *plain text* into a *cipher text* and back is achieved using a certain *key* (not necessarily the same for Alice and Bob), as illustrated in Fig. 1. Here, "secure" means that an eavesdropper, called Eve, has no information on the message. In this chapter we will show that in classical cryptography (using classical signals), security relies on the assumed difficulty to solve certain mathematical tasks, whereas in quantum cryptography (using quantum signals), security arises from the laws of quantum physics.

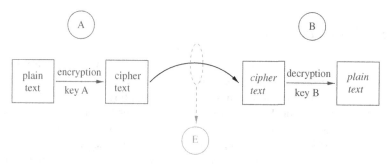

Fig. 1 The principle of cryptography

D. Bruß (✉)
Institut für Theoretische Physik III, Heinrich-Heine-Universität Düsseldorf, Düsseldorf, Germany,
bruss@thphy.uni-duesseldorf.de

T. Meyer (✉)
Institut für Theoretische Physik III, Heinrich-Heine-Universität Düsseldorf, Düsseldorf, Germany,
meyer@thphy.uni-duesseldorf.de

Bruß, D., Meyer, T.: *Quantum Cryptography*. Lect. Notes Phys. **808**, 277–308 (2010)
DOI 10.1007/978-3-642-11914-9_9 © Springer-Verlag Berlin Heidelberg 2010

2 Classical Cryptography

2.1 Archaic Cryptosystems

Two old and simple concepts of classical cryptography are the *transposition* and the *substitution*.

The *transposition* cipher uses a reordering of the letters in the message. As an example, the word "TRIESTE" might get encoded into "TISERET." This method was implemented already in 500 B.C. by the Spartans, using a so-called *Skytale*: The Skytale is a wooden rod, around which a strap of parchment or leather is wrapped. The message is then written on the strap, each letter on a new twist. Unwrapping the strap renders the message unreadable, and it can be transported safely to the receiver, who owns a Skytale of the same diameter to decode the scrambled message.

The *substitution* cipher replaces each letter of the message alphabet with a new one, possibly taken from a new alphabet. For instance, using the encoding A→B, B→C, ..., the message "TRIESTE" gets encoded into "USJFTUF." This cipher, employing a constant shift in the alphabet, is also called "Caesar cipher" as it was used by the Roman general around 50 B.C. for communication with his commanders on the field.

Both the transposition and the substitution cipher are very poor encryption methods, since they can be broken by a frequency analysis of the letters in the ciphertext: In every language, some letters appear more frequently than others, and these frequencies are preserved in the encoding. Thus, by counting the occurrences of all different letters in the ciphertext, it is possible to "guess" the plaintext. Therefore, these two simple cryptosystems are insecure.

2.2 The Vernam Cipher

The *Vernam cipher* was developed in 1926 [38]; it is a much more powerful cryptosystem than the simple ciphers explained above. The main idea of Vernam was to add a *random secret key* to the message. Explicitly, this works as follows: Each letter of the plaintext alphabet is substituted by a number (say, from 1 to 26). Thus, the message consists of numbers m_i. For each i, a random number k_i is chosen, and the ciphertext is calculated as $c_i = m_i \oplus k_i$, where $\oplus$ denotes adding modulo 26. For the decryption, the receiver adds the same key and recovers the message, as $c_i \oplus k_i = m_i$.

Two features of this method are important to note: First, the message and the key need to have the same length, which might become a problem if large messages are to be sent. Second, each key $\{k_i\}$ must only be used *once*. Otherwise, if two messages $\{m_i\}$ and $\{m_i'\}$ are encoded using the same key $\{k_i\}$, yielding $\{c_i\}$ and $\{c_i'\}$, respectively, the XOR of the ciphertexts reveals information about the two messages, since $c_i \oplus c_i' = m_i \oplus m_i'$. For this reason, the Vernam cipher is also called *one-time pad*. Its great advantage is the fact that it is the only cryptosystem which can be proven to be perfectly secure [35], since the ciphertext yields no information about the plaintext. However, this cryptosystem also has a drawback: Since for each message a new key is needed, a large amount of random secret numbers have to be distributed between all parties that wish to communicate.

2.3 RSA

The cryptosystems reviewed so far are *symmetric*, which means that the keys used for encryption and decryption are identical, or at least the latter can be derived from the former in a very simple way. In 1976, Diffie and Hellman [19] proposed the use of one-way functions for constructing an *asymmetric* cryptosystem: It employs two different keys, a public one, which is announced to everybody, and a private one, which is kept secret. The ciphertext is computed from the plaintext by using a one-way function (or more precisely: a trapdoor function), which can be evaluated easily, but which is hard to invert. However, using some additional information (the private key), the inversion is easy, such that the plaintext can be recovered from the ciphertext by the owner of the private key.

This suggestion was utilized by Rivest, Shamir, and Adleman in 1978 [32] who exploited the hardness of factoring large numbers. Their cryptosystem, which became known as the RSA system, is still widely used in everyday life, although there exists no rigorous proof of its security and, more severely, it is endangered by the advent of the quantum computer (see chapter "Quantum Algorithms", Sect. 6.2).

Explicitly, the RSA system works as follows: Choose two prime numbers p_1 and p_2, and compute $N = p_1 p_2$. Let $\phi(N) = (p_1 - 1)(p_2 - 1)$ be the Euler function of N, and choose e such that $1 < e < \phi(N)$ and $\gcd(e, N) = 1$. (Note that finding the greatest common divisor is easy, using the Euclid algorithm.) Finally, calculate d such that $ed = 1 \mod \phi(N)$. The numbers N and e then form the public key, whereas d is the private key. A message M gets encoded into $C = M^e \mod N$. As $a^{ed} = a \mod N$, the decoding is achieved by calculating $C^d \mod N = M$. As an example, take $p_1 = 11$ and $p_2 = 13$, thus $N = 143$ and $\phi(N) = 10 \times 12 = 120$. Choosing $e = 23$ (the public key, together with $N = 143$), we find $d = 47$ (the private key).

3 Quantum Cryptography

In *quantum cryptography*, quantum states are used as information carriers. However, in most cases the term "quantum cryptography" does not refer to quantum cryptosystems, but, somewhat misleadingly, to establishing a random secret key using quantum signals, i.e., implementing the Vernam cipher via *quantum key distribution* (QKD). As described in the previous section, the Vernam cipher is provably secure and thus provides an obvious candidate for a perfect cryptosystem, if the key distribution problem can be solved. Thus, in the following we focus on how Alice and

Bob can create a common random secret bit string, employing the laws of quantum mechanics. There are two different approaches to quantum key distribution: the first possibility is to encode the classical bits in a set of non-orthogonal quantum states. Here, single quantum states are prepared by Alice and measured by Bob, in order to supply them with common random bits. Therefore this scheme is also referred to as "prepare-and-measure" scheme. Non-orthogonal states cannot be distinguished perfectly, and together with the No-Cloning theorem (see below) this leads to the fact that the adversary Eve cannot simply intercept and measure or clone the quantum system underway from Alice to Bob. The protocol provides means to verify that Eve has no (or only very limited) knowledge about the key. The so-called "BB84" protocol (suggested in 1984 by Bennett and Brassard [3]), probably the most famous quantum key distribution protocol, falls into the "prepare-and-measure" category. Other examples are the B92 and the six-state protocol, see below.

The second possibility is to use entangled states that have been distributed between Alice and Bob ("entanglement-based scheme"). This type of protocol is named after A. Ekert [21]. If the entanglement is maximal, simultaneous measurements of Alice and Bob will lead to perfectly correlated secret bits. Any interaction of an eavesdropper necessarily destroys perfect entanglement. However, entangled states are difficult to distribute and store without being affected by noise. We will show later that these two approaches are, to a certain extent, equivalent.

3.1 The No-Cloning Theorem

The No-Cloning theorem, formulated in 1984 by Wootters and Zurek [39], is the underlying fundamental concept that makes quantum key distribution secure. It is based on the fact that the time-evolution of a closed quantum mechanical system is described by a unitary transformation and that this transformation is linear. The No-Cloning theorem states that perfect cloning of an unknown quantum system is impossible. The proof of this theorem is simple (see chapter "Quantum Probability and Quantum Information Theory", Sect. 6.1):

Consider a unitary transformation U that copies the basis states $|0\rangle$ and $|1\rangle$ of a two-level system perfectly, i.e.,

$$U|0\rangle|i\rangle = |0\rangle|0\rangle, \tag{1}$$

$$U|1\rangle|i\rangle = |1\rangle|1\rangle, \tag{2}$$

where $|i\rangle$ is some arbitrary initial state. The action of the linear transformation U on an unknown state $|\psi\rangle = \alpha|0\rangle + \beta|1\rangle$, with $|\alpha|^2 + |\beta|^2 = 1$, is then already fixed by its action on the basis and given by

$$
\begin{aligned}
U|\psi\rangle|i\rangle &= U(\alpha|0\rangle + \beta|1\rangle)|i\rangle, \\
&= \alpha|0\rangle|0\rangle + \beta|1\rangle|1\rangle, \\
&\neq |\psi\rangle|\psi\rangle.
\end{aligned}
\tag{3}
$$

Thus, the unknown state $|\psi\rangle$ is not copied perfectly.

The No-Cloning theorem is a fundamental difference between classical and quantum information theory. For the latter, it is both good and bad news: It prevents a spy from copying quantum signals perfectly without disturbance, hence guaranteeing the security of quantum cryptography. On the other hand, it makes it hard to develop efficient error correction schemes and back-up methods for quantum computers.

3.2 The BB84 Protocol

The main idea of the BB84 protocol, introduced in 1984 by Bennett and Brassard [3], is to generate a secret random key by employing two pairs of orthogonal quantum states, where the classical bit values 0 and 1 are encoded into one pair at a time. The quantum states of one of the pairs are *non-orthogonal* to the ones of the other pair. Explicitly, the two pairs are the eigenstates of the Pauli operators σ_z and σ_x and will be denoted by $|0_z\rangle, |1_z\rangle$ and $|0_x\rangle, |1_x\rangle$, respectively. They have the properties that $|\langle i_x | j_z \rangle| = 1/\sqrt{2}$, for all $i, j \in \{0, 1\}$ and

$$|0_x\rangle = \frac{|0_z\rangle + |1_z\rangle}{\sqrt{2}} = H|0_z\rangle , \quad |1_x\rangle = \frac{|0_z\rangle - |1_z\rangle}{\sqrt{2}} = H|1_z\rangle , \quad (4)$$

where H is the Hadamard transformation, which, in the representation defined by the orthonormal basis $|0_z\rangle, |1_z\rangle$, reads

$$H = \frac{1}{\sqrt{2}} \begin{pmatrix} 1 & 1 \\ 1 & -1 \end{pmatrix} . \tag{5}$$

Alice and Bob are connected via a quantum channel (see chapter "Quantum Entropy and Information", Sect. 2) which is totally insecure, i.e., it can be assumed to be under full control of the eavesdropper. In addition, they have a classical channel that is public but authenticated (i.e., the identity of Alice and Bob is guaranteed by sharing some previous secrecy), which means that Eve cannot send a message via this channel feigning to be Alice or Bob.

The BB84 protocol works as follows (see also Table 1):

(i) *Preparation.* Alice prepares $2n$ qubits, each randomly in one of the four states $|0_z\rangle, |0_x\rangle, |1_z\rangle,$ or $|1_x\rangle$, and sends them along the quantum channel to Bob.

(ii) *Measurement.* For each qubit that Bob receives, he chooses at random one of the two bases (z or x) and measures the qubit with respect to that basis.

(iii) *Sifting.* Alice tells Bob via the classical channel which basis she used for each qubit. They keep the bits where Bob has used the same basis for his measurement as Alice for the preparation. Those n bits are forming the so-called *sifted* key.

(iv) *Parameter estimation.* Alice and Bob choose a subset of the sifted key to esti-
    mate the error rate. They do so by announcing publicly the bit values of the
    subset. If they differ in "too many"[1] cases, they abort the protocol.
(v) *Establishing the secret key.* Finally, Alice and Bob obtain a joint secret key
    from the remaining bits by performing (classical) *error correction* and *privacy
    amplification* [27].

Table 1 The BB84 key distribution protocol. Here, "Y" and "N" stand for "yes" and "no," respectively, and "R" means that Bob obtains a random result

Alice's string	1	1	0	1	0	0	1	0	1	1	1	1	0	0
Alice's basis	+	+	+	×	×	+	×	×	×	×	+	+	+	+
Bob's basis	+	×	+	+	×	+	×	+	×	×	+	+	+	+
Bob's string	1	R	0	R	0	0	1	R	1	1	1	1	0	0
Same basis?	Y	N	Y	N	Y	Y	Y	N	Y	Y	Y	Y	Y	Y
Bits to keep	1		0		0	1			1	1	1	1	0	0
Test	Y		N		N	Y			N	N	N	Y	Y	N
Key			0		0				1		1	1	1		0

Error correction and privacy amplification are purely classical sub-protocols, and
we will only sketch their idea here: *Error correction* is used to eliminate errors in
the sifted key, which may originate from faulty devices, noise, and/or Eve's tamper-
ing with the quantum signals. A simple error correction protocol works as follows:
Alice chooses two bits from the sifted key at random and tells Bob (via the classical
channel) the XOR value of the two bits. Bob tells Alice if he has the same value. In
this case, they keep the first bit and discard the second bit. If their values differ,
they discard both bits. The remaining bits form the key. Once Alice and Bob share
an identical bit string after the error correction, they have to decrease Eve's knowledge about these bits. This is achieved by *privacy amplification*, a simple exemplary
procedure of which works as follows: Alice and Bob agree on pairs of bits of the
(error-corrected) key and replace them by their XOR value. By doing so, they halve
the length of the key, but Eve has less information about this shorter key, even if she
knows the values of the single bits with high probability.

3.2.1 Simple Eavesdropping Strategies and Disturbance Versus Information Gain

We will now sketch a simple eavesdropping strategy and see why the BB84 protocol
is secure. For a full characterization of eavesdropping strategies, see Sect. 4 and for
the rigorous proof of the security of BB84 against all these attacks, see Sect. 5.

The most simple and intuitive attack for Eve is to *intercept and resend* the $2n$
qubits underway from Alice to Bob. Since Eve cannot copy them perfectly, as this is
forbidden by the No-Cloning theorem, an obvious strategy is to measure them. But

---

[1] We elaborate in the next subsection on what "too many" exactly means.

since she does not know the basis in which they were prepared (Alice announces this information only after Bob received all signals), Eve can only guess or just flip a coin for the selection of the measurement basis. In about half of the cases, i.e., for n qubits, she will happen to choose the same basis as Alice and get completely correlated bit values. In the other half, her results will be random and uncorrelated. Then Eve has to send the $2n$ qubits on to Bob, but as she does not know which basis Alice has used, she prepares each qubit in the same basis that she used for the measurement, i.e., only n of the newly created qubits match Alice's bases. After Bob receives Eve's qubits, he measures them, and Alice and Bob apply the sifting (step (iii)). Now, in $n/2$ cases Bob's and Alice's bases are the same, while Eve's basis differs. Since Bob's result is random in those cases, his sifted key will be wrong for $n/4$ bits, i.e., it will contain about 25% errors. In the parameter estimation stage (step (iv)), if Alice and Bob obtain such a high error rate, they abort the protocol.

This simple so-called "intercept-resend" strategy already shows that Eve necessarily creates errors when she learns parts of the key. The following more general (but not most general) eavesdropping method shows the *trade-off* between *disturbance* and *information gain*. Here, Eve uses an additional auxiliary system (ancilla) and a unitary interaction. Let U denote the unitary operation employed by Eve, acting on the qubit sent by Alice and the ancilla system $|E\rangle$ held by Eve. We first consider the case where Eve does not disturb at all the qubits sent by Alice. The action on two non-orthogonal states among the four states used in the BB84 protocol is then given by

$$U|0_z\rangle|E\rangle = |0_z\rangle|E_{0_z}\rangle, \tag{6}$$
$$U|1_x\rangle|E\rangle = |1_x\rangle|E_{1_x}\rangle, \tag{7}$$

with a self-explaining notation for the output states of the ancilla system. Unitary operations preserve the scalar product. The scalar products of the left- and right-hand sides in Eqs. (6) and (7) are

$$\langle 0_z|1_x\rangle\langle E|E\rangle = \langle 0_z|1_x\rangle\langle E_{0_z}|E_{1_x}\rangle, \tag{8}$$

which means that $|E_{0_z}\rangle$ and $|E_{1_x}\rangle$ have to be identical, and therefore Eve can obtain no information when measuring her ancilla.

We now look at the case where Eve's attack does disturb the qubits sent by Alice:

$$U|0_z\rangle|E\rangle = |0'_z\rangle|E_{0_z}\rangle, \tag{9}$$
$$U|1_x\rangle|E\rangle = |1'_x\rangle|E_{1_x}\rangle. \tag{10}$$

Unitarity implies that

$$\langle 0_z|1_x\rangle\langle E|E\rangle = \langle 0'_z|1'_x\rangle\langle E_{0_z}|E_{1_x}\rangle. \tag{11}$$

If Eve wants to obtain information about the states sent by Alice, she needs $|E_{0_z}\rangle$ and $|E_{1_x}\rangle$ to be distinguishable. In order to increase the distinguishability of these two states, their scalar product has to decrease. Due to Eq. (11), this implies that $\langle 0'_z|1'_x\rangle$ has to increase. These simple considerations show that the more information Eve wants to obtain, the more disturbance of the signal she has to introduce.

3.3 The Ekert Protocol

The Ekert protocol, suggested by Ekert in 1991 [21], uses entanglement to create a secret key for Alice and Bob. The idea is to distribute maximally entangled singlet states (see chapter "Bipartite Quantum Entanglement", Sect. 2)

$$|\psi^-\rangle = \frac{1}{\sqrt{2}}(|01\rangle - |10\rangle) \qquad (12)$$

between Alice and Bob and to exploit the fact that a measurement of both qubits in any basis yields correlated results. The remaining problem is how to distribute the qubits such that one can be sure that an eavesdropper can get no (or only very limited) information about the final key. This is done by checking a certain Bell inequality, namely the *CHSH inequality*, as described further below (see also chapter "Quantum Probability and Quantum Information Theory", Sect. 2 and chapter "Photonic Realization of Quantum Information Protocols", Sect. 2.1).

The Ekert protocol works as follows (see also Fig. 2):

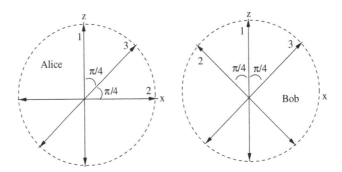

Fig. 2 Ekert protocol: Alice's and Bob's measurement directions (on the Bloch sphere)

(i) *Entanglement distribution.* Alice and Bob distribute a number of singlet states $|\psi^-\rangle$ among them, i.e., the first subsystem belongs to Alice and the second to Bob.

(ii) *Measurements.* For each singlet, Alice and Bob measure an observable randomly chosen from the sets $\{A_i\}$ and $\{B_i\}$, respectively. These observables are spin components, lying in the x–z plane of the Bloch sphere (see chapter "Quantum Probability and Quantum Information Theory", Sect. 4.3) and are given by $A_i = \cos\phi_i^A \sigma_z + \sin\phi_i^A \sigma_x$, $B_i = \cos\phi_i^B \sigma_z + \sin\phi_i^B \sigma_x$, and the angles are $\phi_1^A = 0$, $\phi_2^A = \pi/2$, $\phi_3^A = \pi/4$ for Alice and $\phi_1^B = 0$, $\phi_2^B = -\pi/4$, $\phi_3^B = \pi/4$ for Bob.

(iii) *Announcement of bases.* Alice and Bob announce the directions they chose for each measurement. In the cases where the directions match, i.e., (A_1, B_1) and (A_3, B_3), they get completely anti-correlated results, forming the sifted key by inverting all bits for one party.

(iv) *Bell inequality test.* The results obtained when Alice and Bob measured in the directions (A_1, B_3), (A_1, B_2), (A_2, B_3), and (A_2, B_2) are used to check an CHSH inequality, as described below.

(v) *Establishing the secret key.* Finally, Alice and Bob obtain a joint secret key from the sifted key by performing *error correction* and *privacy amplification*, see above.

The CHSH inequality [15] is an upper bound on a sum of expectation values for certain classical correlations. Consider four classical variables A_1, A_2, B_2, and B_3, which can only take the values $+1$ or -1. It is easy to see that $A_1(B_3 + B_2) + A_2(B_3 - B_2) = \pm 2$ for each assignment to the variables. Taking the average over N assignments to these classical variables, one finds

$$|\langle A_1(B_3 + B_2) + A_2(B_3 - B_2)\rangle| \leq 2, \tag{13}$$

where $\langle A_i B_j \rangle$ denotes the average $\sum_\nu^N A_i^{(\nu)} B_j^{(\nu)}/N$ and $A_i^{(\nu)}$ is assignment number ν to variable A_i. This can be rewritten as the CHSH inequality

$$S := |\langle A_1 B_3\rangle + \langle A_1 B_2\rangle + \langle A_2 B_3\rangle - \langle A_2 B_2\rangle| \leq 2. \tag{14}$$

Let us now consider A_1, A_2, B_2, and B_3 to be quantum mechanical observables and denote the expectation values for their products as

$$\langle A_i B_j \rangle = \mathrm{Tr}(A_i \otimes B_j \rho). \tag{15}$$

Using the measurement directions described in step (ii) of the Ekert protocol, the sum of expectation values S as defined in Eq. (14), with respect to the singlet $\rho = |\psi^-\rangle\langle\psi^-|$, takes the value $2\sqrt{2}$, a violation of the CHSH inequality. Whenever Alice and Bob measure a value of $S = 2\sqrt{2}$, they can be sure to share a maximally entangled state. In this case, Eve has no information on the key, since a maximally entangled state between Alice and Bob cannot be entangled with anything under Eve's control [16]. On the other hand, if Alice and Bob obtain no violation of the CHSH inequality, their measurement results are compatible with a separable state. This means that it is impossible to create a secret key [1, 18].

3.4 The B92 Protocol

Probably the most simple quantum key distribution protocol was suggested in 1992 by Bennett [2]: It uses two non-orthogonal states for encoding the bit values 0 and 1 and relies on the fact that non-orthogonal states cannot be cloned perfectly. The B92 protocol works as follows:

(i) *Preparation.* Alice prepares qubits randomly in one of the two states $|u\rangle$ or $|v\rangle$, with $\langle u|v\rangle \neq 0$, and sends them to Bob via the quantum channel.

(ii) *Measurement.* Bob tries to distinguish these states with a positive operator valued measurement (POVM) [29], given by $\mathcal{M} = \{E_{\neg u}, E_{\neg v}, E_?\}$, where

$$E_{\neg u} = \frac{\mathbb{1} - |u\rangle\langle u|}{1 + |\langle u|v\rangle|}, \tag{16}$$

$$E_{\neg v} = \frac{\mathbb{1} - |v\rangle\langle v|}{1 + |\langle u|v\rangle|}, \tag{17}$$

$$E_? = \mathbb{1} - E_{\neg u} - E_{\neg v} . \tag{18}$$

(iii) *Sifting.* Alice and Bob discard all cases where Bob got an inconclusive result ($E_?$) and obtain the sifted key.

(iv) *Parameter estimation.* Alice and Bob choose a subset of the sifted key to estimate the error rate. They do so by announcing publicly the bit values of the subset. If they differ in too many cases, they abort the protocol.

(v) *Establishing the secret key.* Finally, Alice and Bob obtain a joint secret key from the remaining bits by performing (classical) *error correction* and *privacy amplification.*

3.5 The Six-State Protocol

The six-state protocol [7, 5] employs three pairs of orthogonal states (i.e., six states in total), where the pairs are mutually non-orthogonal. This protocol is obtained from the BB84 protocol by including a third encoding in the eigenstates of σ_y. Thus, here one uses an encoding in three mutually unbiased bases of the two-dimensional Hilbert space: $\{|0_\alpha\rangle, |1_\alpha\rangle\}_{\alpha \in \{x,y,z\}}$, with $|\langle i_\alpha|j_\beta\rangle| = 1/\sqrt{2}$, for $\alpha \neq \beta$ and for all $i, j \in \{0, 1\}$.

The six-state protocol is similar to the BB84 protocol, with only minor modifications: Alice now chooses one out of *three* encodings. Thus, in the sifting procedure, approximately 2/3 of the raw key gets discarded. The advantage of the six-state protocol over the BB84 is that for a given disturbance of the eavesdropper, a higher secret key rate can be extracted. Intuitively speaking, this is due to the fact that the eavesdropper has less prior information, as the six states span the full Bloch sphere, rather than only a great circle (which is spanned by the four states in BB84). We will give a more detailed analysis of the connection between disturbance and secret key generation in Sect. 4.1.

3.6 Protocols with Higher-Dimensional Quantum States

All previously described quantum key distribution protocols use two-dimensional quantum states. One can generalize QKD to higher dimensions, i.e., Alice distributes d-dimensional non-orthogonal quantum states (for $d = 3$, see [6]). Here, Alice and Bob can, e.g., use two bases for a BB84-like protocol, and up to $d + 1$ *mutually unbiased* bases for a tomographically complete protocol (the generalization of the six-state protocol). Mutually unbiased bases are defined via the overlap $|\langle i_\alpha | j_\beta \rangle| = 1/\sqrt{d}$, for $\alpha \neq \beta$ and for all $i, j \in \{0, 1, ..., d - 1\}$. Here, the index α denotes the basis, and i numbers the basis elements. Note, however, that the existence of $d + 1$ mutually unbiased bases is only proven for d being a power of a prime number.

What would be the advantage of a higher-dimensional protocol? It has been shown [9, 12] that for a given disturbance the eavesdropper's information on the key decreases with increasing dimension, see Sect. 4.1.

4 Eavesdropping Strategies

Formally, the set of all possible eavesdropping strategies can be divided into three classes: "individual," "collective," and "coherent" attacks. Individual attacks are the simplest ones, corresponding to an eavesdropper with little power. Coherent attacks are potentially the most powerful, assuming an eavesdropper with unlimited technological power and resources, only being limited by the laws of nature. More concretely, these classes of strategies are defined by how Eve interacts with the quantum signals that are sent from Alice to Bob and how she processes the information she gathers in this way. The most general way to describe how information about a quantum system ρ_A is extracted is the following: Attach an ancilla system in a predefined state $|0\rangle\langle 0|_E$ to ρ_A and perform a (sophisticated) unitary operation U on the composite system $\rho_A \otimes |0\rangle\langle 0|_E$. Then do an (also sophisticated) measurement on the ancilla system $\rho_E := \text{Tr}_A(U^\dagger \rho_A \otimes |0\rangle\langle 0|_E U)$. The measurement is given by a POVM $\mathcal{M} = \{M_j\}$ which yields outcome j with probability $\text{Tr}(M_j\rho)$, when measuring a state ρ. We denote the classical probability distribution which is obtained in this way by $P_{\mathcal{M}}^\rho$, i.e., $P_{\mathcal{M}}^\rho(j) = \text{Tr}(M_j\rho)$.

Consider the case where Alice sends n quantum systems $\rho_A^1, \ldots, \rho_A^n$ to Bob. An *individual* attack is an attack where Eve attaches an ancilla system $|0\rangle\langle 0|_E$ to each state ρ_A^i, applies the same unitary operation U, and measures her part of all the composite systems individually and in the same way. *Collective* attacks are a little more general, as they allow the eavesdropper to measure all ancilla systems collectively. The most general attack is the *coherent* attack, in which it is assumed that Eve attaches one large ancilla system to the state $\rho_A^1 \otimes \cdots \otimes \rho_A^n$ and then performs a *global* unitary transformation U_g and measurement. More formally, the probability distribution that the eavesdropper obtains for each class of attacks is given by:

$$\text{Individual: } P_{\mathcal{M}^1}^{\rho_E^1} \cdots P_{\mathcal{M}^1}^{\rho_E^n}, \ \rho_E^i = \text{Tr}_A(U^\dagger \rho_A^i \otimes |0\rangle\langle0|_E U), \tag{19}$$

$$\text{Collective: } P_{\mathcal{M}^n}^{\rho_E^1 \otimes \cdots \otimes \rho_E^n}, \ \rho_E^i = \text{Tr}_A(U^\dagger \rho_A^i \otimes |0\rangle\langle0|_E U), \tag{20}$$

$$\text{Coherent: } P_{\mathcal{M}^n}^{\rho_E}, \quad \rho_E = \text{Tr}_A(U_g^\dagger (\rho_A^1 \otimes \cdots \otimes \rho_A^n) \otimes |0\rangle\langle0|_E U_g), \tag{21}$$

where $\mathcal{M}^1$ and $\mathcal{M}^n$ denote POVMs on one and n systems, respectively.

4.1 Individual Attacks

Let us analyze the class of individual eavesdropping strategies. Since Eve is restricted to the same interaction with each quantum signal sent from Alice to Bob, all possible attacks can be easily parameterized for a given protocol. In this section, we address the question of how much information the eavesdropper can gain about any single signal, and which restriction on the secret key length this implies. An important result that we will use in this context is a theorem derived in 1978 by Csiszár and Körner [17]. Before stating the theorem, we introduce some definitions: Let A be a random variable with range $\mathcal{A}$ and probability distribution P_A. The *Shannon entropy* (see chapter "Classical Information Theory", Sect. 1.2) of A is defined as $H(A) = -\sum_{a \in \mathcal{A}} P_A(a) \log P_A(a)$. If not stated otherwise, we will use the log to base 2. For two random variables A and B with joint probability distribution P_{AB}, the *mutual information* (see chapter "Classical Information Theory", Sect. 1.2) $I(A : B) = H(A) - H(A|B)$, with $H(A|B) = H(A, B) - H(B)$, is a measure of the amount of information about B contained in A (and vice versa).

We denote by $S(A : B||E)$ the length of the secret key that can be obtained for two parties, each holding a random variable A and B, respectively, if the eavesdropper holds a random variable E, for a given joint probability distribution P_{ABE}. The Csiszár-Körner theorem [17] states that

$$S(A : B||E) \geq \max\{I(A : B) - I(A : E), I(B : A) - I(B : E)\}. \tag{22}$$

This means that whenever Alice or Bob have an advantage over Eve in terms of information about the other party's random variable, they can distill a secret key. Note that this bound is not tight, i.e., there exist probability distributions for which the right-hand side becomes negative, however, a secret key can be distilled. According to the Csiszár–Körner theorem, whenever Eve's maximal mutual information with Alice is smaller than Bob's, the trusted parties can establish a secret key. We will therefore derive eavesdropping strategies that maximize Eve's mutual information.

Let us define the most general individual attack that an eavesdropper can perform on a qubit system, e.g., in the BB84 or six-state protocol: Since every qubit has to be attacked individually and in the same way, the only freedom is in the choice of the unitary transformation. The most general transformation is given by

$$U|0\rangle|X\rangle = \sqrt{F}|0\rangle|A\rangle + \sqrt{1-F}|1\rangle|B\rangle, \tag{23}$$
$$U|1\rangle|X\rangle = \sqrt{F}|1\rangle|C\rangle + \sqrt{1-F}|0\rangle|D\rangle, \tag{24}$$

where $\{|0\rangle, |1\rangle\}$ is the computational basis for Alice's state, $|X\rangle$ is an arbitrary input state of Eve's ancilla, and F is the fidelity (see Chapter "Hilbert Space Methods for Quantum Mechanics", Sect. 3.3) of Bob's state ρ_B with respect to the input state $|\psi_{in}\rangle$ sent by Alice, i.e., $F = \langle\psi_{in}|\rho_B|\psi_{in}\rangle$. It is reasonable to assume that the fidelity is the same for all employed bases (symmetric attack), because otherwise Alice and Bob could detect Eve by comparing the fidelities for different bases. After Eve's attack, she and Bob measure their respective qubits in a randomly chosen basis and obtain (after the sifting) random variables E and B, respectively. Alice's random variable A is determined by the probability that she chooses to send $|0\rangle$ or $|1\rangle$.

It can be shown [22] that for the BB84 protocol, the mutual informations with respect to Alice and Bob and Alice and Eve, in terms of the disturbance $D = 1 - F$, are given by

$$I(A:B) = 1 + D\log D + (1-D)\log(1-D), \tag{25}$$
$$I(A:E) = \frac{1}{2}(1+z)\log(1+z) + \frac{1}{2}(1-z)\log(1-z), \tag{26}$$

where $z = 2\sqrt{D(1-D)}$. The Csiszár-Körner bound (22) gives the maximal disturbance D for which Alice and Bob can surely extract a secret key. The same analysis can be done for the six-state protocol [7], leading to the same expression for $I(A:B)$, but a different one for $I(A:E)$:

$$I(A:E) = 1 + (1-D)\left[f(D)\log f(D) + (1-f(D))\log(1-f(D))\right], \tag{27}$$

with $f(D) = [1 + 1/(1-D)\sqrt{D(2-3D)}]$. When comparing these mutual informations, see Fig. 3, we find that the threshold disturbance, up to which a secret key can still be extracted, is higher for the six state than for the BB84 protocol, making the former more robust against individual attacks.

As mentioned in Sect. 3.6, the six-state protocol can be generalized to d-dimensional quantum systems, which are prepared by Alice in one of $d + 1$ *mutually unbiased bases*. For such a protocol in d dimensions, one finds for the mutual information [9]

$$I(A:B) = 1 + D\log_d\frac{D}{d-1} + (1-D)\log_d(1-D), \tag{28}$$
$$I(A:E) = 1 + (1-D)\left[f_d(D)\log_d f_d(D)\right.$$
$$\left. + (1-f_d(D))\log_d\frac{1-f_d(D)}{d-1}\right], \tag{29}$$

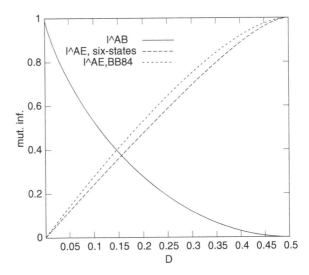

with $f_d(D) = d - 2D + \sqrt{(d-2D)^2 - d^2(1-2D)^2}/(d^2(1-D))$. One finds that
the threshold disturbance is higher for dimension $d = 3$ than for $d = 2$. Moreover,
the mutual information between Alice and Eve decreases as d increases, see Fig. 4.

4.2 Coherent Attacks

Coherent attacks are not only the most general ones but also the most difficult
ones to analyze, due to the high dimension of the global Hilbert space. Coherent
eavesdropping has been studied for BB84 [14] and for the six-state protocol [5].
As a somewhat surprising result, the authors found that coherent eavesdropping
does not increase Eve's Shannon information, however, it does slightly increase the
probability to guess the key.

In many recent security analyses, a stronger security definition is employed
which also covers coherent attacks. We present this definition in detail in Sect. 5.2.

4.3 Eavesdropping Versus Cloning Strategies

As an explanation for the security of quantum cryptography, we have mentioned the impossibility of perfect cloning. However, it is possible to clone an
unknown state *approximately*, i.e., with a fidelity lower than 1. Approximate cloning
has been extensively studied in the literature; for an introduction and overview
see [10, 34].

An obvious eavesdropping strategy is to use the optimal (in terms of fidelity)
cloning transformation, to apply it to the state sent to Bob, to send him an imperfect
clone, and to measure the other imperfect clone. It turns out that for several protocols
the optimal (in terms of mutual information) eavesdropping strategy coincides with

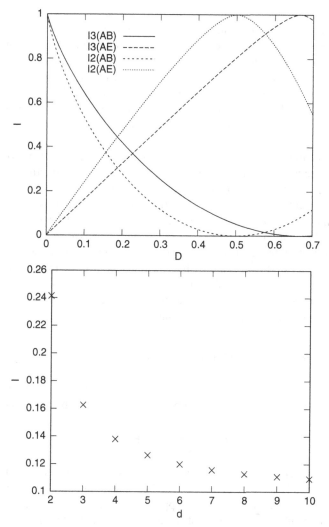

Fig. 4 (a) Comparison of Eve's mutual information in the generalized six-state protocol for dimension $d = 2$ and $d = 3$, as a function of the disturbance. (b) Eve's mutual information at fixed disturbance $D = 0.1$, as a function of the dimension d

optimal cloning. However, in general this is not true (e.g., for B92), and a general proof of the equivalence/non-equivalence for different protocols is still missing.

5 Unconditional Security of BB84

In the previous sections we have analyzed different eavesdropping strategies for various quantum key distribution protocols. An important question that remained open is how secure these protocols are when no restrictions apply to the eavesdropper's strategy. In this section we deal with this issue, the *unconditional security*, for the BB84 protocol, following the ideas of Shor and Preskill [36]. The term "unconditional" means that the eavesdropper is only restricted by the laws of nature and in particular is not limited by any resources such as computing time and/or memory. (Therefore, one has to assume that the eavesdropper holds a quantum system until the very end of the protocol.) A key is defined to be *secure*, if Eve has no "significant" information about it, i.e., if the probability that Alice and Bob agree on a key about which Eve has more than exponentially small information is exponentially small.

The idea of the proof of the unconditional security of the BB84 protocol [36] is to relate its security to entanglement purification and error correcting codes (CSS codes). The proof is divided into three steps: The first step (Sect. 5.1) is to introduce the "entanglement-based" version of the BB84 protocol, which is a reformulation in which Alice does not prepare one of the four states, but rather measures her part of a maximally entangled state that she shares with Bob. We show, after providing a short introduction to classical and quantum error correction (in particular to CSS codes), how Alice and Bob can distribute such a state over an insecure channel. In the second step (Sect. 5.2) we show that the entanglement-based version is secure, and in the third step (Sect. 5.3) we prove the equivalence between the entanglement-based and the original "prepare-and-measure" version.

5.1 Entanglement-Based Version of BB84

The BB84 protocol can be recast into a form that is very similar to the Ekert protocol. The main idea is that Alice and Bob share a maximally entangled state

$$|\phi^+\rangle = \frac{1}{\sqrt{2}} \left(|00\rangle + |11\rangle\right), \tag{30}$$

which leads to perfect correlations when the parties measure in the same basis (chosen from the two bases used in BB84). As this state is pure, it cannot be entangled with anything else, and thus an eavesdropper cannot have any information about the measurement results. Alice's and Bob's aim is consequently to distribute a number m of $|\phi^+\rangle$ states,

$$|\phi^+\rangle^{\otimes m} = |\phi^+\rangle \otimes \cdots \otimes |\phi^+\rangle. \tag{31}$$

Since Alice and Bob have to use an insecure quantum channel, they will not end up with exactly this state, but rather with a mixed state due to noise or interaction of Eve. The next step is thus to correct the errors that occurred during the transmission. Before we explain how this is achieved, we give a short introduction to error correction.

5.1.1 Classical Error Correction

A *classical linear* $[n, k]$ *code* C (encoding a k-bit string into an n-bit string, with $n > k$) is a set of 2^k codewords, where each codeword is a binary vector of length n.

A linear code can be described by an $(n \times k)$-dimensional *generator matrix* G that maps each message a to the encoded message $x = Ga$. (Here, all arithmetic operations are modulo 2.) Thus, the set of codewords is spanned by the columns of G.

Error correction for linear codes can be described by means of a *parity check matrix H*. This is an $((n - k) \times n)$ matrix with the property that $Hx = 0$ for all codewords x. Error correction works as follows: the message a gets encoded as $x = Ga$. If an error e occurs, x evolves to $x' = x + e$, i.e., the 1's in the vector e denote the erroneous bits. Applying the parity check matrix to x' leads to $Hx' = Hx + He = 0 + He = He$, as $Hx = 0$ for all codewords x. It follows that $Hx' = He$, which is called the *(error) syndrome*. If the syndrome is 0, no error has occurred. Otherwise, H is constructed such that the syndrome contains information about the error that allows to correct it. As an example, consider the [3, 1] "repetition code" for 1-bit messages: It has two codewords, $(0, 0, 0)^{\mathrm{T}}$ and $(1, 1, 1)^{\mathrm{T}}$, for 0 and 1, respectively. Its generator matrix is the (3×1) matrix

$$G = \begin{pmatrix} 1 \\ 1 \\ 1 \end{pmatrix}, \tag{32}$$

such that $(0, 0, 0)^{\mathrm{T}} = G \cdot (0)$ and $(1, 1, 1)^{\mathrm{T}} = G \cdot (1)$. To construct H, we choose $n - k = 3 - 1 = 2$ linearly independent vectors orthogonal to the columns of G, e.g., $(1, 0, 1)^{\mathrm{T}}$ and $(0, 1, 1)^{\mathrm{T}}$. The rows of the parity check matrix are then given by these vectors:

$$H = \begin{pmatrix} 1 & 0 & 1 \\ 0 & 1 & 1 \end{pmatrix}. \tag{33}$$

Denote by e_i an error in the ith bit, e.g., $e_2 = (0, 1, 0)^{\mathrm{T}}$. Then for all codewords x, we have $Hx' = H(x + e_i) = He_i$ and we find the syndromes

$$He_1 = \begin{pmatrix} 1 \\ 0 \end{pmatrix}, \quad He_2 = \begin{pmatrix} 0 \\ 1 \end{pmatrix}, \quad He_3 = \begin{pmatrix} 1 \\ 1 \end{pmatrix}, \tag{34}$$

which makes it possible to read off the erroneous bit. This procedure is only successful if one knows that an error has occurred for at most one bit. Therefore, this repetition code only corrects one error.

In general, the properties of an error correcting code depend on the *Hamming distance* of the codewords. The Hamming distance of two binary vectors is the number of positions in which the bit strings differ. If we assume that the error probability is smaller than 1/2, the codeword x that minimizes the Hamming distance between x and the erroneous x', is most likely the correct one. If all codewords of a code have a Hamming distance of at least t, then $\lfloor t/2 \rfloor$ errors can be corrected, where $\lfloor t/2 \rfloor$ denotes the largest integer smaller than or equal to $t/2$.

Finally, we introduce the concept of *duality*: Let C be a linear $[n, k]$ code with generator matrix G and parity check matrix H. Then we can define the dual code $C^{\perp}$ of C, which is the set of all codewords that are orthogonal to each codeword in C. The dual code $C^\perp$ is an $[n-k, n]$ code which is generated by H^T and has a parity check matrix G^T. Dual codes play an important role in the construction of CSS codes, as we will explain below.

5.1.2 Quantum Error Correction

In contrast to a classical bit, where only one type of error is possible (the bit flip), a qubit can undergo three different errors: bit flips, phase errors, and combinations of these two: A bit flip changes $|0\rangle$ into $|1\rangle$ and vice versa. Phase errors transform the state $|1\rangle$ into $-|1\rangle$, but leave $|0\rangle$ unchanged. The combination of the two errors is given by $|0\rangle \rightarrow -|1\rangle$ and $|1\rangle \rightarrow |0\rangle$. These three errors are generated by the Pauli matrices

$$\sigma_x = \begin{pmatrix} 0 & 1 \\ 1 & 0 \end{pmatrix}, \quad \sigma_y = \begin{pmatrix} 0 & -i \\ i & 0 \end{pmatrix}, \quad \sigma_z = \begin{pmatrix} 1 & 0 \\ 0 & -1 \end{pmatrix}, \tag{35}$$

where σ_x corresponds to the bit flip, σ_z to a phase error, and σ_y to the combined phase-bit-error.

A quantum error correction code that corrects bit and phase errors is a CSS code, named after Calderbank, Steane, and Shor [11, 37]. This kind of code has the important property that bit and phase flips are corrected *independently*, by using a quantum version two classical linear codes. This will be described in more detail in this subsection.

Definition 1 Let C_1 and C_2 be classical linear $[n, k_1]$ and $[n, k_2]$ codes, respectively, such that $C_2 \subset C_1$. For each codeword $x \in C_1$, define the quantum state

$$|x + C_2\rangle = \frac{1}{\sqrt{|C_2|}} \sum_{y \in C_2} |x + y\rangle, \tag{36}$$

where $|C_2|$ denotes the cardinality of C_2. The vector space spanned by $\{|x + C_2\rangle\}_{x \in C_1}$ defines a $[n, k_1 - k_2]$ quantum code, a *Calderbank–Shor–Steane* code, or CSS(C_1, C_2) for short.

It is important to realize that two different codewords x and x' in C_1 may lead to identical vectors $|x + C_2\rangle = |x' + C_2\rangle$. This is the case iff $x - x' \in C_2$, or in other words, iff x and x' belong to the same coset of C_1/C_2.[2] Otherwise, the states $|x + C_2\rangle$ and $|x' + C_2\rangle$ are orthogonal. As the number of cosets of C_2 in C_1 is $|C_1|/|C_2|$, the dimension of the space CSS(C_1, C_2) is $|C_1|/|C_2| = 2^{k_1 - k_2}$, thus $m = k_1 - k_2$ qubits can be encoded.

Error correction with CSS codes works as follows. Suppose that C_1 and $C_2^\perp$ both can correct t errors. Moreover, let H_1 be the parity check matrix for C_1 and H_2 that of $C_2^\perp$. Define

---

[2] Let G and H be two groups with $G \subset H$. Then for any $h \in H$, the coset of G in H, determined by h, is defined as $hG = \{h + g : g \in G\}$. The set H/G is the set of all cosets of G in H.

$$\sigma_\alpha^s = \sigma_\alpha^{s_1} \otimes \sigma_\alpha^{s_2} \otimes \cdots \otimes \sigma_\alpha^{s_n}, \tag{37}$$

where $\alpha \in \{x, y, z\}$, $\sigma_\alpha^0 = \mathbb{1}$, and $s = (s_1, s_2, \ldots, s_n)$ is an n bit vector. It can be shown that the syndrome for bit flip errors can be computed by measuring σ_z^s for each row vector s of H_1. Similarly, the syndrome for phase errors can be computed by measuring σ_x^r for each row vector r of H_2. In this way, arbitrary errors on any t qubits can be corrected. For more details, see [28]. An important property of CSS codes is that error correction for phase errors and for bit flips is decoupled from each other.

5.1.3 Entanglement-Based BB84: Step by Step

The last ingredient that we need is the Hadamard transformation (5), which maps the basis $\{|0_z\rangle, |1_z\rangle\}$ to the basis $\{|0_x\rangle, |1_x\rangle\}$ and vice versa, i.e.,

$$H|0, 1_z\rangle = \frac{|0_x\rangle \pm |1_x\rangle}{\sqrt{2}} = |0, 1_x\rangle, \quad H^2 = \mathbb{1}. \tag{38}$$

We are now ready to describe the entanglement-based version of the BB84 protocol:

(i) Alice creates $2n$ qubit pairs in the state $|\phi^+\rangle^{\otimes 2n}$.
(ii) She randomly selects n of those qubits which will later serve as check qubits for the error estimation.
(iii) Alice selects a random $2n$ bit string b and applies the Hadamard transformation (5) to her half of each qubit pair whenever b is "1."
(iv) She sends the other half of all qubit pairs to Bob.
(v) Alice announces b and which qubits are to serve as check qubits.
(vi) Bob performs a Hadamard transformation on those of his qubits where b is "1."
(vii) Alice and Bob measure the check qubits in the $\{|0_z\rangle, |1_z\rangle\}$ basis to estimate the error rate. If more than t results differ, they abort the protocol.
(viii) For the remaining qubits, Alice and Bob measure the syndromes for the codes C_1 and C_2, correct the errors, and obtain $|\phi^+\rangle^{\otimes m}$.
(ix) They measure this state in the $\{|0_z\rangle, |1_z\rangle\}$ basis to obtain a shared secret key.

Note that the application of the Hadamard transformation before and after the qubits are sent through the quantum channel has the same effect as preparing and measuring them in a rotational basis.

5.2 Security of the Entanglement-Based Version of BB84

In this section, we give a more precise argument why measuring $|\phi^+\rangle$ leads to a secure key and how Alice and Bob can reliably estimate the error rate in their measurement data.

First of all, we need to provide a rigorous definition of what "security" means. In the previous section about eavesdropping (Sect. 4) we used the mutual information to quantify the knowledge Eve can have about the key. In their original work [36], Shor and Preskill utilized exactly this security definition, which was accepted to be accurate for a long time. However, it turned out [25] that this measure is not appropriate when one takes into account that the generated key is used in another cryptographic application (e.g., a one-time pad) after the distribution protocol terminates. More concretely, it might happen that the mutual information (or even the *accessible information*, which is the maximal mutual information optimized with respect to all possible measurements) between Eve's data and the key is arbitrary small, yet, upon learning part of the key, Eve can recover the whole key [25]. As a result of this complication, a new security definition was proposed which was shown to provide the *universal composability* which the former definition lacks [4, 30].

Before stating the security definition, we need to introduce some mathematical framework in order to cover *all* possible eavesdropping strategies, in particular those where the adversary keeps a quantum system (containing information about the classical bit strings obtained by Alice and Bob upon measuring their quantum states) until the very end of the protocol. Such a situation, where a quantum system is correlated with classical data, can be described by the so-called *classical-quantum states* (cq-states, for short): Let X be a random variable with range $\mathcal{X}$ and let $\{|x\rangle\}_{x \in \mathcal{X}}$ be some basis of a Hilbert space. Moreover, denote by ρ_E^x the state of the quantum system E conditioned on the value x of the random variable X. Then the overall system can be described by the cq-state

$$\rho_{XE} = \sum_{x \in \mathcal{X}} P_X(x)|x\rangle\langle x| \otimes \rho_E^x. \tag{39}$$

Using this formalism, we can describe all possible states at the end of the protocol: Let S denote the set of all possible keys that can occur. The individual keys held by Alice and Bob can be described by random variables S_A and S_B, respectively, taking values s_A and s_B in S. The adversary holds a quantum system $\rho_E^{s_A s_B}$, which is correlated with those variables, and thus the total system can be described by a classical–classical–quantum (ccq) state:

$$\rho_{S_A S_B E} = \sum_{s_A, s_B \in S} P_{S_A, S_B}(s_A, s_B)|s_A\rangle\langle s_A| \otimes |s_B\rangle\langle s_B| \otimes \rho_E^{s_A s_B}. \tag{40}$$

In the case of a perfect key, Alice's and Bob's random variables are identical and uniformly distributed, i.e., each possible key occurs with equal probability. Moreover, the state of Eve's quantum system is completely independent of the key. Thus, the ideal ccq state is given by

$$\rho_{UU} \otimes \rho_E = \frac{1}{|\mathcal{S}|} \sum_{s \in \mathcal{S}} |s\rangle\langle s| \otimes |s\rangle\langle s| \otimes \rho_E. \tag{41}$$

We are now ready to provide the definition of an unconditionally secure key:

Definition 2 Let $\rho_{S_A S_B E}$, as defined in (40), be the ccq state describing a classical key pair (S_A, S_B) together with an adversary holding a quantum system E. Then (S_A, S_B) is said to be ε-secure with respect to E if and only if

$$\|\rho_{S_A S_B E} - \rho_{UU} \otimes \rho_E\| \leq \varepsilon, \tag{42}$$

where $\rho_{UU} \otimes \rho_E$ is the ideal ccq state, defined in (41).

Here, $\|\rho - \sigma\| = \mathrm{Tr}|\rho - \sigma|/2$ (with $|M| = \sqrt{M^\dagger M}$) denotes the trace distance, which is a proper distance measure in the space of hermitian operators. The above definition of security (Definition 2) has the intuitive interpretation that except with probability ε, the key pair (S_A, S_B) behaves as a perfect key, as described by (41). Moreover, this definition guarantees that the key pair remains secure when used in any cryptographic application.

We can now prove the unconditional security of the entanglement-based version of the BB84 protocol. Recall that the aim of this protocol is to distribute the state $|\phi^+\rangle^{\otimes n}$. Due to noise and/or Eve's interaction, Alice and Bob will in general not end up with *exactly* this state after the distribution; rather, they will hold a state ρ_{AB}, which is (hopefully) similar to $|\phi^+\rangle^{\otimes n}$. "Similarity" with a pure state is measured by means of the so-called *fidelity*, which is defined as $F(\rho, |\psi\rangle) = \langle\psi|\rho|\psi\rangle$. If $F = 1$, the two states are identical. Since we do not make any restrictions about the eavesdropper's strategy, we consider the worst case in which Eve holds a purifying system of ρ_{AB}. This is the state $\rho_E = \mathrm{Tr}_{AB}|\Psi_{ABE}\rangle\langle\Psi_{ABE}|$, where $|\Psi_{ABE}\rangle$ is a pure state (in a higher-dimensional Hilbert space) such that $\rho_{AB} = \mathrm{Tr}_E|\Psi_{ABE}\rangle\langle\Psi_{ABE}|$. This scenario corresponds to the case where the adversary has full control over the quantum channel.

The following lemma relates the fidelity of ρ_{AB} to $|\phi^+\rangle^{\otimes n}$ with the security of the key that is obtained when measuring ρ_{AB}. The proof of this lemma can be found in [25].

Lemma 1 *Let $\varepsilon \geq 0$ and ρ_{AB} be a bipartite quantum state such that*

$$F(\rho_{AB}, |\phi^+\rangle^{\otimes n}) \geq \sqrt{1 - \varepsilon^2}. \tag{43}$$

Then the n-bit strings obtained from measuring ρ_{AB} locally in the $\{|0\rangle, |1\rangle\}$-basis are ε-secure, with respect to an adversary holding the purifying system of ρ_{AB}.

It remains to show how Alice and Bob can estimate this fidelity: Recall that Alice sends the second qubit of the state $|\phi^+\rangle$ to Bob. During the transmission, this qubit may undergo one of the three possible qubit errors introduced in Sect. 5.1 or remain unchanged. Consequently, if a bit flip occurs, Alice and Bob will end up with the state $|\psi^+\rangle$, if a phase flip occurs, they will end up with $|\phi^-\rangle$, and for both errors, they will end up with $|\psi^-\rangle$, where

$$|\phi^\pm\rangle = \frac{1}{\sqrt{2}}(|00\rangle \pm |11\rangle) \qquad (44)$$

and

$$|\psi^\pm\rangle = \frac{1}{\sqrt{2}}(|01\rangle \pm |10\rangle) \qquad (45)$$

are the Bell states (here, we have dropped the index z for the computational basis). In order to determine whether a bit flip occurred, Alice and Bob measure the POVM $\{P_{bf}, \mathbb{1} - P_{bf}\}$, where

$$P_{bf} = P_{|\psi^+\rangle} + P_{|\psi^-\rangle} = |01\rangle\langle 01| + |10\rangle\langle 10|, \qquad (46)$$

and for detecting phase errors, they measure $\{P_{pe}, \mathbb{1} - P_{pe}\}$, where

$$P_{pe} = P_{|\phi^-\rangle} + P_{|\psi^-\rangle} = \frac{1}{2}(\mathbb{1} \otimes \mathbb{1} - \sigma_x \otimes \sigma_x), \qquad (47)$$

on a randomly chosen subset (say, half) of the signals. Note that these POVMs can be implemented locally.

As Eqs. (46) and (47) correspond to measurements in one basis only (namely the Bell basis), their outcomes obey the laws of classical probability theory. Therefore we can apply the following lemma [28], which states that for a sufficiently large bit string the errors of the check bits are representative for the errors of the key bits:

Lemma 2 *Let a random $2n$ bit string that might contain some errors and a random subset of n check bits of that string be given. Then, for any two constants $\delta > 0$ and $\gamma > 0$, the probability of finding less than δn errors on the check bits, and more than $(\delta + \gamma)n$ errors on the remaining bits is less than $e^{-\mathcal{O}(\gamma^2 n)}$, for sufficiently large n.*

Using their estimates about the bit and phase errors, Alice and Bob can finally bound the fidelity of their remaining qubits with respect to the state $|\phi^+\rangle$.

5.3 Equivalence of the Entanglement-Based and Prepare-and-Measure Scheme of BB84

In this section, we show the equivalence of the original version of the BB84 protocol, as presented in Sect. 3.2, and the entanglement-based version, which has been shown to be secure in the previous subsection. We do so by starting with the scheme described in the end of Sect. 5.1 and successively modifying steps to finally arrive at the original version. The main ingredient that we are using are the CSS codes introduced before.

First, we only look at the key qubits, i.e., the qubits that will form the key when measured at the end of the protocol. Note that Alice and Bob ideally share $2n$ maximally entangled pairs, thus measuring one part of a pair leads to a perfect correlation with the other part. Therefore, Alice can equivalently choose $2n$ random bits, prepare $2n$ qubits in the corresponding states, and send them to Bob. Not so obviously, Alice can measure the syndromes for the codes C_1 and C_2 already at the beginning of the protocol, as we will see in the following:

Given a CSS code $\text{CSS}(C_1, C_2)$, we can define a family of equivalent codes $\text{CSS}_{v,w}(C_1, C_2)$, in the sense that they have the same error correcting properties. The codewords of $\text{CSS}_{v,w}(C_1, C_2)$ are given by

$$|x_k, v, w\rangle = \frac{1}{\sqrt{|C_2|}} \sum_{y \in C_2} (-1)^{v \cdot y} |x_k + y + w\rangle, \tag{48}$$

where x_k is one representative of one of the m cosets of C_2 in C_1, and v and w are arbitrary n bit strings. Since the states $\{|x_k, v, w\rangle\}$ form a basis, we can rewrite

$$|\phi^+\rangle^{\otimes n} = \frac{1}{\sqrt{2^n}} \sum_{i=0}^{2^n-1} |i\rangle |i\rangle = \frac{1}{\sqrt{2^n}} \sum_{x_k, v, w} |x_k, v, w\rangle |x_k, v, w\rangle, \tag{49}$$

where i is in binary notation. If now Alice measures the error syndromes, namely σ_z^s for each row vector s of H_1 and σ_x^r for each row vector r of H_2, she obtains a random result for v and w. Finally, if she does a last measurement in the $\{|0\rangle, |1\rangle\}$ basis, she obtains a random string $x_k \in C_1/C_2$. From (49), we see that after these three measurements, Bob's state has collapsed onto $|x_k, v, w\rangle$, which is a random codeword of the randomly chosen code $\text{CSS}_{v,w}(C_1, C_2)$. Here, v and w play the role of the phase and bit flip error correction information, respectively.

Since at the end of the error correction, it is irrelevant whether Alice and Bob obtain the state $|\phi^+\rangle^{\otimes n}$ or $|\phi^-\rangle^{\otimes n}$ (both yield correlated results), they do not need to correct the phase errors, which means that Alice does not need to send the corresponding error correction information v. If she does not reveal this value, she effectively prepares a classical mixture of the states $|x_k, v, w\rangle$ for all possible values of v,

$$\rho_{x_k, w} = \frac{1}{2^n} \sum_v |x_k, v, w\rangle \langle x_k, v, w| = \frac{1}{|C_2|} \sum_{z \in C_2} |x_k + z + w\rangle \langle x_k + z + w|. \tag{50}$$

We see that this state can be equivalently prepared "classically" by choosing a random codeword $z \in C_2$, a random bit string w, and preparing the state $|x_k + z + w\rangle$, where x_k is also chosen randomly, as derived above.

Thus, Alice prepares the state $|x_k + z + w\rangle$ and sends it to Bob. Bob will receive an erroneous state $|x_k + z + w + e\rangle$, where e is the bit string denoting the errors, due channel noise or the adversary's attack. When Bob measures this state in the computational basis, he obtains $x_k + z + w + e$. After learning the error correction information w, which is announced by Alice, he subtracts w from his string and obtains $x_k + z + e$. If e does not contain too many errors, $x_k + z + e$ can be unambiguously corrected to $x_k + z$.

This scheme becomes more simple when Alice instead of choosing $x_k \in C_1/C_2$ chooses $x_k \in C_1$ (randomly), because then w is not needed anymore. Additionally, $x_k + z$ then is a completely random bit string. With this modification, Alice now simply sends the state $|y\rangle$, where y is a randomly chosen bit string, to Bob, instead of sending $|x_k + z + w\rangle$, and Bob receives $|y + e\rangle$. Instead of sending w as error correction information, Alice now sends $y - x_k$, which Bob subtracts from $y + e$ and corrects $x_k + e$ to x_k.

What we have achieved now is that the key qubits are simply prepared in a random state $|y\rangle$, in the same way as the check qubits. The modified protocol thus has the following form:

(i) Alice creates $2n$ random qubits, each in the state $|0\rangle$ or $|1\rangle$, and a random codeword $x_k \in C_1$.

(ii) She randomly selects n positions to be check qubits and the remaining n positions to define $|y\rangle$.

(iii) Alice selects a random $2n$ bit string b and applies the Hadamard transformation to her half of each qubit pair where b is "1."

(iv) She sends the other half of all qubit pairs to Bob.

(v) Alice announces b and $y - x_k$, and which qubits are to serve as check qubits.

(vi) Bob performs a Hadamard transformation on those of his qubits where b is "1."

(vii) Bob measures the check qubits in the $\{|0\rangle, |1\rangle\}$ basis. If he finds more than t results that disagree with Alice's prepared state, they abort the protocol.

(viii) Bob measures the key qubits to get $y + e$, subtracts $y - x_k$, and corrects $x_k + e$ to x_k.

(ix) He calculates the coset to which x_k belongs to get the key k.

Finally, we can remove the Hadamard transformations, i.e., Alice chooses randomly one of the four states in $\{|0\rangle_+, |1\rangle_+, |0\rangle_\times, |1\rangle_\times\}$ for each key and check qubit. Then Bob, instead of waiting for b to be announced, simply chooses one basis at random and measures the arriving qubits. As he will choose the wrong basis in roughly half the cases, Alice should double the number of input qubits to $4n$ to have approximately $2n$ qubits left. After Bob's measurement, Alice announces which basis she used and both discard all instances where they used a different basis. With this last modification, we finally arrived at the prepare-and-measure version of the BB84 protocol.

6 Defense Against Eavesdropping with Photon Number Splitting (PNS) Attacks

So far, we have considered idealized QKD protocols and eavesdropping therein. However, realistic experimental implementations may offer Eve new and more powerful paths. A dangerous eavesdropping strategy becomes obvious when looking at a common experimental implementation for QKD: Qubit systems can be conveniently represented by photons, typically using their polarization as degree of freedom. Ideally each qubit is encoded by exactly one photon. But ideal single-photon sources do not exist; therefore, one often uses weak coherent pulses,

$$|\alpha\rangle = e^{-|\alpha|^2/2} \sum_{n=0}^{\infty} \frac{\alpha^n}{\sqrt{n!}} |n\rangle , \qquad (51)$$

where α is a complex number, and $|n\rangle$ is a Fock state. When the phase $\arg \alpha$ is unknown (or randomized), one arrives at the following mixture of Fock states:

$$\rho = \int \frac{d \arg \alpha}{2\pi} |\alpha\rangle\langle\alpha| = \sum_{n=0}^{\infty} P(n)|n\rangle\langle n| , \qquad (52)$$

where the probability distribution of photons obeys Poissonian statistics,

$$P(n) = e^{-|\alpha|^2} \frac{|\alpha|^{2n}}{n!} . \qquad (53)$$

Here, the mean photon number is $\bar{n} = |\alpha|^2$. A "weak" laser pulse contains no photon in most cases, and more than one photon only with a small probability. A typical weak pulse used for QKD has a mean photon number $\bar{n} = 0.1$. This corresponds to "no photon" with probability $\sim 90.5\%$, "one photon" with $\sim 9\%$, and more than one photon for $\sim 0.5\%$.

Any implementation that uses weak coherent pulses to simulate single photons allows the following eavesdropping attack, which is called *photon number splitting* (PNS) attack: For each pulse sent from Alice to Bob via a lossy fiber, Eve measures its photon number (via a non-demolition measurement). If it is more than one, she splits off one photon, stores it, and sends the remaining photons to Bob through a lossless fiber. If the photon number is one, she blocks the event with a certain probability (in order to maintain the photon statistics that Bob expects to receive) and otherwise may eavesdrop. The vacuum events are simply forwarded to Bob. Now, for all multi-photon events Eve can get full information on the corresponding key bit, if she waits for Alice to announce the bases and then measures her stored photons correspondingly. Thus the protocol looses unconditional security.

A strategy to counter this attack was proposed in [24, 26]: The idea is to introduce a second source in Alice's lab that emits weak coherent pulses with a different photon number distribution (the so-called *decoy states*); namely the decoy source has a much higher mean photon number, but does not differ from the signal source in any other parameter like wavelength, etc. Alice sends decoy pulses to Bob at random, in between the signal pulses. After the distribution of all pulses, she announces which pulses were prepared by which source. If Eve launches the PNS attack described above, then Bob will find an abnormally higher loss than expected for the pulses that were prepared with a lower mean photon number, and thus Eve can be detected.

Another possibility to prevent PNS attacks was proposed by Scarani et al. [33]: The so-called SARG protocol is a variant of the BB84 protocol, where the classical sifting is modified. Instead of announcing the basis in which the signal was prepared, Alice announces a set of two states. This set consists of the signal state and a random state from the second basis. For example, if Alice sent the state $|0\rangle_z$ to Bob, she might announce $\{|0\rangle_z, |0\rangle_x\}$. Bob as usually measures in a randomly chosen basis. In this example, if he measures σ_z, the result is inconclusive. Only if he measures σ_x and obtains $|1_x\rangle$, which happens with probability 1/4, he knows that Alice could not have sent $|0_x\rangle$, thus the state must be $|0\rangle_z$. In this way, the sifted key has only 1/4 of the length of the raw key, in contrast to 1/2 in the case of the BB84 protocol. Although the efficiency is reduced, the security is increased, as multi-photon events do not give Eve the full information. In [33] it has been shown explicitly that for a number of eavesdropping strategies in an implementation with weak coherent pulses, the information extractable by Eve is lower for the SARG protocol than for the BB84 protocol, thus providing a higher security. Moreover, when assuming a realistic channel model, it has been shown that the distance, up to which a secure key can still be generated, becomes also larger for the SARG protocol.

7 Classical Upper Bounds on the Secret Key Rate

It is possible to define a classical measure for the secret key rate at the level were Alice, Bob, and Eve do not hold quantum system anymore, but have carried out their measurements. This situation then can be described purely classically, by random variables A, B, and E, held by the three protagonists, and which are correlated according to a classical probability distribution P_{ABE}.

The easiest way to analyze the protocols introduced so far is by looking at their so-called *entanglement-based* version (cf. Sec. 5.1): Consider a protocol in which Alice prepares some signal states $\{|\phi_i\rangle\}$ with probability p_i and sends them to Bob. This procedure can equivalently be described by preparing an entangled state $|\psi_{\text{eff}}\rangle_{AB} = \sum_i \sqrt{p_i}|i\rangle_A|\phi_i\rangle_B$, where the second half is sent to Bob. At some point, Alice measures her part of the state in the computational basis, obtains the outcome i with probability p_i and effectively prepares $|\phi_i\rangle$ at a distance. Bob does the usual measurement defined in the original version of the protocol, for instance, he measures σ_z or σ_x in the BB84 protocol. The state $|\psi_{\text{eff}}\rangle_{AB}$ is sometimes referred to as "effective state" because it may be used as a concept of thought, in the way explained above.

A general eavesdropping attack can be described by Eve attaching a quantum system to the state underway to Bob and performing an arbitrary unitary operation on this composite system. The result is a tripartite state ρ_{ABE}, where the system in E is under Eve's control. For simplicity, one usually assumes that the state ρ_{ABE} is pure and that the eavesdropper holds the purifying system $\rho_E = \text{Tr}_{AB} \rho_{ABE}$. In order to employ classical methods, one may now assume that each party measures his/her part of the quantum state using some POVM, resulting in the classical variables A, B, E, and probability distribution P_{ABE}. Note that this assumption potentially limits the power of the eavesdropper, as she ultimately has the freedom to keep her quantum system as long as she wishes.

We are now ready to define some classical measures of information, the secret key rate, and bounds thereof.

Definition 3 Let A, B, and E be some random variables with joint probability distribution P_{ABE}. Then the *conditional mutual information* between A and B, given E, is defined as

$$I(A; B|E) = \sum_{e \in \mathcal{E}} P_E(e) \left[H(A|e) + H(B|e) - H(A, B|e) \right]. \tag{54}$$

Here, the conditional Shannon entropy is defined as

$$H(X|e) = - \sum_{x \in \mathcal{X}} P_{X|E}(x, e) \log P_{X|E}(x, e), \tag{55}$$

and we denote by $\mathcal{E}$ and $\mathcal{X}$ the range of E and X, respectively. The conditional mutual information $I(A; B|E)$ quantifies the amount of information revealed about A when one learns B, given the knowledge of E. The following definition incorporates the case where $I(A; B|E)$ is minimized over all possible processings of the variable E:

Definition 4 Let A, B, and E be some random variables with joint probability distribution P_{ABE}. Then the *intrinsic information* between A and B, given E, is defined as

$$I(A; B \downarrow E) = \inf_{E \to \tilde{E}} I(A; B|\tilde{E}), \tag{56}$$

where the infimum is taken over all channels $P_{\tilde{E}|E}$ that can be used to process E.

The final two definitions quantify how many secret bits can be extracted from a given probability distribution and how many are needed to create it.

Definition 5 Let A, B, and E be some random variables with joint probability distribution P_{ABE}. The *secret key rate* $S(A : B\|E)$ is defined as the maximal amount of secret bits that can be extracted asymptotically from P_{ABE}.

The secret key rate is thus a classical analogue of the distillable entanglement E_d (see chapter "Bipartite Quantum Entanglement", Sect. 3.1).

Definition 6 Let A, B, and E be some random variables with joint probability distribution P_{ABE}. The information of formation $I_{form}(A; B|E)$ is defined as the minimal number of secret bits that are needed asymptotically to create P_{ABE}.

The name "information of formation" might be misleading, as it is the classical analogue of the entanglement cost E_c and *not* of the entanglement of formation (see chapter "Bipartite Quantum Entanglement", Quantum entropy and information).

Note that the security definition used here is purely classical. Roughly speaking, it states that the keys extracted by Alice and Bob should be asymptotically equal and uniformly distributed, whereas Eve's conditional mutual information (given the knowledge of all public communication between Alice and Bob) vanishes asymptotically. For a rigorous definition, see, e.g., [31].

Using these quantities, the following bounds of the secret key rate can be derived [31]:

$$S(A; B\|E) \leq I(A; B \downarrow E) \leq I_{form}(A; B|E) , \tag{57}$$

i.e., the secret key rate is bounded from above by the intrinsic information, which is always smaller than or equal to the information of formation.

Remember that we assumed that Eve measures her quantum system before Alice and Bob proceed with the classical post-processing, which limits the validity of the results presented in this section. Moreover, we considered a fixed measurement by Eve, resulting in the classical variable E. A more general approach would be to look for a POVM on Eve's state that maximizes her *accessible information* [28].

8 The Role of Entanglement in QKD

Intuitively, the Ekert protocol and the entanglement-based version of the BB84 protocol suggest that entanglement is a vital requirement for quantum key distribution. In fact, it has been shown by Curty et al. [18] and Acin et al. [1] that there exists a deep relationship between the *provable* entanglement of the state shared by Alice and Bob prior to their measurement and the secret key that can be extracted from the resulting classical data.

8.1 Entanglement and Secret Correlations

In this section, we are investigating the following scenario: Let ρ_{AB} be an (effective) quantum state shared by Alice and Bob, and P_{AB} the probability distribution obtained when measuring some POVM $\{M_a\}$ and $\{M'_b\}$ on Alice's and Bob's side, respectively, i.e., $P_{AB}(a, b) = \text{Tr}(\rho_{AB} M_a \otimes M'_b)$.

It has been shown in [18] that entanglement is a precondition for the creation of a secret key:

Theorem 1 *The correlations in P_{AB} cannot lead to a secret key unless one can show that ρ_{AB} via an entanglement witness $W = \sum_{a,b} c_{ab} M_a \otimes M'_b$, such that $\mathrm{Tr}(W\sigma) \geq 0$ for all separable states σ while $\mathrm{Tr}(W\rho_{AB}) = \sum_{a,b} c_{ab} P_{AB}(a,b) < 0$.*

Proof The proof uses the construction of entanglement witnesses (see chapter "Bipartite Quantum Entanglement", Sect. 3) and can be found in [18].

Note that it is crucial that the measurements $\{M_a\}$ and $\{M'_b\}$ are chosen such that one can construct an entanglement witness out of them. Starting with an entangled state ρ_{AB} is not enough, since Alice and Bob may choose some unclever measurements which do not make use of the quantum correlations in the state ρ_{AB}.

The following theorem [1] provides a necessary and sufficient criterion for the existence of secret correlations (i.e. $I_{\mathrm{form}}(A; B|E) > 0$), based on the possibility of Alice and Bob to detect entanglement in their shared state:

Theorem 2 *For all purifications $|\Psi\rangle_{ABE}$ of ρ_{AB} and all possible measurements by Eve compatible with P_{AB}, it holds $I_{\mathrm{form}}(A; B|E) > 0$ if and only if there exists no separable state which could have produced P_{AB}.*

Proof The proof is also based on entanglement witnesses and can be found in [1].

8.2 Secure Key from Bound Entanglement

In the previous subsection we have shown that provable entanglement is a necessary precondition for secret key generation. An interesting question arises here: does it have to be *free* entanglement or may the entanglement also be *bound* (see chapter "Bipartite Quantum Entanglement", Sect. 3.1)?

Note that bound entanglement (i.e., no singlet states can be distilled from it) is often considered "useless" for practical applications.

The question whether a secret key can be created from bound entanglement was investigated by Horodecki et al. in [23], where the authors introduced the so-called *private states*. The most general form of a private state γ_m, from which one can create m bits of secret key, is given in the following theorem:

Theorem 3 *The most general state shared by Alice and Bob, from which m bits of secret key can be extracted, is given by*

$$\gamma_m = U|\psi_{2^m}^+\rangle\langle\psi_{2^m}^+| \otimes \rho_{A'B'} U^\dagger, \qquad (58)$$

where $|\psi_{2^m}^+\rangle = \sum_{i=0}^{2^m-1} |i\rangle_A |i\rangle_B$ is the maximally entangled state in $\mathbf{C}^{2^m} \otimes \mathbf{C}^{2^m}$, $\rho_{A'B'}$ is some additional state held by Alice and Bob, and

$$U = \sum_{i,j=0}^{2^m-1} |ij\rangle_{AB}\langle ij| \otimes U_{ij}^{A'B'}, \qquad (59)$$

with some unitary operation $U_{ij}^{A'B'}$.

Proof The proof can be found in [23].

The AB part of the state γ_m is called the "key part," whereas the state $\rho_{A'B'}$ is called the "shield," since it shields the key from the eavesdropper, even if she holds the purifying system of the state γ_m. A simple and intuitive example for the state γ_1 is given by

$$\gamma_1 = p|\phi^+\rangle\langle\phi^+| \otimes \rho_s + (1-p)|\phi^-\rangle\langle\phi^-| \otimes \rho_a, \tag{60}$$

where $|\phi^\pm\rangle = (|00\rangle \pm |11\rangle)/\sqrt{2}$ and ρ_s and ρ_a are the projectors onto the symmetric and anti-symmetric subspace of $\mathbf{C}^d \otimes \mathbf{C}^d$, respectively. Surprisingly, for any dimension d and $p = (1+1/d)/2$, one bit of secret key can be extracted from γ_1, but it can be shown that the distillable entanglement E_D is bounded as $E_D \leq \log[(d+1)/d]$. Thus, for increasing d, the amount of distillable entanglement in γ_1 decreases, and in the limit $d \to \infty$ one finds for the distillable key rate $K_D \gg E_D$.

Theorem 3 motivates the following definition [23, 13] of the secret key rate when quantum operations are performed on the state shared by Alice and Bob:

Definition 7 Let ρ_{AB} be a quantum state shared by Alice and Bob and let $\{\Lambda_n\}$ be a sequence of maps such that

$$\Lambda_n(\rho_{AB}^{\otimes n}) = \tilde{\gamma}_m, \tag{61}$$

with

$$\lim_{n\to\infty} \|\tilde{\gamma}_m - \gamma_m\| = 0. \tag{62}$$

Then the *distillable key rate* of the state ρ_{AB} is given by

$$K_D(\rho_{AB}) = \sup_{\{\Lambda_n\}} \lim_{n\to\infty} \frac{m}{n}. \tag{63}$$

It is possible to give an upper bound on $K_D(\rho_{AB})$ in terms of the amount of entanglement contained in ρ_{AB} [23]. The appropriate entanglement measure (see e.g. reference [8]) is an asymptotic version of the *relative entropy of entanglement*

$$E_r(\rho) = \inf_{\sigma \in S} \text{Tr}[\rho(\ln \rho - \ln \sigma)], \tag{64}$$

where S denotes the set of separable states.

Theorem 4 *Let* $E_r^\infty(\rho) = \lim_{n\to\infty} E_r(\rho^{\otimes n})/n$ *be the regularized relative entropy of entanglement. Then*

$$K_d(\rho_{AB}) \leq E_r^\infty(\rho_{AB}). \tag{65}$$

Proof The proof can be found in [23].

Apart from the inequality stated in the above theorem, more relations between entanglement measures and the distillable secret key rate are known. In general, we have that

$$E_D \leq K_D \leq E_r^\infty \leq E_c, \tag{66}$$

where E_c is the entanglement cost. Note that it is possible to have strict inequalities, i.e., there exist examples in which

$$E_D < K_D < E_c. \tag{67}$$

In particular, one can have $K_D > 0$ even for bound entangled states.

In this last section we have shown that secret key distillation and entanglement distillation exhibit certain common features, but are not fully equivalent. Remarkably, concepts and methods developed in *quantum* information theory have been employed as new tools to address problems in *classical* information theory, thus emphasizing the close interplay between classical and quantum information science.

9 Problems/Exercises

(1) Using the measurement directions of the Ekert protocol, show that the CHSH inequality is violated when Alice and Bob share a singlet. Adding white noise to the projector onto the singlet, what is the maximal proportion of noise that still leads to a violation of the CHSH inequality?

(2) Derive Eve's maximal mutual information for the BB84 protocol, see Eq. (26). Start from the general ansatz in Eqs. (23) and (24) and use symmetry of the attack, i.e., equality of the fidelity in the two bases.

(3) Show that the state $|\phi^+\rangle$ in Eq. (30) is form invariant in the Hadamard-rotated basis (eigenbasis of σ_x).

(4) Using the definition in Eq. (36), show that $|x + C_2\rangle = |x' + C_2\rangle$ if and only if $x - x' \in C_2$.

(5) Consider the following probability distribution, which arises from measuring a certain bound entangled state [20] in the computational basis. Show that this

A B E	P_{ABE}
0 0 0	1/6
0 0 1	1/6
0 1 0	1/6
1 0 1	1/6
1 1 0	1/6
1 1 1	1/6

distribution has positive intrinsic information, $I(A; B \downarrow E) > 0$. *Hint:* Show that for all channels $E \rightarrow \tilde{E}$ one has $I(A; B|\tilde{E}) > 0$. It suffices to consider only alphabets of $\tilde{E}$ not larger than the size of E.

References

1. Acin, A., Gisin, N.: Phys. Rev. Lett. **94**, 020501 (2005)
2. Bennett, C.: Phys. Rev. Lett. **68**, 3121 (1992)
3. Bennett, C.H., Brassard, G.: Quantum cryptography: Public key distribution and coin tossing. In: Proceedings of the IEEE International Conference on Computers, Systems, and Signal Processing, Bangalore, India, pp. 175–179. IEEE, New York (1985)
4. Ben-Or, M., Horodecki, M., Leung, D.W., et al.: The universal composable security of quantum key distribution. In: Kilian, J. (ed.) Second Theory of Cryptography Conference, TCC 2005. Lecture Notes in Computer Science, vol. 3378, pp. 386–406. Springer, New York (2005)
5. Bechmann-Pasquinucci, H., Gisin, N.: Phys. Rev. A **59**, 4238 (1999)
6. Bechmann-Pasquinucci, H., Peres, A.: Phys. Rev. Lett. **85**, 3313 (2000)
7. Bruß, D.: Phys. Rev. Lett. **81**, 3018 (1998)
8. Bruß, D. Leuchs, G.: Lectures on Quantum Information. Wiley-VCH, Weinheim (2006)
9. Bruß, D., Macchiavello, C.: Phys. Rev. Lett. **88**, 127901 (2002)
10. Bruß, D., Macchiavello, C.: Approximate quantum cloning. In: Bruß, D., Leuchs, G. (eds.) Lectures on Quantum Information, pp. 53–72. Wiley-VCH, Weimheim (2006)
11. Calderbank, A.R., Shor, P.W.: Phys. Rev. A **54**, 1098 (1996)
12. Cerf, N.J., Bourennane, M., Karlsson, A., et al.: Phys. Rev. Lett. **88**, 127902 (2002)
13. Christandl, M., Ekert, A., Horodecki, M., et al.: Unifying classical and quantum key distillation. In: Theory of Cryptography. Lecture Notes in Computer Science, vol. 4392, pp. 456–478. Springer, Heidelberg (2007)
14. Cirac, I., Gisin, N.: Phys. Lett. A **229**, 1 (1997)
15. Clauser, J., Horne, M., Shimony, A., et al.: Phys. Rev. Lett. **23**, 880 (1969)
16. Coffman, V., Kundu, J., Wootters, W.K.: Phys. Rev. A **61**, 052306 (2000)
17. Csiszár, I., Körner, J.: IEEE Trans. Inf. Theory **24**, 339 (1978)
18. Curty, M., Lewenstein, M., Lütkenhaus, N.: Phys. Rev. Lett. **92**, 217903 (2004)
19. Diffie, W., Hellmann, M.: IEEE Trans. Inf. Theory **22**, 644 (1976)
20. Dür, W., Cirac, J.I., Tarrach, R.: Phys. Rev. Lett. **83**, 3562 (1999)
21. Ekert, A.: Phys. Rev. Lett. **67**, 661 (1991)
22. Fuchs, C.A., Gisin, N., Griffiths, R.B., et al.: Phys. Rev. A **56**, 1163 (1997)
23. Horodecki, K., Horodecki, M., Horodecki, P., et al.: Phys. Rev. Lett. **94**, 160502 (2005)
24. Hwang, W.-Y.: Phys. Rev. Lett. **91**, 057901 (2003)
25. König, R., Renner, R., Bariska, A., et al.: Phys. Rev. Lett. **98**, 140502 (2007)
26. Lo, H.-K., Ma, X., Chen, K.: Phys. Rev. Lett. **94**, 230504 (2005)
27. Maurer, U.M.: IEEE Trans. Inf. Theory **39**, 733 (1993)
28. Nielsen, M.A., Chuang, I.C.: Quantum Computation and Quantum Information. Cambridge University Press, Cambridge (2000)
29. Peres, A.: Quantum Theory: Concepts and Methods. Kluwer Academic Publishers, Dordrecht (1995)
30. Renner, R., König, R.: Universally composable privacy amplification against quantum adversaries. In: Kilian, J. (ed.) Second Theory of Cryptography Conference, TCC 2005. Lecture Notes in Computer Science, vol. 3378, pp. 407–425. Springer, New York (2005)
31. Renner, R., Wolf, S.: New bounds in secret-key agreement: The gap between formation and secrecy extraction. In: Biham, E. (ed.) Advances in Cryptology — EUROCRYPT 2003. Lecture Notes in Computer Science, vol. 2656, pp. 562–577. Springer, New York (2003)
32. Rivest, R., Shamir, A., Adleman, L.: Commun. ACM **21**, 120 (1978)
33. Scarani, V., Acin, A., Ribordy, G., et al.: Phys. Rev. Lett. **92**, 057901 (2004)
34. Scarani, V., Iblisdir, S., Gisin, N., et al.: Rev. Mod. Phys. **77**, 1225 (2005)
35. Shannon, C.: Bell Syst. Tech. J. **28**, 656 (1949)
36. Shor, P.W., Preskill, J.: Phys. Rev. Lett. **85**, 441 (2000)
37. Steane, A.M.: Proc. R. Soc. Lond. A **452**, 2551 (1996)
38. Vernam, G.S.: J. IEEE **55**, 109 (1926)
39. Wootters, W.K., Zurek, W.H.: Nature **299**, 802 (1982)

Quantum Algorithms

J. Kempe and T. Vidick

The idea to put computing machines on a physical footing and to use the laws of physics as the basis of a computer already dates back several decades. In the 1980s, Feynman [24, 25] was the first to consider quantum mechanics from a computational point of view by observing that the simulation of quantum mechanical systems on a classical computer seemed to require an increase in complexity exponential in the size of the system. He asked whether this exponential overhead was inevitable, and if it was possible to design a universal *quantum* computer, which could simulate any quantum system without the exponential overhead. In 1985 Deutsch [17] defined the model of the quantum Turing machine, generalizing the classical Turing machine to follow the laws of quantum mechanics. Yao later showed that it was equivalent to the quantum circuit model, also defined by Deutsch.

Deutsch was the first to exhibit a concrete (albeit artificial) computational task which admitted a quantum algorithm strictly more efficient than the best classical algorithm solving the same problem (see Sect. 3). Although this algorithm seems rather simple (it involves only two quantum bits - qubits), it carries the main ingredients of later quantum algorithms and is a nice toy model for understanding why and how quantum algorithms work.

The most striking demonstration of the computational power of quantum computers was given by Shor in 1994 [44], who exhibited efficient quantum algorithms for factoring and the discrete logarithm. There are no known efficient classical algorithms for these problems, and in fact the security of many cryptographic systems (such as some primitives for secure authentication over the internet like RSA) rely on their supposed hardness. Shor thus demonstrated that, should a quantum computer be built, most current encryption schemes will be broken immediately.

We begin this chapter by describing some useful notations in Sect. 1. We then describe the quantum circuit model of computation, introduced by Deutsch, in

J. Kempe (✉)
CNRS & LRI, University of Paris-Sud, Orsay, France; School of Computer Science,
Tel Aviv University, Tel Aviv, Israel, kempe@post.tau.ac.il

T. Vidick (✉)
DI, École Normale Supérieure, Paris, France, vidick@eecs.berkeley.edu

Kempe, J., Vidick, T.: *Quantum Algorithms*. Lect. Notes Phys. **808**, 309–342 (2010)
DOI 10.1007/978-3-642-11914-9_10   © Springer-Verlag Berlin Heidelberg 2010

Sect. 2. In Sect. 3 we give a detailed account of Deutsch's algorithm. We then outline the quantum Fourier Transform (Sect. 4), which is a crucial ingredient in many
quantum algorithms. As applications, we describe the algorithms of Deutsch–Josza
and Simon in Sect. 5 and give in detail Shor's algorithm for factoring in Sect. 6.
These algorithms can be seen as special cases of a more general problem, called
the Hidden Subgroup Problem, which we present in Sect. 7. In Sect. 8 we give the
details of another notable quantum algorithm: Grover's 1996 algorithm for unstructured search. We finally end this chapter by giving an overview of new algorithms
and algorithmic techniques in Sect. 9.

Those who are interested to know more will find detailed expositions of the clas-
sic quantum algorithms in the literature (in particular [39, 34]) and of more recent
developments in the reference list.

1 Notations

In this chapter we will make use of the following conventions and notations:

A *bit* b is either 0 or 1, i.e., $b \in \{0, 1\}$. An *n-bit string* is a sequence of n bits,
i.e., it is an element of $\{0, 1\}^n$. We often work in arithmetic over the 2-element field
$GF(2)$. In this case we denote addition by $\oplus$: $0 \oplus 0 = 1 \oplus 1 = 0$ and $0 \oplus 1 = 1 \oplus$
$0 = 1$, this is just addition mod 2. We will also use $\oplus$ to denote bit-wise addition
of bit strings mod 2, e.g., $10 \oplus 11 = 01$. For multiplication of 2 bits b_1 and b_2 we
will write either $b_1 b_2$ or $b_1 \cdot b_2$. The *inner product* of 2 bit strings of same length n
will also be denoted by "$\cdot$" and is defined for $x = x_1 x_2 \ldots x_n$ and $y = y_1 y_2 \ldots y_n$
as $x \cdot y = x_1 y_1 + x_2 y_2 + \ldots + x_n y_n$ mod 2. We also require a little bit of group
theory. A group G is a set of elements with a binary operation defined on them,
often denoted by "$+$," which is associative, has an identity element and is such that
every element in G has an inverse for this operation. The group is called *Abelian*
if for all elements $g, h \in G$ we have $g + h = h + g$. An example of a family of
Abelian groups are the *cyclic* groups of M elements, denoted as $\mathbb{Z}_M$. The elements
are $\mathbb{Z}_M = \{0, 1, \ldots, M - 1\}$ and the group operation is simply addition modulo M.
The simplest such group is the group $(\mathbb{Z}_2, \oplus)$: its elements satisfy $0 \oplus 0 = 1 \oplus 1 = 0$
and $0 \oplus 1 = 1 \oplus 0 = 1$.

A qubit is a two-dimensional quantum system. We will denote the two basis
states by $|0\rangle$ and $|1\rangle$. The state of several qubits is spanned by the *tensor product*
of individual qubits. If the first qubit is in the state $|\phi\rangle$ and the second qubit is
in the state $|\psi\rangle$, then we will write the state of the joint system of two qubits as
$|\phi\rangle \otimes |\psi\rangle$. In case there is no ambiguity we will omit the tensor product $\otimes$. The
tensor product space of n qubits is spanned by the 2^n states $|00 \ldots 00\rangle$, $|00 \ldots 01\rangle$,
$\ldots$, $|11 \ldots 11\rangle$. These states are also called *computational basis* states, as they
correspond to the classical configurations of n bits. A *measurement* in the computational basis projects onto these basis states. Its outcomes are hence described by an
n-bit string. If the system is in a state $|\phi\rangle$, then the probability that a measurement
in the computational basis gives a bit string x is given by the inner product squared

$|\langle x|\phi\rangle|^2$. Sometimes we will only perform a *partial* measurement on a state. This means we only measure a subset of the qubits in the computational basis. The state collapses to a state that is consistent with the measurement we have performed. For instance if we have the state $\frac{1}{\sqrt{3}}|00\rangle + \frac{1}{\sqrt{3}}|01\rangle + \frac{1}{\sqrt{3}}|11\rangle$ and we measure the second qubit in the computational basis, then the state of the first qubit will be $|0\rangle$ if the outcome is 0 (which happens with probability $\frac{1}{3}$) and $\frac{1}{\sqrt{2}}(|0\rangle + |1\rangle)$ if we get outcome 1 (with probability $\frac{2}{3}$). This becomes more obvious if we rewrite the state as

$$\frac{1}{\sqrt{3}}|00\rangle + \frac{\sqrt{2}}{\sqrt{3}}\frac{|0\rangle + |1\rangle}{\sqrt{2}} \otimes |1\rangle. \tag{1}$$

To analyze the algorithms that follow, we will need some standard notation to describe the asymptotic behavior of such functions as the running time of an algorithm as a function of the input size. We will use the following:

- $f = O(g)$ (f is "big-O" of g) if there exist positive constants C and k such that $|f(x)| \leq C|g(x)|$ for all $x > k$.
- $f = \Omega(g)$ (f is "big-Omega" of g) if there exist positive constants C and k such that $|g(x)| \leq C|f(x)|$ for all $x > k$.
- $f = \Theta(g)$ (f is "big-Theta" of g) if there exist positive constants C_1, C_2, and k such that $C_1|g(x)| \leq |f(x)| \leq C_2|g(x)|$ for all $x > k$.

In complexity theory problems are often classified according to the resources required by the best algorithm that solves them. Here the notion of *efficiently solvable* is used. An algorithm is considered *efficient* if the resources it uses (time and space) scale like a polynomial in the number of input bits. In other words, if the input can be described by n bits, then an algorithm is *efficient* if its resources scale as $O(n^\alpha)$ for some constant α, which means that there exists some polynomial in n that bounds the amount of resources used. This gives rise to the complexity class P, which is the class of problems for which there is an efficient algorithm (note that if the algorithm runs in time polynomial in n, it also uses only polynomial space, as it takes at least a unit of time to use a unit of memory). Another widely used complexity class is the class NP. Problems in this class are such that, given a solution to the problem, there is an algorithm that can *efficiently verify* if this solution is correct. However, in general these problems are not known to have an efficient algorithm that finds the solution. In particular we have $P \subseteq NP$ and it is a famous open question whether $P \neq NP$. Equality of these classes would imply that it is as easy to find a solution to a problem as it is to verify a candidate solution that has been given. One important property of the class NP is that it contains *complete* problems. A problem is NP-complete if any other problem in NP can be reduced to it, whereby a reduction from a problem A to a problem B we mean a polynomial-time algorithm which, given access to solutions of problem B, can solve problem A efficiently. Many of the hardest algorithmic problems, like constraint satisfaction problems, or

the traveling salesman problem are *NP*-complete and finding an efficient algorithm
to one of them would allow to solve all problems in *NP* efficiently.

2 The Quantum Circuit Model

A classical computer can be described by a circuit. The input is a string of bits
($\{0, 1\}^n$). The input is processed by a succession of logical gates like NOT, OR,
AND, or NAND, which transforms the input to the output. In general the output bits
are Boolean functions $f : \{0, 1\}^n \rightarrow \{0, 1\}$ of the input bits. A schematic view is
depicted in Fig. 1.

The way Feynman put it, a quantum computer is a machine that obeys the laws
of quantum mechanics, rather than Newtonian classical physics. In the context of
computation this has two important consequences, which define the two aspects in
which a quantum computer differs from its classical counterpart. First, the *states*
describing the machine in time are quantum mechanical wave functions. Each basic
unit of computation – the *qubit* – can be thought of as a two-dimensional complex
vector of norm 1 in some Hilbert space with basis $|0\rangle$ and $|1\rangle$. We can think of the
basis states as corresponding to the states of a classical bit, 0 and 1. And second,
the *dynamics* that governs the evolution of the state in time is *unitary*, i.e., described
by a unitary matrix that transforms the state at a certain time to the state at some
later time. A second dynamical ingredient is the *measurement*. In quantum mechanics observing the system changes it. In the more restricted setting of a quantum
algorithm a measurement can be thought of as a projection on the basis states.

With these notions in place, we can describe a general quantum circuit. In the
beginning the qubits are initialized to some known classical state (a basis state of
$|0\rangle$ and $|1\rangle$, also called the *computational basis*), like for instance $|00\ldots0\rangle$. Then
a unitary transformation U is performed on the qubits. In the end the qubits are
measured in the computational basis and the result is processed classically. The
general setting can be seen in Fig. 2.

Given this model, it is not even clear if such a quantum computer is able to
perform classical computations. After all, a unitary matrix is invertible and hence
a quantum computation is necessarily *reversible*. Classical computation given by
some circuit with elementary gates, like the AND and NOT gates, is not reversible,

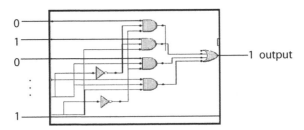

Fig. 1 Schematic representation of a classical circuit computing a Boolean one-bit function

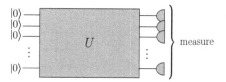

Fig. 2 A general quantum circuit consisting of three steps: initialization of the qubits, unitary transformation, and measurement in the computational basis

let alone because a gate like the AND gate has two inputs and only one output. However, the question of reversibility of classical computation has been studied in the context of energy dissipation by Bennett in the 1970s [9] (see also (Toffoli, 1980, Automata, Languages and Programming, unpublished)), who established that classical computation can be made reversible with only a polynomial overhead in the number of bits and gates used.

2.1 Universal Reversible Classical Computation

One way to make classical computation reversible is to simulate any classical circuit with AND and NOT operations by a circuit consisting entirely of what is called a Toffoli gate, which acts as depicted in Fig. 3.

When viewed as a matrix over the eight basis states 000, 001, 010, 011, 100, 101, 110, 111 the Toffoli gate acts as

$$T = \begin{pmatrix} 1 & 0 & 0 & 0 & 0 & 0 & 0 & 0 \\ 0 & 1 & 0 & 0 & 0 & 0 & 0 & 0 \\ 0 & 0 & 1 & 0 & 0 & 0 & 0 & 0 \\ 0 & 0 & 0 & 1 & 0 & 0 & 0 & 0 \\ 0 & 0 & 0 & 0 & 1 & 0 & 0 & 0 \\ 0 & 0 & 0 & 0 & 0 & 1 & 0 & 0 \\ 0 & 0 & 0 & 0 & 0 & 0 & 0 & 1 \\ 0 & 0 & 0 & 0 & 0 & 0 & 1 & 0 \end{pmatrix} . \tag{2}$$

Clearly, the Toffoli gate is reversible (it is its own inverse and implements a permutation of the 8 bit strings). Note also that this gate is unitary.

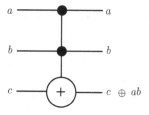

Fig. 3 The Toffoli gate flips the last bit if the first two bits are in the state 11

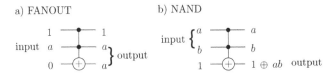

Fig. 4 Implementing the FANOUT (**a**) and the NAND gate (**b**) with the Toffoli gate

To show how a circuit consisting of Toffoli gates can simulate *any* classical circuit we will assume here that the classical circuit is made of NAND gates and FANOUT gates. It is known (and not hard to show) that any Boolean function can be computed by a circuit of NAND and FANOUT gates. The FANOUT gate simply copies 1 bit into 2 bits. The NAND gate outputs 0 if and only if both input bits are 1; otherwise it outputs 1. In other words, it acts as NAND $(a, b) = 1 \oplus ab$. Figure 4 shows how the Toffoli gate can implement the FANOUT and the NAND gate with some extra ancillary bits. Replacing every occurrence of a FANOUT or a NAND gate in a circuit by the corresponding Toffoli gate supplemented with ancillary bits, we can simulate the whole circuit in a reversible fashion. Since classical reversible computation is just a permutation on the bit strings of its input, it is in particular unitary.

As a result, quantum computation is at least as strong as classical computation. To implement a classical subroutine with a quantum machine, all one has to do is to make the classical circuit reversible (using, e.g., Toffoli gates and auxiliary qubits) and implement the Toffoli gates via quantum gates.

Some care has to be taken with the ancilla qubits when a classical computation is made reversible and then used as part of a quantum circuit. Since the input to the circuit can now be a quantum state, it is important to make sure that the state of the ancilla qubits at the end of the computation does not depend on the input to the circuit: if it did, any interference effect (constructive or destructive) that we would like to use in the algorithm would be prevented by the ancilla qubits. This can be dealt with by copying the result of the computation onto some clean bits and then undoing any operation on the ancilla qubits, a procedure usually referred to as *garbage removal*. See Exercise 1 for one effect quantum states can have with classical gates.

The next important question is whether it is possible to build a universal quantum machine (rather than special purpose computers). In other words, is there a small set of operations that implements any unitary transformation? Classically, it is well known that any Boolean function can be computed with a small set of *universal* gates, like AND and NOT, or NAND, as we have seen above. Moreover each of the elementary gates in a classical circuit only operates on few (1 or 2) bits at a time. Fortunately, it turns out that a similar statement is true for the quantum world; it was shown [18, 20] that there are sets of universal quantum gates on at most 2 qubits. In particular it has been shown that any unitary transformation can be decomposed into a sequence of gates consisting of one-qubit gates and CNOTs. We can describe the

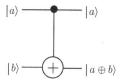

Fig. 5 The CNOT gate flips the second qubit if the first qubit is in the state $|1\rangle$

action of the CNOT gate by showing how it transforms the basis states $|00\rangle$, $|01\rangle$, $|10\rangle$, and $|11\rangle$: see Fig. 5.

However, the set of gates consisting of CNOT and single qubit gates is a continuous gate set. Sometimes, in particular in the context of fault-tolerant computation, it is convenient to work with a small finite set of universal gates. One such universal gate set is the set of three gates $\left\{\frac{\pi}{8}, H, CNOT\right\}$, where

$$\frac{\pi}{8} = \begin{pmatrix} e^{-\frac{i\pi}{8}} & 0 \\ 0 & e^{\frac{i\pi}{8}} \end{pmatrix}, \quad H = \frac{1}{\sqrt{2}} \begin{pmatrix} 1 & 1 \\ 1 & -1 \end{pmatrix} \tag{3}$$

$\frac{\pi}{8}$ is a gate that multiplies the basis state $|1\rangle$ with $e^{i\frac{\pi}{4}}$ (up to a global, unobservable phase), the Hadamard gate H maps $|0\rangle \rightarrow \frac{1}{\sqrt{2}}(|0\rangle + |1\rangle)$ and $|1\rangle \rightarrow \frac{1}{\sqrt{2}}(|0\rangle - |1\rangle)$ and the controlled-NOT gate (CNOT) is as defined above. This gate set is universal in the sense that any unitary transformation can be *approximated* to arbitrary precision by a sequence of gates from this set.

Exercise 1 (Inverted CNOT) *Study the following mini two-qubit quantum circuit: apply a Hadamard transform on each of the two qubits, then apply a CNOT as in Fig. 5 followed by another Hadamard on each of the qubits. What is the resulting action of this circuit? (Hint: This circuit has an equivalent description with just one two-qubit gate.) This little circuit illustrates how a classical gate can change behavior unexpectedly when given a quantum superposition as input (the state after the first two Hadamard transforms).*

Exercise 2 (*n*-fold Controlled Operations) *Let V be a one-qubit gate. Let C(V) be the two-qubit gate that implements V on the second qubit if the first qubit is in the state $|1\rangle$. For instance $CNOT = C(X)$, where X is the one-qubit gate that flips a bit, i.e., $X|0\rangle = |1\rangle$ and $X|1\rangle = |0\rangle$. Let $C^2(V)$ be the three-qubit gate that implements V on the third qubit if the first two qubits are in the state $|11\rangle$. For instance Toffoli $= C^2(X)$. Similarly one can define $C^3(V)$, $C^4(V)$ and so on.*

1. *Find the matrix representations for $C(V)$ and $C^2(V)$ from the one of V.*
2. *Show how to implement $C^n(X)$ using $2n - 3$ Toffoli gates (make sure that any ancilla qubits used are returned to their initial state).*
3. *Let $R' = C^{n-1}(Z)$, where Z is the one-qubit unitary that flips the sign of $|1\rangle$ and leaves $|0\rangle$ invariant, i.e., $Z|0\rangle = |0\rangle$ and $Z|1\rangle = -|1\rangle$. Find a circuit with*

elementary (or small) gates to implement R'. Define R to be the transformation on n qubits that flips the sign of all basis states except for the state $|0\rangle^{\otimes n}$, which is left invariant. Build R from R'.

It turns out that there are many other choices of finite universal gate sets acting on few qubits only. In principle the complexity of a unitary transformation, when counted as the number of elementary gates needed to implement it, might depend on the choice of universal gate set. However, Solovay and Kitaev have shown (see e.g. [39]) that all finite universal gate sets are equivalent in the sense that if we can approximate an n-qubit unitary with $poly(n)$ gates from one set, we can also approximate it with $poly(n)$ gates from the other. This means that the notion of an *efficient* algorithm does not depend on the particular choice of gate set.

When we wish to solve a problem (like factor a large number) we are often content with a circuit that solves the problem *with some probability* p, which is not too small (i.e., constant). If we have such a circuit and we can verify rapidly if the circuit gave the correct answer (e.g., in the case of factoring we just need to check if the numbers that are output are indeed factors of our original number), we can repeat it several times to boost the success probability. Repeating the circuit r times boosts the success probability up to $1 - (1 - p)^r$ which is exponentially close to 1 if $r = n$.

In complexity theoretic terms, a problem that can be solved by a quantum circuit with $poly(n)$ elementary gates (from some fixed universal gate set) with high probability (e.g., constant), where n is size of the input, is said to be in the complexity class BQP (which stands for bounded quantum polynomial time).

3 First Algorithms

The first truly quantum algorithm was proposed by Deutsch [17]. Albeit quite simple, it contains in essence most of the future developments of quantum algorithms. Deutsch's algorithm provides an improvement over the best classical algorithm solving the same problem in terms of its *black-box* or *query* complexity, which is defined below, and will be one of the complexity measures that we use in comparing the performance of different algorithms.

3.1 The Black-Box Model

Suppose we are given a function $f : \{0, 1\}^n \rightarrow \{0, 1\}^m$, and we want to obtain some information about f. For example, we might want to compute a specific value $f(x)$, or find out whether f ever takes the value $y \in \{0, 1\}^m$, or if f is a constant function, etc. The complexity of these tasks will depend on the format in which f is given to us: is it given by a table of values, or by some little program computing f,

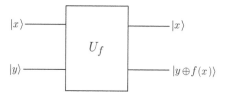

Fig. 6 The gate U_f

or by characterizing it in some mathematical way (e.g., as the solution of an implicit equation)?

To abstract away any special features that f might have, we place ourselves in the *black-box*, or *oracle*, model of computation. In this model, f is given as a special gate that we can use in our circuit. This gate basically takes an input $x \in \{0, 1\}^n$ and outputs $f(x)$. However, such a transformation is not reversible (consider for instance the all-0 function). To make it reversible, and thus implementable as a quantum circuit, we introduce another input $y \in \{0, 1\}^m$. The box U_f outputs x and $f(x) \oplus y$ (see Fig. 6). U_f is then a permutation of all $(n + m)$-bit strings, and therefore reversible.

Exercise 3 *What is the inverse of U_f? Suppose f is invertible with inverse f^{-1}. Is $U_{f^{-1}}$ the inverse of U_f?*

In the black-box model, the complexity of a given circuit is the number of gates U_f that it uses. The reasoning behind this is that in general it is expensive in terms of resources to apply U_f and so we focus on counting these applications. We will not count other gates, such as the CNOT or Hadamard gates, however, we will still be careful to use as few as possible of these gates (since later in the context of factoring we will replace the black-box by an actual function that we evaluate in a subroutine and we will count *all* the gates in the circuit). The *query complexity* of an algorithm computing some information on a function f given as a black-box (U_f) will thus be given by the minimal number of evaluations of (queries to) f (U_f) that it needs to perform its task, in the worst case.

3.2 Deutsch's Algorithm

Deutsch was the first to exploit the unique features of quantum computation to devise a quantum algorithm that had a strictly lower query complexity than the best classical deterministic algorithm. The problem he considered is the following:

Input: A Boolean function $f : \{0, 1\} \to \{0, 1\}$ (given as a black-box).
Output: Constant (if $f(0) = f(1)$) or balanced (if $f(0) \neq f(1)$).

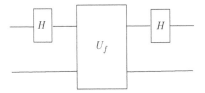

Fig. 7 The transformation D_f

Any classical deterministic algorithm for this problem needs at least two queries to f, since it needs both $f(0)$ and $f(1)$ to decide whether f is constant or balanced. Deutsch showed how this problem could be solved by a quantum circuit using only one query to f. The main idea is that, since we are interested in a *global* property of the function f, we might be able to get sufficient information by querying f *in superposition* and measuring the resulting quantum state in order to gather relevant information on f. Toward this aim, Deutsch introduced the following unitary transformation D_f that uses only one unitary U_f and two Hadamard gates.

What is the effect of D_f when applied on a basis state? Suppose, for example, that we start in the $|0\rangle \otimes |0\rangle$ state. We first apply a Hadamard transformation to the first qubit

$$
\begin{aligned}
H \otimes \mathbb{1} : |0\rangle \otimes |0\rangle &\mapsto \frac{1}{\sqrt{2}}(|0\rangle + |1\rangle) \otimes |0\rangle \\
&= \frac{1}{\sqrt{2}}(|0\rangle \otimes |0\rangle + |1\rangle \otimes |0\rangle).
\end{aligned}
\tag{4}
$$

We then apply U_f, which gives us the state

$$
\frac{1}{\sqrt{2}}(|0\rangle \otimes |f(0)\rangle + |1\rangle \otimes |f(1)\rangle)
\tag{5}
$$

and the final Hadamard transform applied on the first qubit then yields, as described in Fig. 7

$$
D_f|00\rangle = \frac{1}{2}\Big(|0\rangle \otimes (|f(0)\rangle + |f(1)\rangle) + |1\rangle \otimes (|f(0)\rangle - |f(1)\rangle)\Big).
\tag{6}
$$

Exercise 4 *Compute the result of the D_f transformation on the other three basis states.*

Deutsch's algorithm uses this transformation D_f together with another Hadamard gate on the second qubit. It is described in Fig. 8.
The input to D_f is $(\mathbb{1} \otimes H)|01\rangle = |0\rangle \otimes \frac{1}{\sqrt{2}}(|0\rangle - |1\rangle)$. Since

$$
D_f|01\rangle = \frac{1}{2}\Big(|0\rangle \otimes (|f(0) \oplus 1\rangle + |f(1) \oplus 1\rangle) + |1\rangle \otimes (|f(0) \oplus 1\rangle - |f(1) \oplus 1\rangle)\Big),
\tag{7}
$$

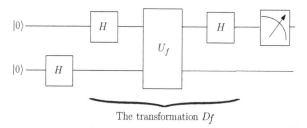

The transformation D_f

Fig. 8 Deutsch's circuit

(see Exercise 4) by linearity we can write the state $|\Psi\rangle$ after application of D_f as

$$
|\Psi\rangle = \frac{1}{2\sqrt{2}}\Big(|0\rangle \otimes \big(|f(0)\rangle - |f(0) \oplus 1\rangle + |f(1)\rangle - |f(1) \oplus 1\rangle\big)
$$
$$
+ |1\rangle \otimes \big(|f(0)\rangle - |f(0) \oplus 1\rangle - (|f(1)\rangle - |f(1) \oplus 1\rangle)\big)\Big).
$$
$$(8)$$

For any $x \in \{0, 1\}$, $|f(x)\rangle - |f(x) \oplus 1\rangle$ is equal to $(-1)^{f(x)}(|0\rangle - |1\rangle)$. Hence $|\Psi\rangle$ can be rewritten as

$$
|\Psi\rangle = \frac{1}{2}\big(((-1)^{f(0)}+(-1)^{f(1)})|0\rangle + ((-1)^{f(0)}-(-1)^{f(1)})|1\rangle)\big) \otimes \frac{1}{\sqrt{2}}(|0\rangle - |1\rangle) .
$$
$$(9)$$

We have therefore managed to transfer the values $f(0)$ and $f(1)$ to local phases in front of the $|0\rangle$ and $|1\rangle$ states.

If f is balanced, then $|\Psi\rangle = \pm\frac{1}{\sqrt{2}}|1\rangle \otimes (|0\rangle - |1\rangle)$ and if it is constant, then $|\Psi\rangle = \pm\frac{1}{\sqrt{2}}|0\rangle \otimes (|0\rangle - |1\rangle)$. A measurement of the first qubit of $|\Psi\rangle$ in the computational basis thus gives the answer to our problem with certainty. Furthermore, the construction of $|\Psi\rangle$ only requires one gate U_f (note that $|\Psi\rangle = D_f|0\rangle \otimes (\frac{1}{\sqrt{2}}(|0\rangle - |1\rangle)) = D_f \cdot (\mathbb{1} \otimes H) \cdot (|0\rangle \otimes |1\rangle)))$, so that one can determine if f is balanced or constant with only one call to the black-box U_f. This is a strict improvement over the best classical algorithm.

4 The Quantum Fourier Transform

In this section we briefly recall some facts about the classical Fourier Transform on Abelian groups, and we define the quantum Fourier Transform, a key ingredient in many quantum algorithms. In fact we will see that Deutsch's algorithm, which we studied in the previous section, used the quantum Fourier Transform over the group $(\mathbb{Z}_2, \oplus)$.

Let $(G, +)$ be a finite Abelian group and $(\mathcal{U}, \times)$ the multiplicative group of complex numbers with modulus 1.

Definition 1 A character on G is a function $\chi : G \to \mathcal{U}$ such that

$$\forall g, h \in G \qquad \chi(g + h) = \chi(g)\chi(h). \tag{10}$$

It is known that in the Abelian case the set of all characters on G forms a group $\hat{G}$ isomorphic to G (in particular there are as many characters as there are group elements). We fix an isomorphism $\phi : G \to \hat{G}$ and denote by χ_g the image $\phi(g)$ of g by ϕ. The only fact we will need on characters is the following lemma.

Lemma 1 *(Schur's orthogonality lemma) For every $h, h' \in G$,*

$$\frac{1}{|G|} \sum_{g \in G} \chi_h(g)\overline{\chi_{h'}(g)} = \delta_{hh'}. \tag{11}$$

Example 1 Let $G = (\mathbb{Z}_2, \oplus)$. Then a character $\chi : G \to \mathcal{U}$ must be such that $\chi(0) = \chi(0 \oplus 0) = \chi(0)\chi(0)$, and $\chi(0) \in \mathcal{U}$, so necessarily $\chi(0) = 1$. Moreover, in $\mathbb{Z}_2$, $1 \oplus 1 = 0$, so $\chi(1 \oplus 1) = \chi(1)^2 = \chi(0) = 1$, and there are two possibilities: either $\chi(1) = 1$ (this gives us the *trivial* character, which is constant and equal to 1, and is denoted by χ_0) or $\chi(1) = -1$, in which case we get χ_1. The group $\hat{G}$ is $\hat{G} = \{\chi_0, \chi_1\}$, and it can be checked that $\phi : G \to \hat{G}$ that sends 0 to χ_0 and 1 to χ_1 is indeed a group isomorphism.

Exercise 5 *Let $G = (\mathbb{Z}_2^n, \oplus)$: elements of G are n-dimensional vectors of 0s and 1s, and the group law is componentwise addition modulo 2. Show that the characters of G are the maps*

$$\chi_y : x \in \mathbb{Z}_2^n \mapsto (-1)^{x \cdot y} \quad \forall y \in \mathbb{Z}_2^n. \tag{12}$$

Let $f : G \to X$ be a function defined on G and taking its values in some set X (usually $X = G, \mathbb{Z}$ or $\mathbb{R}$).

Definition 2 (Abelian Fourier Transform) The Fourier coefficients of f are

$$\hat{f}(x) = \frac{1}{\sqrt{|G|}} \sum_{y \in G} \chi_x(y)f(y). \tag{13}$$

The Fourier transform of f is the function

$$\hat{f} : x \mapsto \frac{1}{\sqrt{|G|}} \sum_{y \in G} f(y)\chi_x(y). \tag{14}$$

Example 2 Let $f : \mathbb{Z}_2 \to \{0, 1\}$ be a Boolean function. If f is constant, then

$$\hat{f}(0) = \frac{\chi_0(0) f(0) + \chi_0(1) f(1)}{\sqrt{2}} = \sqrt{2} f(0), \tag{15}$$

$$\hat{f}(1) = \frac{\chi_1(0) f(0) + \chi_1(1) f(1)}{\sqrt{2}} = \frac{f(0) - f(1)}{\sqrt{2}} = 0. \tag{16}$$

If f is balanced, then

$$\hat{f}(0) = \frac{f(0) + f(1)}{\sqrt{2}} = \frac{1}{\sqrt{2}}, \quad \hat{f}(1) = \frac{f(0) - f(1)}{\sqrt{2}} = \pm\frac{1}{\sqrt{2}}. \tag{17}$$

Label the elements of G as $e_1, \ldots, e_M$, where $M = |G|$. Then the classical Fourier Transform can be viewed as a matrix multiplication, multiplying the vector $\mathbf{f} = (f(e_1), \ldots, f(e_M))$, by the Fourier transform matrix

$$F = \frac{1}{\sqrt{|G|}} \begin{pmatrix} \chi_1(e_1) & \chi_1(e_2) & \cdots & \chi_1(e_M) \\ \chi_2(e_1) & \chi_2(e_2) & \cdots & \chi_2(e_M) \\ \vdots & \vdots & & \vdots \\ \chi_M(e_1) & \chi_M(e_2) & \cdots & \chi_M(e_M) \end{pmatrix}, \tag{18}$$

where the $\chi_i = \chi_{e_i}$ are the characters of $\hat{G}$. Note that from Schur's orthogonality lemma (Lemma 1) it follows that the Fourier Transform matrix is unitary. From the definition of the Fourier coefficients of f, we have that $\hat{\mathbf{f}} = F\mathbf{f}$, where $\hat{\mathbf{f}}$ is the vector $(\hat{f}(e_1), \ldots, \hat{f}(e_M))$. If computed naively by matrix multiplication, this operation would take $O(M^2)$ elementary operations. However, it is well known it can be sped up to $O(M \log M)$ elementary operations, using the fast Fourier transform (FFT).

Exercise 6 *In Exercise 5 you showed that the characters of $G = (\mathbb{Z}_2^n, \oplus)$ are the maps $\chi_y : x \in \mathbb{Z}_2^n \mapsto (-1)^{x \cdot y}$ for all $y \in \mathbb{Z}_2^n$. Write down the Fourier matrix F in this case. If $n = 1$, can you relate F to the Hadamard transform? And in general?*

Definition 3 The quantum Fourier transform (QFT) over an Abelian group G of cardinality M is the unitary operation

$$QFT : |x\rangle \mapsto \frac{1}{\sqrt{M}} \sum_{y \in G} \chi_y(x) |y\rangle. \tag{19}$$

With any function $f : G \to X$ we associate the state

$$|f\rangle = \frac{1}{\sqrt{M}} \sum_{x \in G} |x\rangle |f(x)\rangle. \tag{20}$$

The state associated with the Fourier Transform $\hat{f}$ of f is then

$$|\hat{f}\rangle = \frac{1}{M} \sum_{x,y\in G} \chi_y(x)|y\rangle|f(x)\rangle. \qquad (21)$$

We say that $|\hat{f}\rangle$ is obtained by quantum Fourier sampling (QFS).

Note that quantum Fourier sampling produces a *superposition* over the Fourier coefficients of f and that a measurement of state $|\hat{f}\rangle$ outputs $y \in G$ and z in X with probability proportional to $|\sum_{x\in G, f(x)=z} \chi_y(x)|^2$. We can then estimate these amplitudes by repeated sampling. $|\hat{f}\rangle$ thus gives us some information on the function f through its Fourier coefficients and will be best used to distinguish functions with very different Fourier coefficients.

Exercise 7 *What is the state $|\hat{f}\rangle$ obtained by quantum Fourier sampling when $f : \mathbb{Z}_2 \to \{0, 1\}$ is a constant function? And when f is balanced? Compare it to the state produced in Deutsch's algorithm. Show that there is a measurement such that, if you are given only one copy of $|\hat{f}\rangle$, this measurement enables you to distinguish with certainty between the case where f is constant and the case where f is balanced.*

We end this section by showing how the black-box U_f can be combined with the Hadamard transform and the QFT to construct the state $|\hat{f}\rangle$. Sampling from this state will be the basis of several algorithms, and in particular of Shor's algorithm for factoring, in the next section.

For ease of exposition assume that $|G| = M = 2^n$ and $|X| = 2^p$ are powers of 2, and suppose that we input the state $|0^{\otimes n}\rangle|0^{\otimes p}\rangle$ to the circuit of Fig. 9. Since $H|0\rangle = 1/\sqrt{2}(|0\rangle+|1\rangle)$ (we will investigate the exact behavior of $H^{\otimes n}$ in Sect. 5.1), after the Hadamard transform, the state is

$$\begin{aligned}\left(H^{\otimes n}|0^{\otimes n}\rangle\right)|0^{\otimes p}\rangle &= \left(\frac{1}{\sqrt{2}}(|0\rangle + |1\rangle)\right)^{\otimes n}|0^{\otimes p}\rangle \\ &= \frac{1}{\sqrt{2^n}} \sum_{x\in\{0,1\}^n} |x\rangle|0^{\otimes p}\rangle.\end{aligned} \qquad (22)$$

We then apply U_f, resulting in $\sum_{x\in\{0,1\}^n} |x\rangle|f(x)\rangle$. Finally, after applying the QFT to finish the quantum Fourier sampling routine, this state becomes

$$|\hat{f}\rangle = \sum_{x,y\in G} \chi_y(x)|y\rangle|f(x)\rangle. \qquad (23)$$

It is therefore possible to construct the state $|\hat{f}\rangle$ using only *one* gate U_f. In the next section we will show how it is possible to *efficiently* implement the QFT over some groups G. Since we will reuse this circuit repeatedly, we state this as a lemma.

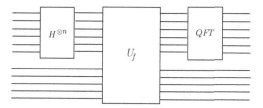

Fig. 9 The transformation D_f (quantum Fourier sampling) in the general case

Lemma 2 *Let $f : G \to X$ be any function and $M = |G|$. There is a circuit that, when given input $|0^{\otimes n+p}\rangle$, where $n = \lceil \log M \rceil$ and $p = \lceil \log |X| \rceil$, outputs the state $|\hat{f}\rangle$.[1] Moreover this circuit can be implemented by using only one gate U_f, one set of gates implementing the QFT over G, and n Hadamard gates.*

5 Deutsch–Josza and Simon's Algorithms

In this section we will present two algorithms that make more elaborate use of the QFT over the group $\mathbb{Z}_2^n$ illustrating the applicability of Lemma 2.

5.1 The QFT over $\mathbb{Z}_2^n$

Let $G = (\mathbb{Z}_2^n, \oplus)$. We are going to show that the quantum Fourier transform over G is simply the tensor product of Hadamard gates $H^{\otimes n}$, so that it can be implemented very efficiently (using $n = \log |G|$ gates).

Let us examine the action of $H^{\otimes n}$ on an arbitrary n-qubit basis state x. The transformation induced by a single Hadamard gate on a qubit i in the basis state $|x_i\rangle$ can be written as

$$H|x_i\rangle = \frac{1}{\sqrt{2}}(|0\rangle + (-1)^{x_i}|1\rangle) = \sum_{y_i \in \{0,1\}} (-1)^{x_i \cdot y_i}|y_i\rangle. \qquad (24)$$

Applying this to $H^{\otimes n}$ with $|x\rangle = |x_1 \ldots x_n\rangle$ we get

$$
\begin{aligned}
& H^{\otimes n}|x_1 \ldots x_n\rangle \\
&= \frac{1}{\sqrt{2^n}} \left(\sum_{y_1 \in \{0,1\}} (-1)^{x_1 \cdot y_1}|y_1\rangle \right) \otimes \ldots \otimes \left(\sum_{y_n \in \{0,1\}} (-1)^{x_n \cdot y_n}|y_n\rangle \right) \\
&= \frac{1}{\sqrt{2^n}} \sum_{y \in \{0,1\}^n} (-1)^{x_1 \cdot y_1 + \ldots + x_n \cdot y_n}|y\rangle = \frac{1}{\sqrt{2^n}} \sum_{y \in \{0,1\}^n} (-1)^{x \cdot y}|y\rangle,
\end{aligned}
\qquad (25)
$$

---

[1] $\lceil x \rceil$ denotes the smallest integer larger than $x \in \mathbb{R}$.

where $x \cdot y$ is the inner product of the vectors x and y mod 2. Since the characters of $\mathbb{Z}_2^n$ are simply the maps $\chi_y(x) = (-1)^{x \cdot y}$ for $y \in \mathbb{Z}_2^n$ (cf. Exercise 5), the gate $H^{\otimes n}$ exactly implements the QFT over $\mathbb{Z}_2^n$.

5.2 The Deutsch–Josza Algorithm

Deutsch and Josza [19] introduced the following problem:

Input: An integer n and a Boolean function $f : \{0, 1\}^n \to \{0, 1\}$.
Promise: f is either constant or balanced ($|\{x : f(x) = 0\}| = |\{x : f(x) = 1\}| = 2^{n-1}$).
Output: Constant or balanced.

This is a *promise* problem: we only need to answer correctly on inputs that satisfy the promise of being either constant *or* balanced. There is no requirement on the output of the algorithm if the input is neither constant nor balanced.

In the worst case, a classical deterministic algorithm will need at least 2^{n-1} calls to f to answer this problem, since for any 2^{n-1} $x_i \in \{0, 1\}^n$, there exists a constant function that evaluates to 0 on all the x_i's, as well as a balanced function that evaluates to 0 on all x_i's.

However, this problem can be solved exactly using only *one* query to U_f by using the circuit D_f described in Sect. 4! To see this, we will introduce the *complement* f^* of f: for all x, $f^*(x) = f(x) \oplus 1$. If f is a constant function, then

$$|\hat{f}\rangle = \frac{1}{2^n} \sum_{x,y} (-1)^{x \cdot y} |y\rangle |f(x)\rangle$$
$$= \frac{1}{2^n} \sum_y \left(\sum_x (-1)^{x \cdot y} \right) |y\rangle |f(0)\rangle \qquad (26)$$
$$= |0^{\otimes n}\rangle |f(0)\rangle$$

and similarly, $|\hat{f}^*\rangle = |0^{\otimes n}\rangle |f(0) \oplus 1\rangle$. But if f is a balanced function, then

$$|\hat{f}\rangle = \frac{1}{2^n} \sum_{x,y} (-1)^{x \cdot y} |y\rangle |f(x)\rangle$$
$$= \frac{1}{2^n} \left(|0^{\otimes n}\rangle \left(\sum_x |f(x)\rangle \right) + \sum_{x,y \neq 0} (-1)^{x \cdot y} |y\rangle |f(x)\rangle \right) \qquad (27)$$

and similarly,

$$|\hat{f}^*\rangle = \frac{1}{2^n} \left(|0^{\otimes n}\rangle \left(\sum_x |f(x) \oplus 1\rangle \right) + \sum_{x,y \neq 0} (-1)^{x \cdot y} |y\rangle |f(x) \oplus 1\rangle \right). \qquad (28)$$

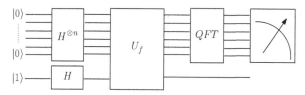

Fig. 10 The Deutsch–Josza circuit

Since f is balanced, $\sum_x |f(x)\rangle = \sum_x |f(x) \oplus 1\rangle$, so $|\hat{f}\rangle - |\hat{f}^*\rangle$ has zero amplitude on all states that have their first register in the basis state $|0^{\otimes n}\rangle$. We saw that in case f was constant, $|\hat{f}\rangle - |\hat{f}^*\rangle = |0^{\otimes n}\rangle (|f(0)\rangle - |f(0) \oplus 1\rangle)$. It is therefore possible to distinguish with certainty between the case when f is balanced and the case when it is constant by measuring the first register of $|\hat{f}\rangle - |\hat{f}^*\rangle$: if the outcome is $|0^{\otimes n}\rangle$, then the function is constant; otherwise it is balanced.

How can we construct $|\hat{f}\rangle - |\hat{f}^*\rangle$ efficiently? Recall that $U_f |x\rangle |a\rangle = |x\rangle |f(x) \oplus a\rangle$. So, when given as input the state $|0^{\otimes n}\rangle |1\rangle$, the circuit D_f outputs exactly $|\hat{f}^*\rangle$. By linearity, $|\hat{f}\rangle - |\hat{f}^*\rangle$ can be constructed using only one call to D_f: it suffices to input the state $|0^{\otimes n}\rangle (|0\rangle - |1\rangle)/\sqrt{2}$ (which can be produced from $|0^{\otimes n}\rangle \otimes |1\rangle$ by applying a Hadamard gate on the last qubit) to the circuit D_f. The final circuit for solving the Deutsch–Josza problem is described in Fig. 10.

Exercise 8 *For any $s \in \mathbb{Z}_2^n$ let $f_s : x \in \mathbb{Z}_2^n \mapsto s \cdot x$. Show that the quantum states $|\hat{f}_s\rangle$ and $|\hat{f}_{s'}\rangle$ are orthogonal if $s \neq s'$. Use this to devise a quantum algorithm for the following problem, due to Bernstein and Vazirani* [11]:

Input: An integer n and a Boolean function $f : \mathbb{Z}_2^n \to \{0, 1\}$.
Promise: There is an $s \in \mathbb{Z}_2^n$ such that $f = f_s$.
Output: s.

What is the query complexity of your algorithm? Give a classical algorithm that solves this problem by using n queries to f. Using coding theory, it can be shown that this is essentially optimal, since any query to f gives us at most 1 bit of information (f takes its values in $\{0, 1\}$).

5.3 Simon's Algorithm

Simon's algorithm [46] studies the periodicity of functions defined on $\mathbb{Z}_2^n$. This is the first quantum algorithm that has an exponential advantage when compared to the best classical *probabilistic* algorithm. It solves the following promise problem:

Input: A function $f : \mathbb{Z}_2^n \to \{0, 1\}^n$.
Promise: f is periodic : there exists $a \in \mathbb{Z}_2^n$ such that for all $x, y \in \mathbb{Z}_2^n$, $f(x) = f(y)$ if and only if $y = x \oplus a$.
Output: a.

Since f has no particular structure apart from its periodicity, the best any classical probabilistic algorithm can do is query elements at random until a pair $x \neq y$ such that $f(x) = f(y)$ is found, and then output $a = x \oplus y$. Using the birthday paradox, the expected number of queries necessary before such a collision is found is of order $\Omega(2^{n/2})$.

We will solve this problem using almost exactly the same quantum circuit as we used in the Deutsch–Josza algorithm. The periodicity of f means that there exists a partition of the 2^n input strings into two sets X and $\overline{X} = \{x \oplus a | x \in X\}$ with $|X| = |\overline{X}| = 2^{n-1}$, such that all the values $f(x)$ are distinct for $x \in X$ and similarly for $\overline{X}$. Now let us take a look at state $|\hat{f}\rangle$ obtained by quantum Fourier sampling (see Definition 3):

$$
\begin{aligned}
|\hat{f}\rangle &= \frac{1}{2^n} \sum_{x,y \in \mathbb{Z}_2^n} (-1)^{x \cdot y} |y\rangle |f(x)\rangle \\
&= \frac{1}{2^n} \sum_{x \in X, \, y \in \mathbb{Z}_2^n} ((-1)^{x \cdot y} + (-1)^{(x \oplus a) \cdot y}) |y\rangle |f(x)\rangle \\
&= \frac{1}{2^{n-1}} \sum_{\substack{x \in X, \, y \in \mathbb{Z}_2^n \\ y \cdot a = 0}} (-1)^{x \cdot y} |y\rangle |f(x)\rangle.
\end{aligned}
\tag{29}
$$

By measuring the first register in the computational basis, we obtain a random $y_1 = y \in \mathbb{Z}_2^n$ such that $y_1 \cdot a = 0$, thereby gaining 1 bit of information about the period a. We can now repeat this algorithm to obtain y_2 such that $y_2 \cdot a = 0$, y_3 such that $y_3 \cdot a = 0$, and so on. These y_i form a subspace of the n-dimensional vector space of all n-bit strings (over $\mathbb{Z}_2$). If among the y_i there are $n - 1$ vectors that are linearly independent (i.e., such that they span a space of dimension $n-1$), then the equations $a \cdot y_i = 0$ completely determine $a \neq 0$. But for each set of y_i that do not yet span a space of dimension $n - 1$ the probability that the next y will be outside the space is at least $1/2$, because the space spanned by them contains at most 2^{n-2} out of the 2^{n-1} possible y's. Hence after $O(n)$ repetitions of the algorithm with a probability exponentially close to 1 we will have enough information to determine a.

Since the construction of $|\hat{f}\rangle$ requires only one black-box query to f (cf. Lemma 2), this gives an algorithm that solves Simon's problem with probability exponentially close to 1 using only $O(n)$ queries to f, which constitutes an exponential speed-up compared to the $\Omega(2^{n/2})$ queries required by any classical probabilistic algorithm.

6 Factoring in Polynomial Time

The algorithms in the previous section have laid the foundation for one of the major breakthroughs in quantum computation: Shor's 1994 algorithm for factoring [44]. The basic ingredients of this algorithm can already be found in Deutsch's and

Simon's algorithms. However, some of the ingredients are much more sophisticated, and we will detail them in this section one by one.

6.1 The QFT over $\mathbb{Z}_M$

Let M be a positive integer and $\mathbb{Z}_M = \{0, \ldots, M - 1\}$ the cyclic group of integers modulo M. Let $\omega = e^{2i\pi/M}$ be a primitive Mth root of unity. The group of characters of $\mathbb{Z}_M$ is constituted of the M maps $\chi_i : \mathbb{Z}_M \to \mathbb{U}$, $\chi_i(k) = \omega^{i \cdot k}$, for $i = 0, \ldots, M - 1$. So, in the case of the cyclic group $\mathbb{Z}_M$, the QFT is the unitary operation:

$$QFT : |x\rangle \mapsto \frac{1}{\sqrt{M}} \sum_{y \in \mathbb{Z}_M} \omega^{x \cdot y} |y\rangle. \tag{30}$$

For the QFT over $\mathbb{Z}_M$ to be used in an efficient algorithm, we need to show that it can be efficiently implemented using only standard one or two-qubit gates. We will see that $O(\log^2 M)$ such gates suffice.

For ease of presentation let us assume that $M = 2^n$. To evaluate $\omega^{x \cdot y} = \exp(\frac{2\pi i x \cdot y}{2^n})$, decompose y as $y = y_0 + 2y_1 + \ldots + 2^{n-1} y_{n-1}$, so that

$$\omega^{x \cdot y} = \omega^{x y_0} \cdot \omega^{2 x y_1} \ldots \omega^{2^{n-1} x y_{n-1}} \tag{31}$$

and

$$\sum_{y \in \mathbb{Z}_M} \omega^{x \cdot y} |y\rangle = \sum_{y_0 \in \{0,1\}, \ldots, y_{n-1} \in \{0,1\}} \omega^{x y_0} \cdot \omega^{2 x y_1} \ldots \omega^{2^{n-1} x y_{n-1}} |y_0, \ldots, y_{n-1}\rangle,$$

$$= \bigotimes_{i=0}^{n-1} \left(\sum_{y_i \in \{0,1\}} \omega^{2^i x y_i} |y_i\rangle \right),$$

$$= \bigotimes_{i=0}^{n-1} \left(|0\rangle + \omega^{2^i x} |1\rangle \right). \tag{32}$$

Note that $\omega^{2^i x} = e^{2 i \pi 2^{i-n} x}$ only depends on the low-order bits of x: to reflect this we introduce the notation $.x_i x_{i-1} \ldots x_0 = x_i/2 + x_{i-1}/4 + \ldots + x_0/2^{i+1}$, so that $\omega^{2^i x} = e^{2 i \pi .x_{n-1-i} \ldots x_0}$ and finally

$$\frac{1}{\sqrt{M}} \sum_{y \in \mathbb{Z}_M} \omega^{x \cdot y} |y\rangle = \frac{1}{\sqrt{2^n}} \left(|0\rangle + e^{2\pi i (.x_{n-1} \ldots x_1 x_0)} |1\rangle \right) \otimes$$

$$\cdots \otimes \left(|0\rangle + e^{2\pi i (.x_1 x_0)} |1\rangle \right) \otimes \left(|0\rangle + e^{2\pi i (.x_0)} |1\rangle \right). \tag{33}$$

Since the Hadamard on qubit x_i can be thought of as performing $|x_i\rangle \rightarrow (|0\rangle + e^{2\pi i (.x_i)}|1\rangle)$, if we apply a Hadamard to the first qubit of $|x\rangle = |x_{n-1} \dots x_0\rangle$ (notice that we have written x in reverse, so that the registers containing x have to be swapped at the beginning of the circuit), then the resulting state is

$$1/\sqrt{2} \left(|0\rangle + e^{2\pi i (.x_{n-1})}|1\rangle \right) |x_1 \dots x_{n-1}\rangle. \tag{34}$$

Now, define the rotations

$$R_d = \begin{pmatrix} 1 & 0 \\ 0 & e^{-\frac{2i\pi}{2^d}} \end{pmatrix}. \tag{35}$$

Then, applying R_2 controlled on qubit 2, R_3 on qubit 3, up to R_n controlled on qubit n (see Exercise 2 on how to implement this) yields the state

$$\frac{|0\rangle + e^{2\pi i (.x_{n-1}\cdots x_1 x_0)}|1\rangle}{\sqrt{2}} |x_{n-2} \dots x_0\rangle. \tag{36}$$

We have thus produced the first qubit of $QFT|x\rangle$. Using the same idea, we apply a Hadamard gate on the second qubit of $|x\rangle$ (which was left unchanged by the previous operations) to get $1/\sqrt{2} \left(|0\rangle + e^{2\pi i (.x_{n-2})}|1\rangle \right) |x_{n-3} \dots x_0\rangle$, and applying R_2 controlled on qubit 3, R_3 controlled on qubit 4, up to R_{n-1} controlled on qubit n, we get $1/\sqrt{2} \left(|0\rangle + e^{2\pi i (.x_{n-2}\cdots x_1 x_0)}|1\rangle \right) |x_{n-3} \dots x_0\rangle$. Repeating this operation with the third, etc., qubits gives $QFT|x\rangle$ (see Fig. 11). This circuit uses $\frac{1}{2}n(n + 1) = O(\log^2 M)$ gates.

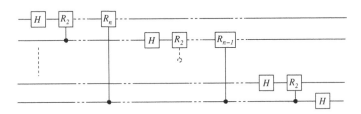

Fig. 11 The QFT over $\mathbb{Z}_M$

6.2 Shor's Algorithm

The fact that factoring big integers is hard is a cornerstone of many cryptographic systems used today (e.g., the RSA cryptosystem). Factoring clearly is in *NP*, since it is easy to check whether the product of two integers equals a given value, but it is not believed to be *NP*-complete. Currently the best classical algorithms for factoring are probabilistic and take sub-exponential time (in the input size : the number of bits required to represent the number that we want to factor) to complete.

So it was an important breakthrough when Shor published a polynomial-time quantum algorithm for factoring in 1994. His algorithm follows the basic scheme underlying the previous algorithms that we have studied. Its heart is an efficient quantum algorithm for the following problem, which is similar to Simon's period-finding problem, but over another group.

Input: An integer $N \in \mathbb{N}^*$ and a function $f : \mathbb{Z} \to X$.
Promise: There is an $a < N$ such that f is a-periodic:

$$\forall x, y \in \mathbb{Z} \qquad f(x) = f(y) \iff \exists k \in \mathbb{Z} \quad x - y = ka . \qquad (37)$$

Output: a.

We first outline a classical polynomial-time reduction from factoring to the period-finding problem and then we show how period finding can be efficiently solved by a quantum algorithm.

Let N be the integer that we would like to factor. Our aim is to find a non-trivial factor q of N. If we are able to do so, then we can run our algorithm recursively on q and N/q to completely factor N. Suppose we are given an x such that $x^2 = 1$ mod N, and $x \neq \pm 1$ mod N. Then $N|(x^2 - 1) = (x - 1)(x + 1)$, so N has a common factor with $x + 1$ or $x - 1$, and by hypothesis this factor is different from N. By computing the greatest common divisor $\gcd(N, x+1)$ and $\gcd(N, x-1)$ we then find non-trivial factors of N, and this can be done efficiently in $O(\log N)$ operations using Euclid's algorithm. Now let us see how we can find an x such that $x^2 = 1$ mod N. Let y be any integer modulo N. If y is not co-prime to N, then by Euclid's algorithm we have already found a divisor of N. Otherwise, the function $f : \mathbb{Z} \to \mathbb{Z}_N$, $f : x \mapsto y^x$ is periodic of period a, where a is the *order* of y: the smallest positive integer such that $y^a = 1$ mod N. Now if a is even, then $(y^{a/2})^2 = 1$ mod N and we can choose $x = y^{a/2}$. By using the Chinese remainder theorem, it can be shown (see, e.g., [34, 39, 40]) that a y taken at random in $\{0, \ldots N - 1\}$ will have even order modulo N and will be such that $y^{a/2} \neq \pm 1$ mod N with probability at least a constant (except if N is even, which immediately gives a factor 2 or if N is a power of a prime, which can be tested efficiently beforehand). So all we have to do is pick a y at random, compute its order, and if it is even then compute $\gcd(y^{a/2} - 1, N)$ and $\gcd(y^{a/2} + 1, N)$ to find a non-trivial factor of N.

But how do we compute the order of a given y? Let $f : \mathbb{Z} \to \mathbb{Z}_N$ be the function that maps an integer x to y^x mod N. Then f is periodic, and its period is exactly the order of y. So we have reduced factoring to finding the period of a given function defined over the integers.

Before devising a quantum algorithm for the general case, let's first restrict ourselves to the case where $f : \mathbb{Z}_M \to X$ is a periodic function of period $a|M$. Then $|\hat{f}\rangle$ is given by

$$|\hat{f}\rangle = \frac{1}{M} \sum_{x,y \in \mathbb{Z}_M} \omega^{xy} |y\rangle |f(x)\rangle,$$

$$= \frac{1}{M} \sum_{x \in \mathbb{Z}_a, y \in \mathbb{Z}_M} \left(\sum_{j=0}^{M/a-1} \omega^{(x+ja)y} \right) |y\rangle |f(x)\rangle. \tag{38}$$

Whenever $ay = 0 \mod M$, the geometric sum $\sum_{j=0}^{M/a-1} \omega^{(x+ja)y}$ equals $\omega^{xy} M/a$, and otherwise it is 0, so

$$|\hat{f}\rangle = \frac{1}{a} \sum_{x \in \mathbb{Z}_a} \left(\sum_{c \in \mathbb{Z}_a} \omega^{x \, cM/a} |cM/a\rangle \right) |f(x)\rangle. \tag{39}$$

A measurement of the first register of $|\hat{f}\rangle$ will thus give a uniformly random $y = cM/a$. To get a, we need to solve $y/M = c/a$, where y and M are known. Decomposing a as a product of primes, it is easily seen that at most $\lceil \log_2 a \rceil$ elements of $\mathbb{Z}_a$ are not co-prime to a, and since c is taken uniformly at random in $\mathbb{Z}_a$, after $\lceil \log_2 a \rceil = O(\log N)$ repetitions of this procedure, we obtain a c that is co-prime with a with high probability. In this case, a can be found easily by expressing y/M as a reduced fraction: the denominator then gives a.

So $O(\log N)$ copies of the state $|\hat{f}\rangle$ obtained from quantum Fourier Sampling suffice to determine a with high probability, and we have already shown that $|\hat{f}\rangle$ could be produced by a circuit using only one query to f (Lemma 2), $O(\log M)$ Hadamard gates and $O(\log^2 M)$ two-qubit gates implementing the QFT over $\mathbb{Z}_M$ (Sect. 6.1). Notice that here f is the exponentiation $x \mapsto y^x$. This function can be implemented very efficiently, using only $O(\log N)$ gates, by using the classical fast exponentiation algorithm, which is based on repeated squaring.

In the general case, f is a function from $\mathbb{Z}$ to X. This poses two added difficulties. First we need to cut off the domain of f. To this end we first pick an integer $M > N$ and consider the restriction f_M of f to $\mathbb{Z}_M$. Picking M large enough (see below) guarantees that there is sufficient periodicity in the domain we have chosen. The second difficulty is that we are unlikely to pick M to be exactly a multiple of the unknown period a. Hence, since f_M is not perfectly periodic, there will be some extra terms in the superposition (39). However, these terms are of small norm compared to the relevant terms. More precisely, Exercise 9 shows that there is a constant probability of measuring a y satisfying

$$\frac{c}{a} - \frac{1}{2M} \leq \frac{y}{M} \leq \frac{c}{a} + \frac{1}{2M}, \tag{40}$$

where $c \in \{0, \ldots, a-1\}$. But two distinct fractions with denominator at most N must be at least $1/N^2$ apart, so if we choose $M > N^2$ then c/a is the unique fraction with denominator at most N within distance $1/2M$ from y/M and can be determined with the continued fraction expansion. The total run-time of this algorithm is still $O(poly(\log N))$.

Exercise 9 *The aim of this exercise is to show that if a does not divide M, then a measurement of (39) still yields an integer y such that y/M is close to c/a, where c is an unknown integer, with good probability.*

1. *Show that there $-a/2 \leq ay \mod M \leq a/2$.*
2. *Prove the following inequalities:*

$$\forall \theta \in [-\pi, \pi] \qquad \frac{2|\theta|}{\pi} \leq |1 - e^{i\theta}| \leq |\theta|. \qquad (41)$$

3. *Let $A = \lfloor M/a \rfloor$ and $\omega = e^{2i\pi/M}$.[2] Show that if y is such that $-a/2 \leq ay \mod M \leq a/2$, then*

$$\left| \frac{\omega^{iAay} - 1}{\omega^{ay} - 1} \right| \geq \frac{2}{\pi} A - \left(1 + \frac{2}{\pi}\right). \qquad (42)$$

4. *By bounding the geometric series $\sum_{j=0}^{\lfloor M/a \rfloor - 1} \omega^{(x+ja)y}$, conclude that when measuring the first register of the resulting superposition (corresponding to (39)), we have constant probability of measuring an integer y such that there exists an integer c such that*

$$\frac{c}{a} - \frac{1}{2M} \leq \frac{y}{M} \leq \frac{c}{a} + \frac{1}{2M}. \qquad (43)$$

7 The Hidden Subgroup Problem

The algorithms of Simon and Shor can be seen as solving special instances of a more general problem, called the *hidden subgroup problem* (HSP), as we show in this section. With very similar methods to the ones in the previous section the HSP over *Abelian* groups can be solved efficiently by a quantum algorithm. Other Abelian Hidden Subgroup problems that can be solved this way are discrete logarithm [45], the hardness of which is the basis of some classical cryptosystems, and finding solutions to Pell's equation [29], a number theoretic problem known to be at least as hard as factoring. As another application of the HSP, Friedl et al. [26] solve the hidden translation problem: given two functions f and g defined over some group $\mathbb{Z}_p^n$ such that $f(x) = g(x + t)$ for some hidden translation t, find t.
Formally, the HSP is the following promise problem:

Input: A finite group G, a set S, and a function $f : G \rightarrow S$ (given as a black-box).
Promise: f is constant on (left) cosets of some unknown subgroup H of G, and f
     takes different values on distinct cosets.
Output: H (a set of generators).

---

[2] $\lfloor x \rfloor$ denotes the largest integer smaller than $x \in \mathbb{R}$.

Example 3 Take $G = \mathbb{Z}_2^n$, and $H = \{0, a\} \subset G$. Then the cosets of H in G are all sets of the form $\{x, x \oplus a\}$ for $x \in \mathbb{Z}_2^n$. A function f is constant on every coset and takes different values on distinct cosets, if and only if it is periodic of period a. Simon's algorithm shows that in this case, we can efficiently (with $O(\log |G|)$ queries to the black-box implementing f) find a, and thus reconstruct the subgroup H, which is generated by a.

In case G is an Abelian group, the HSP can be efficiently solved by a quantum algorithm that follows the basic lines of Simon's and Shor's algorithms as follows.

1. *Prepare a random coset state.* Use the circuit D_f to prepare the state

$$|\hat{f}\rangle = \frac{1}{|G|} \sum_{x, y \in G} \chi_y(x)|y\rangle|f(x)\rangle = \frac{1}{|G|} \sum_{\substack{c \in G/H \\ y \in G}} \left(\sum_{x \in H} \chi_y(cx)\right)|y\rangle|f(cH)\rangle. \quad (44)$$

Measure the last register. This selects a coset cH of H in G at random and projects the state onto

$$|\hat{f}'\rangle = \frac{1}{\sqrt{|G| \cdot |H|}} \sum_{y \in G} \left(\sum_{x \in H} \chi_y(cx)\right)|y\rangle. \quad (45)$$

2. *Fourier sampling.* Measure the first register. Get $y \in G$ with probability

$$\left|\sum_{x \in H} \chi_y(cx)\right|^2 = \left|\chi_y(c) \sum_{x \in H} \chi_y(x)\right|^2 = \left|\sum_{x \in H} \chi_y(x)\right|^2. \quad (46)$$

Exercise 10 *Show that, for any $y \in G$, $\sum_{h \in H} \chi_y(h)$ is nonzero if and only if χ_y is the identity on H. (Hint: if there is an h such that $\chi_y(h) \neq 1$, multiply the sum by $\chi_y(h)$, use the multiplicativity of characters and the fact that $h' \in H \mapsto h \cdot h' \in H$ is an isomorphism). Deduce that $|\hat{f}'\rangle$ is a uniform superposition over all $y \in G$ such that $\chi_y(h) = 1$ for all $h \in H$.*

Exercise 10 shows that at Step 2 of the algorithm we get a random $y \in H^\perp = \{y \in G, \chi_y(h) = 1 \quad \forall h \in H\}$. Such a y can be seen as giving a linear constraint on H, and with $O(\log |H|)$ independent such constraints we can efficiently reconstruct H. Hence, by repeating the quantum Fourier sampling a sufficient number of times, we can find a generating set for H.

This algorithm does not extend to the non-Abelian case, because characters have a much more complicated structure (in fact for non-Abelian groups one looks at the so-called *irreducible representations*, which take their values in the set of unitary matrices $U_k(\mathbb{C})$ rather than simply $\mathbb{U} = U_1(\mathbb{C})$). Giving an efficient algorithm solving the HSP for non-Abelian groups is an active area of research, and such an algorithm would have interesting consequences. For instance, an efficient solution for the symmetric group S_n (permutations of n elements) would give an efficient algorithm for the graph isomorphism problem: to determine whether two given graphs are equal up to permutation of the vertices (see Exercise 11). This is another NP-problem that is not known to be NP-complete, similar to the factoring problem. It is still possible to define, and sometimes efficiently compute, the Fourier transform and the QFT over non-Abelian groups: for example, it has been shown by Beals [8] that the QFT over the symmetric group can be implemented efficiently; the difficulty is to extract enough information about the hidden subgroup from the corresponding coset states. In fact a recent series of papers [31, 27, 33, 38, 30] showed that this approach to the problem cannot work (in the case of measurements on one, two, or even $\log n$ copies of the state obtained from quantum Fourier sampling).

Another important problem is the HSP over the dihedral group D_N (the group of symmetries of a regular N-gon). Again the QFT can be implemented efficiently and the difficulty is posed by the subsequent extraction of the necessary information from the resulting states. A solution in this case would give an algorithm for the shortest vector problem in a lattice; this reduction was shown by Regev [41]. The shortest vector problem is at the base of several classical cryptographic schemes designed as an alternative to those based on factoring or discrete logarithm. For the HSP over the dihedral group D_{2^n} Kuperberg [35] gives a quantum algorithm that runs in time $2^{O(\sqrt{n})}$, a quadratic improvement in the exponent over any known classical algorithm.

Exercise 11 (Reduction from graph isomorphism to HSP) *Let G_1 and G_2 be two connected graphs on n vertices and define $G = G_1 \cup G_2$.*

Let $f : S_{2n} \rightarrow \{permutations of G\}$ be defined as $f(\pi) = \pi(G)$. We decide to represent graphs by the ordered list of their edges (v_i, v_j), where (v_i, v_j) is an ordered pair of vertices. In this representation, f is efficiently computable, and it is easy to check whether $f(\pi_1) = f(\pi_2)$ for any two permutations π_1 and π_2. Let $H = \{\pi \in S_{2n}, f(\pi) = G\}$.

1. *Show that H is a subgroup of S_{2n}.*
2. *Show that $f(\pi_1) = f(\pi_2) \iff \pi_1 H = \pi_2 H$. Conclude that f is constant on H and different on different cosets of H.*
3. *Let $\pi : G \simeq G$ be an isomorphism. Show that if π cannot be decomposed as (π_1, π_2) where π_i is an automorphism of G_i, then G_1 and G_2 are isomorphic. Prove that given a set of generators for H, it is easy to check if H contains any π that does not admit such a decomposition.*
4. *Conclude that if you had access to an efficient quantum algorithm for solving the HSP over S_{2n}, then you could efficiently decide whether G_1 and G_2 are isomorphic.*

8 Grover's Algorithm for Unstructured Search

The problem of unstructured search is one of the most frequently encountered fundamental problems in information processing. Imagine you have access to a database that contains N items and such that one of these items has a special mark. This mark is modeled by a function $f : \{1, \ldots, N\} \rightarrow \{0, 1\}$ such that $f(i) = 1$ for exactly one i in $\{1, \ldots, N\}$, the marked item. Your goal is to find the marked item using the least amount of queries to f possible. It is assumed that the database has no particular structure (for instance, its items are not sorted), so f is simply given as a black-box and has no special properties that could be used to speed up the search. In some sense this problem is a black-box version of NP-complete problems: when we strip off the structure of the problem, what we want is to search for a solution, which, once found, can be verified efficiently (i.e., the function $f(x) = 1$ if and only if x is a valid solution of the problem can be computed efficiently when x is given).

Any deterministic or randomized algorithm that succeeds to solve this problem with constant probability will have to make $\Omega(N)$ queries to f on average, since any query has probability $1/N$ of giving the marked item, and otherwise gives us no information at all about its location. In 1996 Grover [28] found a quantum algorithm that could find the marked item with very high probability by using only $O(\sqrt{N})$ queries to f. Moreover, a matching lower bound was proved in [10], showing that Grover's algorithm is essentially optimal, and also showing that quantum algorithms cannot solve NP-complete problems efficiently in a black-box setting, i.e., without using some structure of the problem.

The heart of the algorithm is the repeated application of a special gate D, called the *diffusion transform*, that can be seen as an *inversion about the mean* or *the average*. Consider an arbitrary superposition $|\Psi\rangle = \sum_{y \in \{0,1\}^n} \alpha_y |y\rangle$ on n qubits (for simplicity we assume that $N = 2^n$ is a power of 2), where $\alpha_y \in \mathbf{C}$ are the amplitudes: $\sum_y |\alpha_y|^2 = 1$. Let $\alpha = \frac{1}{N} \sum_y \alpha_y$ be the average of the amplitudes of all basis states in $|\Psi\rangle$. The transformation D is such that, after applying D, the amplitude of a basis state is transformed into its symmetric around α, i.e., $\alpha_y \mapsto 2\alpha - \alpha_y$. Figure 12 illustrates the transformation D.

To see how to implement D, note that it can be defined as $D = H^{\otimes n} R H^{\otimes n}$, where R is the diagonal unitary matrix that sends the basis state $|0^{\otimes n}\rangle$ to itself and flips the sign of all other basis states (see Exercise 2 for implementing R with elementary gates). Indeed,

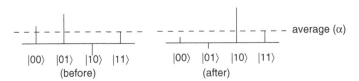

Fig. 12 The diffusion transform D

$$H^{\otimes n} R H^{\otimes n} |\Psi\rangle = \frac{1}{\sqrt{2^n}} H^{\otimes n} R \sum_{x,y} (-1)^{x \cdot y} \alpha_y |x\rangle,$$

$$= -\frac{1}{\sqrt{2^n}} H^{\otimes n} \left(\sum_{x,y} (-1)^{x \cdot y} \alpha_y |x\rangle \right) + \frac{2}{\sqrt{2^n}} \left(\sum_y \alpha_y \right) H^{\otimes n} |0\rangle,$$

$$= -\sum_y \alpha_y |y\rangle + 2\alpha \sum_x |x\rangle,$$

$$= \sum_y (2\alpha - \alpha_y)|y\rangle. \tag{47}$$

Note that the uniform superposition $\frac{1}{\sqrt{2^n}} \sum_y |y\rangle$ is a fixed point of D.

In Sect. 4 we saw how the gate

$$S_f = (I^{\otimes n} \otimes H) \cdot U_f \cdot (I^{\otimes n} \otimes H) \tag{48}$$

can be used to introduce a phase $-1 = e^{i\pi}$ for every state $|y\rangle$ such that $f(y) = 1$, by using an ancilla qubit initialised in $|1\rangle$.

Exercise 12 *Check that if y is such that $f(y) = 1$, then $S_f|y\rangle \otimes |1\rangle = -|y\rangle \otimes |1\rangle$, and if $f(y) = 0$, then $S_f|y\rangle \otimes |1\rangle = +|y\rangle \otimes |1\rangle$.*

Suppose we first prepare a uniform superposition over all $N = 2^n$ basis states on n qubits. Applying the transformation S_f, we can change the sign of the basis state corresponding to the marked item to its opposite, while leaving all other phases the same. So now the amplitudes of all unmarked items are still equal to the same value, $1/\sqrt{N}$, while the sign of the marked item's amplitude is flipped, $-1/\sqrt{N}$. The average amplitude α is thus close to $1/\sqrt{N}$. If we apply D, then the amplitudes of unmarked items, being close to the average, will remain roughly unchanged, whereas the amplitude of the marked item will go from $-1/\sqrt{N}$ to $2/\sqrt{N} - (-1/\sqrt{N}) = 3/\sqrt{N}$. Repeating this process $c\sqrt{N}$ times for some constant c to be determined later will result in the marked item's amplitude being close to 1, and thus a measurement in the computational basis will give us the marked state with good probability.

More precisely, Grover's algorithm proceeds as follows (see Fig. 13):

1. Using the Hadamard transform, prepare a uniform superposition over all N basis states and initialise an ancilla qubit to state $|1\rangle$. The state of the whole system is then $|\Psi_0\rangle = \left(\frac{1}{\sqrt{2^n}} \sum_{x \in \{0,1\}^n} |x\rangle \right) \otimes |1\rangle$.

2. Repeat the following $c\sqrt{N}$ times

   (a) Apply the unitary S_f to the whole system

   (b) Apply the diffusion transform D to the first register

3. Measure the resulting state in the computational basis

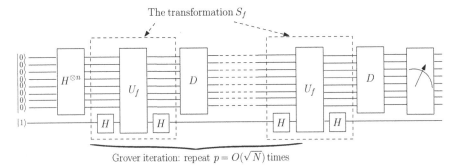

Grover iteration: repeat $p = O(\sqrt{N})$ times

Fig. 13 The circuit for Grover's algorithm

Number the inner loops from 1 to $p = c\sqrt{N}$, where c is a constant to be determined later (in fact the exact value for p is very important, as we will see below). Let α_i denote the amplitude of any unmarked state at the beginning of the ith inner loop (note that by symmetry all unmarked states are treated in the same way and hence have the same amplitudes throughout the algorithm), and β_i the amplitude of the marked state, so that $\alpha_0 = \beta_0 = 1/\sqrt{2^n}$. We would like to have $|\beta_p|^2 \geq 1/2$, so that the measurement gives the marked state with good probability.

Lemma 3 *For all* $i = 0, \ldots, p - 1$,

$$\alpha_{i+1} = \left(1 - \frac{2}{N}\right)\alpha_i - \frac{2}{N}\beta_i \quad and \quad \beta_{i+1} = 2\left(1 - \frac{1}{N}\right)\alpha_i + \left(1 - \frac{2}{N}\right)\beta_i. \tag{49}$$

Proof Let x_0 be the marked state and $|\Psi_i\rangle = \left(\sum_{x \neq x_0} \alpha_i |x\rangle + \beta_i |x_0\rangle\right) \otimes |1\rangle$ the state of the system at the beginning of the ith iteration of the loop. Then, by Exercise 12, after applying S_f we have $|\Psi_i'\rangle = S_f |\Psi_i\rangle = \left(\sum_{x \neq x_0} \alpha_i |x\rangle - \beta_i |x_0\rangle\right) \otimes |1\rangle$. Here $\alpha = 1/N\left((N-1)\alpha_i - \beta_i\right)$, so after applying D we get

$$\begin{aligned}
|\Psi_{i+1}\rangle &= D \otimes \mathbb{1}|\Psi_i'\rangle \\
&= \left(\sum_{x \neq x_0} (2\alpha - \alpha_i)|x\rangle + (2\alpha + \beta_i)|x_0\rangle\right) \otimes |1\rangle \\
&= \left(\sum_{x \neq x_0} \left(\left(1 - \frac{2}{N}\right)\alpha_i - \frac{2}{N}\beta_i\right)|x\rangle \right. \\
&\quad \left. + \left(2\left(1 - \frac{1}{N}\right)\alpha_i + \left(1 - \frac{2}{N}\right)\beta_i\right)|x_0\rangle\right) \otimes |1\rangle ;
\end{aligned} \tag{50}$$

which proves the lemma. $\square$

Lemma 4 *For all* $i = 1, \ldots, p$,

$$\alpha_i = \frac{1}{\sqrt{N-1}} \cos((2i+1)\theta) \qquad and \qquad \beta_i = \sin((2i+1)\theta) \qquad (51)$$

where θ is such that $\sin^2 \theta = 1/N$.

Proof [Exercise] Hint: prove it by induction using Lemma 3 and the trigonometric formulas $\cos(a+b) = \cos a \cos b - \sin a \sin b$, $\cos 2\theta = 1 - 2\sin^2 \theta = 1 - 2/N$, and $\cos \theta = \sqrt{1 - \sin^2 \theta} = 2\sqrt{N-1}/N$.

Take $p = \lceil \frac{\pi}{8\theta} \rceil$ (i.e., choose c appropriately). Since for large N, $\theta \simeq \sin \theta \simeq 1/\sqrt{N}$, we have $p = O(\sqrt{N})$. By Lemma 4, $\beta_p = \sin((2p+1)\theta) = \sin(\lceil \pi/4 \rceil + \theta)$. Assuming N is big enough so that $\theta < \pi/4$, we have $1/\sqrt{2} < \beta_p \leq 1$. This means that, after exactly p iterations of the inner loop, the amplitude of the marked state will be bigger than $1/\sqrt{2}$, and so this state will be measured with probability greater than $1/2$ in the last part of the algorithm.

Lemma 4 shows that the amplitude of the marked state oscillates as a sine function when we go through the inner loop of Grover's algorithm. It is thus important to know precisely the number of iterations of the inner loop that are necessary: for instance if we make $4p$ iterations, then we will have $\beta_i \approx \sin(\pi + \theta) \approx 1/\sqrt{N}$, and there will be a negligible probability of measuring the marked state in the last step of the algorithm (see, however, Exercise 14 on how to circumvent knowing p exactly). $\qquad \square$

Exercise 13 (Extension to t marked items) *Assume that the database has exactly t marked items, and your goal is to return any one of these marked items. By proving a closed form formula similar to the one in Lemma 4, determine the amplitudes of the marked and unmarked basis states after i iterations of the inner loop. Use this to show that $O(\sqrt{N/t})$ iterations of the inner loop suffice to have probability greater than $1/2$ of measuring a marked item.*

Exercise 14 (Unknown number of marked states) *Assume that there are t marked items, but t is unknown. A priori we do not know for how long to run the iteration in Grover's algorithm to measure a marked state with good probability. Devise an algorithm that runs in time $O(\sqrt{N})$ and finds a marked item with high probability in this case. Hint: Run Grover's algorithm for $p = 1, 2, 4, \ldots, 2^{\log N/2}$ checking if the resulting measurement gives a marked item or not. Show that with good probability one of these runs succeeds to find a marked item.*

Several quantum algorithms that use Grover's search as a subroutine have been found and shown to provide a polynomial speed up over their classical counterparts. For example, Brassard et al. [12] give a quantum algorithm for the problem of finding collisions in a k-to-1 function. For a k-to-1 black-box function f the task is to find a collision, i.e., two inputs $x \neq y$ such that $f(x) = f(y)$. The idea is to first classically query a set K of size $|K| = (N/k)^{1/3}$ and check it for collisions, which can be done with $O((N/k)^{1/3})$ queries. If a collision is found the algorithm outputs it and stops, otherwise we set up a Grover search for a function f defined outside K that is 1 if and only if there is a collision with an element in K. In that case there are $(k-1)|K| \approx k^{2/3}N^{1/3}$ "marked items" and Grover's search runs in time $\sqrt{N/(k^{2/3}N^{1/3})} = (N/k)^{1/3}$, by Exercise 13. So the total number of queries of this algorithm is $O((N/k)^{1/3})$, better than any classical algorithm.

Other applications of Grover's algorithm include deciding whether all elements in the image of a function on N inputs are distinct [13], which can be done in time $O(N^{3/4})$ with Grover's algorithm as a subroutine. Note that recently a better quantum algorithm based on quantum walks has been given for this problem [5] (see next Sect.). In [21] optimal quantum algorithms for graph problems such as (strong) connectivity, minimum spanning tree, and shortest path are given using Grover's search.

9 Developments

We have seen two types of quantum algorithms: those that implement a version of the hidden subgroup problem or use the QFT and those that use a version of Grover's search. For some time these algorithms have dominated the field. Recently, two alternative trends have emerged, which we briefly outline in this section.

9.1 Quantum Walks

One of the biggest breakthroughs in classical algorithm design was the introduction of randomness and the notion of a probabilistic algorithm. Many problems have good algorithms that use a *random walk* as a subroutine: for instance Aleliunas et al. [3] give a very space-efficient classical probabilistic algorithm for the undirected graph connectivity problem based on random walks. With this motivation in mind, quantum analogues of random walks have been introduced. There exist two different models of a quantum walk: the continuous-time model introduced in [23] and the discrete-time model of [1, 6]. The continuous model gives a unitary transformation directly on the space on which the walk takes place. The discrete model is a direct analogue of the classical random walk: it introduces an extra coin register which is used as a "quantum coin flip," and performs a walk step controlled on the state of the coin.

The two main quantities that characterize a random walk's performance are

- its mixing time: the time it takes for the walk to have a close to uniform probability of having reached any point in the domain,
- its hitting time: the expected time it takes to hit a certain point in the domain

These quantities have been analyzed for several graphs in both the continuous and the discrete model. It turns out that a quantum walk's mixing time cannot be more than quadratically faster than its classical counterpart's [1]; so the classical and quantum performances are polynomially related. However, a quantum walk can provide an exponential improvement in terms of its hitting time: it has been shown that there are graphs and two vertices in them such that the classical hitting time from one vertex to the other is polynomial in the number of vertices of the graph, whereas the quantum walk is exponentially faster. This idea was used in [15] to devise an (artificial) problem for which a quantum walk-based algorithm gives a provable exponential speed-up over any classical probabilistic algorithm. It is open whether quantum hitting times can be used to speed up classical algorithms for relevant problems.

Following this work a quantum walk algorithm has been introduced in [43] for the problem of finding a marked vertex in a graph. The idea is very simple: the algorithm starts in the uniform superposition over all vertices. At each step it performs a quantum walk that basically moves randomly over the graph, except when it hits a marked vertex, at which point it stays there. It can be shown that after a sufficient time the amplitude of the state concentrates in the marked item(s); a measurement then finds a marked item with high probability.

This algorithm solves Grover's problem on a graph. Why do we need a quantum walk search if we have Grover's algorithm? It turns out that there are situations when the diffusion step D of Grover's algorithm is too costly to implement (because the local topology of the database does not allow for it, because of limitations on the quantum gates or because it is too costly in a query setting). A quantum walk only makes local transitions and can be more advantageous. One example is the search for a marked item in a 2-dimensional database. In this case Grover's algorithm requires $\sqrt{N}$ queries, but to shift amplitude from one item of the database to another on the grid takes an additional $\sqrt{N}$ steps on average per query. The net complexity of the algorithm becomes $\sqrt{N} \cdot \sqrt{N} = N$ and the quantum advantage is lost. The quantum walk algorithm has been shown to find a marked item in time $O(\sqrt{N} \log N)$ [7].

Another example of the superiority of the quantum walk search over Grover's algorithm has been given in [5] by Ambainis, who uses a quantum walk to give an improved algorithm for element distinctness, which runs in optimal time $O(N^{2/3})$; thus improving over Grover-based algorithms for this problem (which runs in time $O(N^{3/4})$, see Sect. 8). Several new quantum walk-based algorithms with polynomial improvements over Grover-based algorithms have followed suit (see [37, 14, 36]). For references on quantum walks, see [32, 4].

9.2 Adiabatic Quantum Algorithms

Adiabatic quantum algorithms were introduced by Farhi et al. [22] as an interesting alternative to the use of quantum gates and circuits for the design of algorithms. The idea is to describe a specific hermitian matrix H (called Hamiltonian in the context of physical systems) that concentrates all the constraints that we want to impose on a solution to our problem through a decomposition $H = \sum_i H_i$, where each H_i represents a local constraint. By definition, the ground state (lowest eigenstate) of H violates the smallest number of such constraints and represents the desired optimal solution, so our aim is to obtain this state.

Consider a time-dependent Hamiltonian $H = H(t)$, and let state $|\Psi(t)\rangle$ evolve over time according to $H(t)$. The *adiabatic theorem* says that, if $|\Psi(0)\rangle$ is the ground state of $H(0)$, and if $H(t)$'s evolution is sufficiently slow, then $|\Psi(t)\rangle$ will be the ground state of $H(t)$ at all times t. The bounds on the evolution speed are determined by the energy differences between $H(t)$'s ground-state and first excited state, i.e., the difference between its smallest and second smallest eigenvalue when it is viewed as a matrix. Therefore, in order to construct the ground state of a specific Hamiltonian H, we first start by devising another Hamiltonian H' whose ground state $|\Psi'\rangle$ is easy to prepare. We then slowly evolve our system's Hamiltonian from H' to H, usually in a linear fashion over time, by letting $H(t) = (1 - t/T)H' + (t/T)H$, where T is the total run-time of the algorithm. If T is large enough, then the adiabatic theorem guarantees that the state at time t will be the ground state of $H(t)$, leading to the solution, the ground state of H, at time T. For the algorithm to be efficient, one has to pick H and H' such that the gap of $H(t)$ at all times t is at least inverse polynomial in the size of the problem, so that the total running time T will be polynomial.

Farhi et al. set up adiabatic algorithms for NP-complete problems [22]. It has been impossible so far to determine the gap analytically and the number of qubits in numerical simulations is limited. However, this approach seems promising, even though there is now mounting evidence that an adiabatic algorithm cannot solve NP-complete problems efficiently. For instance, quantum unstructured search has been implemented adiabatically and shown to have the same run-time as Grover's algorithm [42, 16]. It is not hard to see that an adiabatic algorithm can be simulated efficiently with a quantum circuit [22] – one needs to implement a time-dependent unitary that is given by a set of local Hamiltonians, each one acting only on a few qubits. Recently it has been shown [2] that also any quantum circuit can be simulated efficiently by an appropriate adiabatic algorithm; hence these two models of computation are essentially equivalent. This means that a quantum algorithm can be designed in each of the two models. The advantage of the adiabatic model is that it deals with gaps of Hermitian matrices, an area that has been widely studied both by solid-state physicists and probabilists. Hopefully this new toolbox will yield new algorithms.

References

1. Aharonov, D., Ambainis, A., Kempe, J., et al.: Quantum walks on graphs. In: Proceedings of the 33th ACM Symposium on Theory of Computing (STOC), pp. 50–59. ACM Press, New York (2001)
2. Aharonov, D., van Dam, W., Kempe, J., et al.: Adiabatic quantum computation is equivalent to standard quantum computation. In: Proceedings of the 45th Annual IEEE Symposium on

Foundations of Computer Science (FOCS), pp. 42–51. IEEE Computer Society, Washington (2004)

3. Aleliunas, R., Karp, R., Lipton, R., et al.: Random walks, universal traversal sequences, and the time complexity of maze problems. In: Proceedings of the 20th Annual IEEE Symposium on Foundations of Computer Science (FOCS), pp. 218–223. IEEE Computer Society, Washington (1979)

4. Ambainis, A.: SIGACT News **35**, 22 (2004)

5. Ambainis, A.: Quantum walk algorithm for element distinctness. In: Proceedings of the 45th Annual IEEE Symposium on Foundations of Computer Science (FOCS), pp. 22–31. IEEE Computer Society, Washington (2004)

6. Ambainis, A., Bach, E., Nayak, A., et al.: One-dimensional quantum walks. In: Proceedings of the 33th Annual ACM Symposium on Theory of Computing (STOC), pp. 60–69. ACM Press, New York (2001)

7. Ambainis, A., Kempe, J., Rivosh, A.: Coins make quantum walks faster. In: Proceedings of the 16th ACM-SIAM Symposium on Discrete Algorithms (SODA), pp. 1099–1108. ACM Press, New York (2005)

8. Beals, R.: Quantum computation of Fourier transforms over symmetric groups. In: Proceedings of the 29th Annual ACM Symposium on Theory of Computing (STOC), pp. 48–53. ACM Press, New York (1997)

9. Bennett, Ch.: IBM J. Res. Dev. **17**, 5225 (1973)

10. Bennett, C.H., Bernstein, E., Brassard, G., et al.: SIAM J. Sci. Comput. **26**, 1510 (1997)

11. Bernstein, E., Vazirani, U.: SIAM J. Sci. Comput. **26**, 1411 (1997)

12. Brassard, G., Hoyer, P., Tapp, A.: Quantum cryptanalysis of hash and claw-free functions. In: Proceedings of the 3rd Latin American Symposium on Theoretical Informatics (LATIN'98). Lecture Notes in Computer Science, vol. 1380, p. 163. Springer, Heidelberg (1998)

13. Buhrman, H., Dürr, C., Heiligman, M., et al: SIAM J. Sci. Comput. **34**, 1324 (2005)

14. Buhrman, H., Spalek, R.: Quantum verification of matrix products. In: Proceedings of the 17th ACM-SIAM Symposium on Discrete Algorithms (SODA), pp. 880–889. ACM Press, Miami (2006)

15. Childs, A.M., Cleve, R., Deotto, E., et al.: Exponential algorithmic speedup by a quantum walk. In: Proceedings of the 35th Annual ACM Symposium on the Theory of Computing (STOC), pp. 59–68. ACM Press, New York (2003)

16. van Dam, W., Mosca, M., Vazirani, U.: How powerful is adiabatic quantum computation? In: Proceedings of the 42nd Annual IEEE Symposium on Foundations of Computer Science (FOCS), IEEE Computer Society, Washington, pp. 279–287 (2001)

17. Deutsch, D.: Proc. R. Soc. Lond. A **400**, 97 (1985)

18. Deutsch, D., Barenco, A., Ekert, A.: Proc. R. Soc. Lond. A **449**, 669 (1995)

19. Deutsch, D., Jozsa, R.: Proc. R. Soc. Lond. A **439**, 553 (1992)

20. Di Vincenzo, D.P.: Phys. Rev. A **51**, 1015 (1995)

21. Dürr, C., Heiligman, M., Høyer, P., et al.: Quantum query complexity of some graph problems. In: Proceedings of the 31st International Colloquium on Automata, Languages, and Programming (ICALP). Lecture Notes in Computer Science, vol. 3142, p. 481. Springer, Heidelberg (2004)

22. Farhi, E., Goldstone, J., Gutmann, S., et al.: Science **292**, 472 (2001)

23. Farhi, E., Gutmann, S.: Phys. Rev. A **58**, 915 (1998)

24. Feynman, R.: Int. J. Theor. Phys. **21**, 467 (1982)

25. Feynman, R.: Opt. News **11**, 11 (1985)

26. Friedl, K., Ivanyos, G., Magniez, F., et al.: Hidden translation and orbit coset in quantum computing. In: Proceedings of the 35th ACM Symposium on Theory of Computing (STOC), pp. 1–9. ACM Press, New York (2003)

27. Grigni, M., Schulman, L., Vazirani, M., et al.: Quantum mechanical algorithms for the non-abelian hidden subgroup problem. In: Proceedings of the 33rd ACM Symposium on the Theory of Computing (STOC), pp. 68–74. ACM Press, New York (2001)

28. Grover, L.: A fast quantum mechanical algorithm for database search. In: Proceedings of the 28th ACM Symposium on the Theory of Computing (STOC), pp. 212–219. ACM Press, Philadelphia (1996)

29. Hallgren, S.: Polynomial-time quantum algorithms for Pell's equation and the principal ideal problem. In: Proceedings of the 34th ACM Symposium on the Theory of Computing (STOC), pp. 653–658. ACM Press, New York. (2002)

30. Hallgren, S., Moore, C., Rötteler, M., et al.: Limitations of quantum coset states for graph isomorphism. In: Proceedings of the 38th ACM Symposium on the Theory of Computing (STOC), pp. 604–617. ACM Press, New York (2006)

31. Hallgren, S., Russell, A., Ta-Shma, A.: Normal subgroup reconstruction and quantum computation using group representations. In: Proceedings of the 32nd ACM Symposium on the Theory of Computing (STOC), pp. 627–635. ACM Press, New York (2000)

32. Kempe, J.: Contemp. Phys. **44**, 302 (2003)

33. Kempe, J., Shalev, A.: The hidden subgroup problem and permutation group theory. In: Proceedings of the 16th ACM-SIAM Symposium on Discrete Algorithms (SODA), pp. 1118–1125. ACM Press, New York (2005)

34. Kitaev, A.Y., Shen, A.H., Vyalyi, M.N.: Classical and Quantum Computation. Graduate Series in Mathematics, vol. 47. AMS, Providence (2002)

35. Kuperberg, G.: SIAM J. Sci. Comput. **35**, 170 (2005)

36. Magniez, F., Nayak, A.: Quantum complexity of testing group commutativity. In: Proceedings of the 32nd International Colloquium on Automata, Languages, and Programming (ICALP), pp. 1312–1325. Lecture notes in computer science 3580, Springer-Verlag, Berlin (2005)

37. Magniez, F., Santha, M., Szegedy, M.: Quantum algorithms for the triangle problem. In: Proceedings of the 16th ACM-SIAM Symposium on Discrete Algorithms (SODA), pp. 1109–1117. ACM Press, New York (2005)

38. Moore, C., Russell, A., Schulman, L.: The symmetric group defies strong Fourier sampling. In: Proceedings of the 46th IEEE Symposium on Foundations of Computer Science (FOCS). IEEE Computer Society, Washington (2005)

39. Nielsen, M.A., Chuang, I.L.: Quantum Computation and Quantum Information. Cambridge University Press, Cambridge (2000)

40. Preskill, J.: Quantum information and computation. Lecture Notes http://www.theory.caltech.edu/people/preskill/ph229/

41. Regev, O.: Quantum computation and lattice problems. In: Proceedings of the 43rd Annual IEEE Symposium on Foundations of Computer Science (FOCS), pp. 520–529. IEEE Computer Society Press, Los Alamitos (2002)

42. Roland, J., Cerf, N.: Phys. Rev. A **65**, 042308 (2002)

43. Shenvi, N., Kempe, J., Whaley, K.B.: Phys. Rev. A **67**, 052307 (2003)

44. Shor, P.W.: Algorithms for quantum computation: Discrete log and factoring. In: Proceedings of the 35th IEEE Annual Symposium on the Foundations of Computer Science (FOCS), pp. 124–134. IEEE Computer Society, Los Alamitos (1994)

45. Shor, P.W.: SIAM J. Comput. **26**, 1484 (1997)

46. Simon, D.: SIAM J. Comput. **26**, 1474 (1997)

Index

F. Benatti et al.: *Index*. Lect. Notes Phys. **808**, 343–350 (2010)
DOI 10.1007/978-3-642-11914-9